Problems and Methods of Optimal Control

Mathematics and Its Applications

Managing Editor:

Volume 286

Problems and Methods of Optimal Control

by

Leonid D. Akulenko
Institute for Problems in Mechanics,
Russian Academy of Sciences,
Moscow, Russia

SPRINGER-SCIENCE+BUSINESS MEDIA, B.V.

A C.I.P. Catalogue record for this book is available from the Library of Congress

ISBN 978-94-010-4522-3 ISBN 978-94-011-1194-2 (eBook)
DOI 10.1007/978-94-011-1194-2

This is an updated and revised translation
of the original Russian work *Asymptotic Methods of Optimal Control*
Translated by N. N. Bolotnik.
Published by Nauka, Moscow © 1987

Printed on acid-free paper

Contents

Preface

The numerous applications of optimal control theory have given an incentive to the development of approximate techniques aimed at the construction of control laws and the optimization of dynamical systems. These constructive approaches rely on small parameter methods (averaging, regular and singular perturbations), which are well-known and have been proven to be efficient in nonlinear mechanics and optimal control theory (maximum principle, variational calculus and dynamic programming). An essential feature of the procedures for solving optimal control problems consists in the necessity for dealing with two-point boundary-value problems for nonlinear and, as a rule, nonsmooth multi-dimensional sets of differential equations. This circumstance complicates direct applications of the above-mentioned perturbation methods which have been developed mostly for investigating initial-value (Cauchy) problems.

There is now a need for a systematic presentation of constructive analytical perturbation methods relevant to optimal control problems for nonlinear systems. The purpose of this book is to meet this need in the English language scientific literature and to present consistently small parameter techniques relating to the constructive investigation of some classes of optimal control problems which often arise in practice. This book is based on a revised and modified version of the monograph: L. D. Akulenko "Asymptotic methods in optimal control". Moscow: Nauka, 366 p. (in Russian).

Chapter 1 deals with the procedure of the averaging method for optimal control problems of quasilinear oscillatory systems with slowly varying parameters. We consider the means for introducing a small parameter, and describe some ways of reducing original equations to the standard form of a controlled system having a rotating phase. We have constructed algorithms for approximately solving optimal control problems with a fixed asymptotically large time interval, as well as time-minimum-type problems. The algorithms are illustrated by solving (to a first approximation in terms of a small parameter) a number of problems dealing with the optimal control of mechanical systems. In Chapter 2, asymptotic methods for solving boundary-value problems of the maximum principle and optimal control problems are considered. The facilities and features of using the averaging method for more general systems, as well as in critical (nonstandard) situations, are also discussed here.

The averaging method for problems of the optimal control of essentially nonlinear rotatory-oscillatory systems is developed in Chapters 3 and 4. In Chapter 3, we consider the methods used for reducing general mechanical systems to the standard form of a controlled system having a rotating phase. We describe the procedures of constructing the solution to a first approximation for single-frequency optimal control problems on an asymptotically large time interval in the cases of both fixed and free instants of termination. The schemes of averaging the maximum principle equations, based on integrals of motion of the original unperturbed system, are developed.

In Chapter 4, we discuss the basis of methods for the canonical separation of motions in nonlinear Hamiltonian systems of the maximum principle. The method of constructing higher approximations on the basis of Whittaker's equations is developed; we describe a two-parameter scheme of canonical averaging which leads to fewer manipulations. We consider multifrequency, and more general, systems in the vicinity of a stable local integral manifold, averaging schemes for a relatively short time interval, and discuss features for applying the averaging method to nonlinear systems. General theoretical results are illustrated by solving model examples admitting interpretation in a mechanics context.

In Chapters 5 and 6, the asymptotic methods developed in Chapters 1–4 are used for investigating, to a first approximation in terms of a small parameter, optimal control problems for some mechanical systems of practical interest. Chapter 5 deals with the results of solving different control problems for "pendulum-type" oscillatory systems. Spatial and plane models, the cases of a velocity-controlled and acceleration-controlled suspension point, and the variable or controlled length of a pendulum, are considered. In Chapter 6, we study problems of spacecraft orbital motions and the optimal control of rotations by means of "low thrust". Quasioptimal and locally optimal modes of braking rigid body rotations are also considered.

Small parameter techniques for the synthesis of regularly perturbed systems having an invariant norm and equivalent systems are developed in Chapter 7. Perturbations from a comparatively wide class are considered, which is sufficient for satisfying many applied goals. We describe a method of the successive approximations of control vector domains and the numerical implementation of algorithms for the approximate solution of feedback control problems. Solutions of various problems of spacecraft stabilization and orientation are given, including models having internal degrees of freedom.

In Chapter 8, we propose new statements of problems and study some other aspects of perturbation theory consistent with the analysis of controlled systems. We construct procedures for the approximate and asymptotic solution of regularly and singularly perturbed problems of control and optimization, carry out decomposition of weakly coupled subsystems on the basis of dynamic programming and the canonical separation of motions. Perturbation techniques for constructing switching curves for bang-bang controls for nonoscillating systems are developed. Approaches to the solution of complex control problems for hybrid oscillatory systems which are of considerable interest in modern engineering, are suggested.

This book has been written for those interested in applied problems of dynamic systems control. It is also intended to be useful for graduate and postgraduate students acquainted with the fundamentals of control theory (the maximum principle and dynamic programming) and small parameter techniques (averaging, regular and singular perturbations), since it contains solutions to a number of problems having a clear mechanical interpretation. These problems will help students develop their imagination and intuition.

The analytic content of the results presented and perspectives for developing further the methods of constructing solutions for sophisticated problems of the control of dynamic systems will also be of interest to researchers. The direct numerical solution of these problems causes significant, often insuperable, difficulties, due to rapid oscillations and the degeneration of boundary-value problems. Therefore, a preliminary analytical treatment of a given control problem seems to be necessary.

Finally, the approach of this the book will be of interest to engineers and designers, since it allows them to study basic properties of controlled processes and to estimate efficiently the influence of basic parameters and the essential characteristics on the performance quality of a controlled system.

The formulae numbering in this book uses three numbers: the first two indicate the section number, and the third one is the formula number within a given section. When referring to a formula within a given section, we use only the formula number. Reference to a formula from another section within a given chapter has two numbers: the number of the section (the second figure in the triple numeration) and the formula number. Reference to a formula from another chapter uses the full triple numeration.

Lemmas, theorems and figures have double numeration: the chapter number and the ordinal number.

Acknowledgments

I would like to express my acknowledgement to Kluwer Academic Publishers and "Nauka" Publishing House. Their interaction and cooperation have provided the possibility of publishing my book in English. I also wish to thank charming Ms. M. Deignan and Mrs. M. Ermolova, who both contributed a lot in providing specific advice and help. I also thank Dr. N. Bolotnik for his hard work on the translation and Mrs. R. Soldatova for her significant work on the preparationon of the typescript.

CHAPTER 1

Averaging Method in Optimal Control Problems for Quasilinear Oscillatory Systems

We consider here a wide class of control problems for quasilinear oscillatory and rotatory systems on an asymptotically large time interval by applying a small control force. We develop an approximate solution technique based on the combination of Pontryagin's maximum principle [131, 40] and the asymptotic methods of nonlinear mechanics [38, 149, 150, 35, 36, 45, 74, 98, 119, 137]. Some problems which can be interpreted in terms of mechanics are investigated to a first approximation.

1.1. Controlled Quasilinear Oscillatory Systems

1.1.1. GENERAL MULTI-FREQUENCY CONTROLLED SYSTEM

Consider a system containing slow, x, and fast, y, variables [38, 150]:

$$\begin{aligned} \dot{x} &= \varepsilon X(\tau, x, y, u), \qquad x(t_0) = x^0, \\ \dot{y} &= Y_0(\tau, x, y) + \varepsilon Y(\tau, x, y, u), \qquad y(t_0) = y^0. \end{aligned} \tag{1.1.1}$$

Here, $t \in [t_0, T]$ denotes time, $T = \Theta\varepsilon^{-1}$, $\Theta = \text{const}$, $x \in D_x$ is an n_x-vector, $y \in D_y$ is an n_y-vector ($n_y \geqslant 1$, $n_x + n_y \geqslant 2$); $u \in U$ is a control vector-valued function of dimension $m \geqslant 1$, $\tau = \varepsilon t$ is the slow time, and $\varepsilon \in (0, \varepsilon_0]$ is a small parameter. The functions X, Y_0, and Y are assumed to be sufficiently smooth in some domains; the smoothness properties are specified below. The domains D_x and D_y are open and connected. The domain D_x is bounded and D_y is unbounded. The set U, as a rule, is connected and fixed. The constants t_0, x^0, and y^0 are the initial data, $x^0 \in D_x$, $y^0 \in D_y$. The parameters t_0, x^0, and y^0, as well as the functions X and Y, may continuously depend on ε; however, for the sake of compactness, this dependence is not indicated.

System (1) reduces to the standard form, provided that, when $\varepsilon = 0$ and

$$\dot{y} = Y_0(\tau, x, y), \qquad x, \tau = \text{const}, \tag{1.1.2}$$

system (1) has a general multi-frequency solution of the kind:

$$\begin{aligned}
&y = y_0(\psi, c, x, \tau) = P(\psi) + \varphi(\psi, c, x, \tau),\\
&P = (P_1, \ldots, P_{n_\psi}), \qquad P_i = \Pi_i \psi_i/(2\pi), \quad i = 1, \ldots, n_\psi,\\
&\psi = \nu(\tau)(t - t_0) + \psi_0, \qquad c, \psi_0 = \text{const},\\
&Y_0(\tau, x, y + \Pi) \equiv Y_0(\tau, x, y).
\end{aligned} \tag{1.1.3}$$

Here, Π is a constant n_y-vector with zero components for oscillating variables and nonzero components for rotating variables; φ is a 2π-periodic function of the vector phase ψ; $\nu(\tau)$ is a vector frequency, $\nu_i(\tau) \geqslant \nu_{i0} > 0$, $\nu_{i0} = \text{const}$; (c, ψ_0) is a n_y-vector of the constants of integration: $c \in D_c$, $|\psi_0| < \infty$, $n_\psi + n_c = n_y \geqslant 1$.

When $\varepsilon > 0$, the relationships (3) are used for the change of variables $(x, y, \tau) \to (x, c, \psi, \tau)$. Having differentiated (3) according to Equations (1) and using the identity $(\partial y_0/\partial \psi)\nu \equiv Y_0(\tau, x, y_0)$, we obtain the following system with the initial conditions:

$$\begin{aligned}
&\dot{x} = \varepsilon X(\tau, x, y_0(\psi, c, x, \tau), u), \qquad x(t_0) = x^0,\\
&\dot{c} = \varepsilon C(\tau, x, c, \psi, u), \qquad c(t_0) = c^0,\\
&\dot{\psi} = \nu(\tau) + \varepsilon \Psi(\tau, x, c, \psi, u), \qquad \psi(t_0) = \psi^0.
\end{aligned} \tag{1.1.4}$$

Here, the functions C and Ψ are defined as the unique solution of the linear algebraic system

$$\begin{aligned}
&\frac{\partial y_0}{\partial \psi}\Psi + \frac{\partial y_0}{\partial c}C = Y(\tau, x, y_0, u) - \frac{\partial y_0}{\partial \tau} - \frac{\partial y_0}{\partial x}X(\tau, x, y_0, u),\\
&\det\left(\frac{\partial y_0}{\partial \psi}, \frac{\partial y_0}{\partial c}\right) \neq 0.
\end{aligned} \tag{1.1.5}$$

The right-hand side of system (4) is 2π-periodic in ψ if the functions X and Y are Π-periodic in y. The initial values for c and ψ are calculated according to (3). Having constructed a combined n_a-vector a from x and c, we represent the controlled system (4) in a standard form with rotating phase:

$$\begin{aligned}
&\dot{a} = \varepsilon f(\tau, a, \psi, u), \qquad a(t_0) = a^0,\\
&\dot{\psi} = \nu(\tau) + \varepsilon F(\tau, a, \psi, u), \qquad \psi(t_0) = \psi^0.
\end{aligned} \tag{1.1.6}$$

Here, f and F are sufficiently smooth functions of $a \in D_a$ ($D_a = D_x \times D_c$), $\tau \in [\tau_0, \Theta]$, $u \in U$. In the theory of nonlinear oscillations, systems of the form (6) without u are usually called quasilinear [1–3, 15, 24, 35, 36, 38, 46, 74, 120, 150]. Analogously, we call system (6) a controlled quasilinear system in the standard form with rotating phase [1, 46].

It is worth noting that the dependence $\nu(\tau)$ can be specified in a more sophisticated manner, for instance, as

$$\nu(\tau) = \Omega(z), \qquad \dot{z} = \varepsilon Z(z), \qquad z(t_0) = z^0, \tag{1.1.7}$$

where z is a slow vector of arbitrary dimension $n_z \geqslant 0$.

To define the motion completely one has to specify the control u in (6) as a function of time t, or of time t and the state variables a and ψ. We describe admissible controls as those piecewise continuous functions $u(t)$ or $u(t, a, \psi)$, $u \in U$, whose substitution into (6) gives a set of equations whose solution exists and, as a rule, is unique for $t \in [t_0, T]$, $T = \Theta\varepsilon^{-1}$. The motion is considered on an asymptotically large time interval $t - t_0 \sim \varepsilon^{-1}$, where the slow variable a receives, generally speaking, an increment $a - a_0 \sim 1$ and the phase ψ takes the increment $\psi - \psi_0 \sim \varepsilon^{-1}$. Additional requirements imposed on the control u follow from the statement of the optimal control problem and are specified below.

It is worth noting that the representation of the original system in the form (1) is, in a sense, a tribute to tradition. Without loss of generality (under the assumptions imposed on the generating system (2)), the vector x can be included in the vector y. A more general situation, when the dimension of the vector c in (3) is less than $n_y - n_\psi$, is treated in Section 2.5.

1.1.2. MULTI-FREQUENCY QUASILINEAR SYSTEM

Consider an oscillatory system with many degrees of freedom and slowly changing parameters, reduced to the normal form:

$$\begin{aligned} &\ddot{y}_i + \nu_i^2(\tau) y_i = Y_i(\tau, x) + \varepsilon f_i(\tau, x, y, \dot{y}, u), \\ &y_i(t_0) = y_i^0, \qquad \dot{y}_i(t_0) = \dot{y}_i^0, \quad \nu_i^2 \geqslant \nu_{i0}^2 > 0, \quad i = 1, \ldots, n_y; \\ &\dot{x} = \varepsilon X(\tau, x, y, \dot{y}, u), \qquad x(t_0) = x^0. \end{aligned} \tag{1.1.8}$$

When $\varepsilon = 0$, the system oscillates about the shifted equilibrium position

$$\bar{y}_i = Y_i(\tau, x)\nu_i^{-2}(\tau).$$

We can apply a transformation similar to (3) to each component $y_i, \dot{y}_i$:

$$y_i = a_i \sin\psi_i + \bar{y}_i(\tau, x), \qquad \dot{y}_i = a_i \nu_i \cos\psi_i, \tag{1.1.9}$$

where a_i are amplitudes of partial oscillations, and ψ_i are phases. Differentiation of (9) according to (8) gives rise to the system with n_y phases:

$$\dot{a}_i = \varepsilon\frac{f_i}{\nu_i}\cos\psi_i - \varepsilon\frac{\nu_i'}{\nu_i}a_i\cos^2\psi_i - \varepsilon y_i'\sin\psi_i,$$

$$\dot{\psi}_i = \nu_i - \varepsilon\frac{f_i}{\nu_i a_i}\sin\psi_i + \varepsilon\frac{\nu_i'}{\nu_i}\cos\psi_i\sin\psi_i - \varepsilon\frac{y_1'}{a_i}\cos\psi_i,$$

$$\dot{x} = \varepsilon X, \qquad x(t_0) = x^0 \qquad \left(\bar{y}_i' = \frac{\partial\bar{y}_i}{\partial\tau} + \left(\frac{\partial\bar{y}_i}{\partial x}\cdot X\right)\right), \tag{1.1.10}$$

$$a_i^0 = \left[(y_i^0 - \bar{y}_i^0)^2 + \left(\frac{\dot{y}_{i0}^0}{\nu_i^0}\right)^2\right]^{1/2},$$

$$\sin\psi^0 = \frac{y_i^0 - \bar{y}_i^0}{a_i^0}, \qquad \cos\psi^0 = \frac{\dot{y}_i^0}{a_i^0\nu_i^0},$$

the right-hand side of which is 2π-periodic in each phase ψ_i.

Note that the application of asymptotic methods to the investigation of uncontrolled systems of the form (10) meets significant difficulties due to the "resonance problem". These difficulties are aggravated for variable frequencies $\nu_i(\tau)$. In the case when the right-hand side of the system is represented, for example, as a "truncated" multiple Fourier series, one can use the method of averaging. In applications, system (10) often appears to be actually single-frequency, i.e. $\nu_i = \nu_*(\tau) + O(\varepsilon)$. For example, one can reduce to the class of single-frequency systems the equations of the perturbed motion of a particle in a central gravitational or linear field, the equations of pendulum oscillations or rigid body rotations with respect to a fixed point, etc., see Section 1.1.3 below. Separating one of the phases $\psi_{i*} = \psi_*$ and introducing off-resonance amounts θ_i by the formulae $\psi_i = \psi_* + \theta_i$ lead to (6) with $n_\psi = 1$.

By means of a change of variables

$$\psi_i = \varphi_i + \theta_i, \qquad \varphi_i = \int \nu_i(\tau)\,dt, \quad i = 1, \ldots, n_\varphi \tag{1.1.11}$$

both single-frequency and multi-frequency systems can be represented in the standard form

$$\dot{\chi} = \varepsilon X(t, \chi, u, \varepsilon), \qquad \chi(t_0) = \chi^0, \quad \chi = (a, x, \theta). \tag{1.1.12}$$

Here, X is a complicated function of t; the analysis of the existence of the uniform average for this function, as well as the construction of this average, are difficult problems. System (8) can also be reduced to the form (12) by using a change of variables of the Van der Pohl type [35, 36, 38]

$$\begin{aligned} y_i &= a_i \sin\varphi_i + b_i \cos\varphi_i + \bar{y}_i, \\ \dot{y}_i &= \nu_i(a_i \cos\varphi_i - b_i \sin\varphi_i), \qquad i = 1,\dots,n_y, \end{aligned} \tag{1.1.13}$$

where a_i and b_i are oscillating (slow) variables. For the unknowns a_i and b_i (13), the following system is obtained:

$$\begin{aligned} \dot{a}_i &= \varepsilon\frac{f_i}{\nu_i}\cos\varphi_i + \varepsilon\frac{\nu_i'}{\nu_i}(a_i\cos\varphi_i - b_i\sin\varphi_i)\cos\varphi_i - \varepsilon\bar{y}_i'\sin\varphi_i, \\ \dot{b}_i &= -\varepsilon\frac{f_i}{\nu}\sin\varphi_i - \varepsilon\frac{\nu_i'}{\nu_i}(a_i\cos\varphi_i - b_i\sin\varphi_i)\sin\varphi_i - \varepsilon\bar{y}_i'\cos\varphi_i, \\ a_i^0 &= \frac{\dot{y}_i^0}{\nu_i^0}, \qquad b_i^0 = y_i^0 - \bar{y}_i^0, \qquad \varphi_i = \int_{t_0}^{t}\nu_i(\varepsilon s)\,ds. \end{aligned} \tag{1.1.14}$$

The general multi-frequency system (1), where $y = y_0(\psi, a, \tau)$ is a multi-phase function (ψ is a vector), can be reduced to the form (6) by using the transformation described in Section 1.1.1, and, by using the change (11), this can be reduced to the form (12). The standard form of a controlled system has to be chosen for reasons of convenience in each specific case. It is worth adding that, in general, the changes (9) are singular if some amplitudes a_i vanish and Equations (10) for phases ψ_i have singular right-hand sides. Van der Pohl's substitutions (13) are nonsingular, provided $\nu_i \geqslant \nu_{i0} > 0$, and this is the basic advantage of such changes. The convenience of variables of amplitude-phase type (a_i, ψ_i) is that averaging over phases results in reducing the order of the system by at least unity. The order of the system for the averaged variables (a_i, b_i) is conserved and equals $2n_y$ in the general case.

Below, we consider and study control problems for quasilinear oscillatory systems, reduced to the standard form (6) with a scalar phase ψ, $n_\psi = 1$. We also consider the facilities for the application of asymptotic (averaging) methods to multifrequency systems of the type (6) (Sections 1.4, 2.5, 4.4, and 4.5).

1.1.3. EXAMPLES OF SINGLE-FREQUENCY CONTROLLED SYSTEMS. WAYS OF INTRODUCING A SMALL PARAMETER

(1) *Model of oscillations of a lift.* Consider the vertical oscillations of a body (load) with mass m suspended by a uniform elastic thread in the gravity field (Fig. 1.1). If

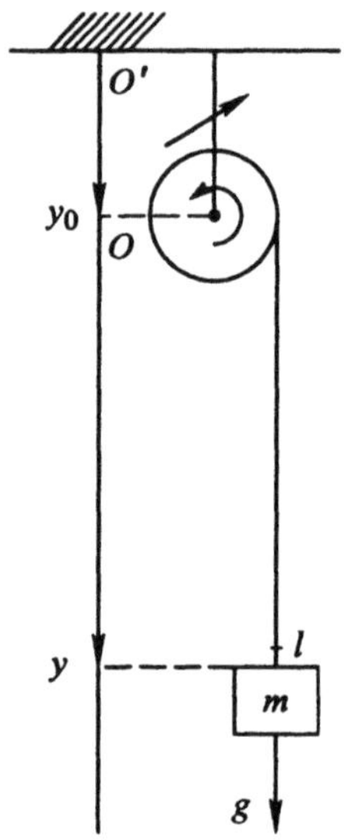

Fig. 1.1.

the thread is unextended, its length would change slowly, governed by the law $l = l(\tau)$, $\tau = \varepsilon t$. The equation of motion for this system in the quasistatic approximation has the form [98, 145, 146]:

$$m\ddot{y} + ESl^{-1}(y - l) = mg - N, \qquad N = m\ddot{y}_0. \tag{1.1.15}$$

Here, y is the relative coordinate of the load, E is Young's modulus of the material of the thread, S is the cross-sectional area of the thread, g is the acceleration due to gravity; N describes the kinematic control action and is assumed to be small. Let l_* be a characteristic length of the thread. Then $T_* = 2\pi(ml_*/(ES))^{1/2}$ is the characteristic period of free oscillations. The slowness of changing l means that $\dot{l}T_*/l_* \sim \varepsilon \ll 1$. Having replaced the variables in (15) by

$$t_* = \nu_* t, \qquad \nu_* = 2\pi/T_*, \qquad y_* = y/l_*,$$
$$\sigma = l/l_*, \qquad G = g/(l_*\nu_*^2), \qquad \varepsilon u = N/(ml_*\nu_*^2),$$

we obtain an equation of the type (8) in dimensionless form

$$\ddot{y} + \nu^2(\tau)y = 1 + G - \varepsilon u, \qquad \nu^2 = 1/\sigma. \tag{1.1.16}$$

Using the approach described in Section 1.1.2, one can reduce Equation (16) to the standard form (10) or (14). The control u, in particular, can be intended for quenching the oscillations caused by perturbations.

(2) *Plane pendulum with controlled suspension point.* Consider planar oscillations of a pendulum with variable length $l(\tau)$ in the quasilinear approximation. Its suspension point P can move in a vertical plane Oxy with the acceleration (w_x, w_y), see Fig. 1.2. We assume that the constraint force is positive during the motion. The equations of motion are written in the form

$$\ddot{\varphi} + \frac{g}{l}\sin\varphi = -\frac{1}{l}(w_x \cos\varphi + w_y \sin\varphi) - 2\frac{\dot{l}}{l}\dot{\varphi},$$
$$\dot{v}_x = w_x, \qquad \dot{v}_y = w_y. \tag{1.1.17}$$

Here, φ is the angle of the deviation of the pendulum from the vertical, w_x and w_y are control variables. Based on the characteristic quantities l_* and $\nu_* = (g/l_*)^{1/2}$ of

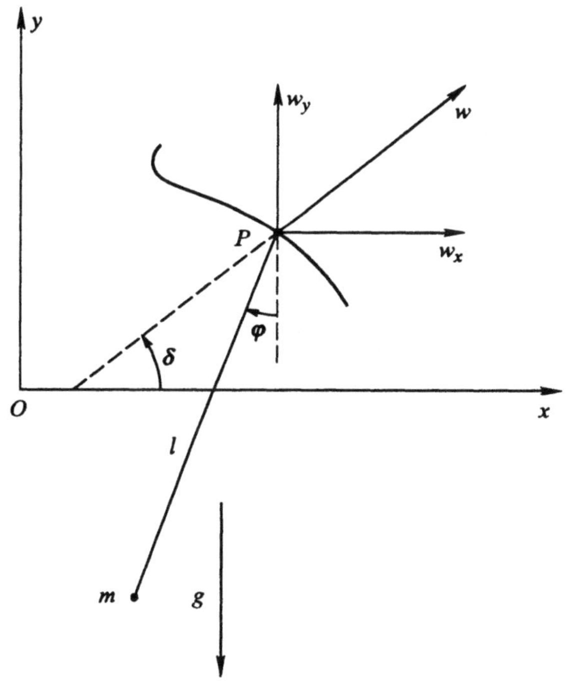

Fig. 1.2.

system (17), dimensionless variables are introduced. As a result, we have

$$\ddot{\varphi} + \sigma^{-1}\sin\varphi = -\sigma^{-1}(W_x \cos\varphi + W_y \sin\varphi) - 2\dot{\sigma}\sigma^{-1}\dot{\varphi}$$
$$(t_* = \nu_* t,\ W_{x,y} = w_{x,y} g^{-1},\ \sigma = l l_*^{-1}). \tag{1.1.18}$$

We suppose that $\sigma = \sigma(\tau)$, $\tau = \varepsilon t$, the amplitude of oscillations and controlled variables being assumed small, i.e.

$$\varphi, \dot{\varphi} \sim \varepsilon^{\alpha}, \quad \alpha > 0; \qquad W_{x,y} \sim \varepsilon^{\alpha_{x,y}}, \quad \alpha_{x,y} > 0.$$

For different values of the exponents α and $\alpha_{x,y}$, one can obtain the quasilinear model of the pendulum oscillations

$$\ddot{\varphi} + \nu^2(\tau)\varphi = \varepsilon f(\varphi, \dot{\varphi}, \sigma, \sigma', u_x, u_y) + \varepsilon^2 + \cdots,$$
$$\nu^2 = \sigma^{-1} > 0, \tag{1.1.19}$$

with different functions f, for example,

$$\text{a) } \alpha = 1, \quad \alpha_x = 2, \quad \alpha_y = 1;$$
$$f = -\sigma^{-1}(u_x + \varphi u_y) - 2\sigma'\sigma^{-1}\dot{\varphi};$$
$$\text{b) } \alpha = 1/2, \quad \alpha_x = 3/2, \quad \alpha_y = 1; \tag{1.1.20}$$
$$f = (1/6)\sigma^{-1}\varphi^3 - \sigma^{-1}(u_x + \varphi u_y) - 2\sigma'\sigma^{-1}\dot{\varphi}.$$

If the suspension point moves along a straight line inclined at the angle δ to the horizon ($W_x = W\cos\delta$, $W_y = W\sin\delta$), see Fig. 1.2, then, according to (18)–(20),

$$\ddot{\varphi} + \sigma^{-1}\sin\varphi = -\sigma^{-1}W\cos(\varphi - \delta) - 2\dot{\sigma}\sigma^{-1}\dot{\varphi};$$
$$\delta \neq \pm\pi/2: \quad \text{a) } \alpha_\delta = 2, \quad \text{b) } \alpha_\delta = 3/2; \tag{1.1.21}$$
$$\delta = \pm\pi/2: \quad \text{a) } \alpha_\delta = 1, \quad \text{b) } \alpha_\delta = 1.$$

The purposes of control may be the damping of oscillations ($\varphi = \dot{\varphi} \equiv 0$), the acceleration of the pendulum to a prescribed velocity v_x^*, v_y^* with the damping of relative oscillations, and others (see Chapter 5).

Note that, in the above examples (1) and (2), the length is the parameter slowly varying according to a prescribed law $l(\tau)$ of the type (7). However, in the quasilinear consideration, l cannot be treated as a variable depending on the controlled coordinates. Otherwise, the system is essentially nonlinear. Such systems are studied in Chapters 3–5.

(3) *Spatial oscillator.* Equations of controlled motion for a quasilinear oscillator can be written in the vector form

$$\dot{\mathbf{r}} = \mathbf{v}, \qquad m\dot{\mathbf{v}} = -k\mathbf{r} + \mathbf{f}(\mathbf{r}, \mathbf{v}) + \mathbf{F}, \qquad \mathbf{r}(t_0) = \mathbf{r}^0, \quad \mathbf{v}(t_0) = \mathbf{v}^0. \tag{1.1.22}$$

Here, $\mathbf{r}$ is a radius-vector of the particle of mass m, $\mathbf{v}$ is its velocity, k is a scalar coefficient of the restoring force, $\mathbf{f}$ is a perturbing force, and $\mathbf{F}$ is a control. If $\mathbf{f} = \mathbf{F} = \mathbf{0}$, then the system performs harmonic oscillations in a plane, and the trajectory of the particle is an ellipse, see Fig. 1.3. Let us introduce, in (22), the dimensionless variables by the formulae

$$t_* = \nu t, \qquad \nu^2 = k/m, \qquad \mathbf{r}_* = \mathbf{r}/r_*, \qquad \mathbf{v}_* = \mathbf{v}/(r_*\nu),$$
$$\varepsilon f_*(\mathbf{r}_*, \mathbf{v}_*) \equiv \mathbf{f}/(mr_*\nu^2), \qquad \varepsilon\mathbf{u} = \mathbf{F}/(mr_*\nu^2),$$

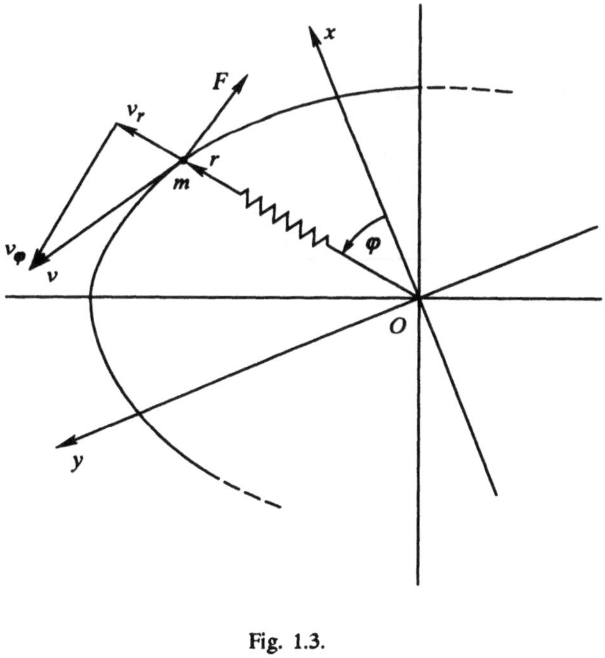

Fig. 1.3.

where r_* is a characteristic linear dimension of the ellipse. Then we obtain a single-frequency system of the special form (8): $n_x = 0$, $\nu_i = 1$. This system is reduced to the form (1.4) with $\varphi_i = t - t_0$ by the change

$$\mathbf{r} = \mathbf{a}\sin(t - t_0) + \mathbf{b}\cos(t - t_0), \qquad \mathbf{v} = \mathbf{a}\cos(t - t_0) - \mathbf{b}\sin(t - t_0),$$

where **a** and **b** are osculating variables. For the obtained system, it is possible to state various problems of control of slow variables by means of small control forces, see Section 1.4 below and Chapter 5.

(4) *"Small thrust" control in Kepler's problem.* The equations of motion for a particle in a gravitational field, provided the orbit is close to being circular (Fig. 1.4), are reduced to the standard form (14). The controlled motion in the orbital plane can be described by the equations [71, 72]

$$\begin{aligned} \dot r &= v_r, & m\dot v_r &= mv_\varphi^2 r^{-1} - mkr^{-2} + W_r + P_r, \\ \dot\varphi &= v_\varphi r^{-1}, & m\dot v_\varphi &= -mv_r v_\varphi r^{-1} + W_\varphi + P_\varphi. \end{aligned} \tag{1.1.23}$$

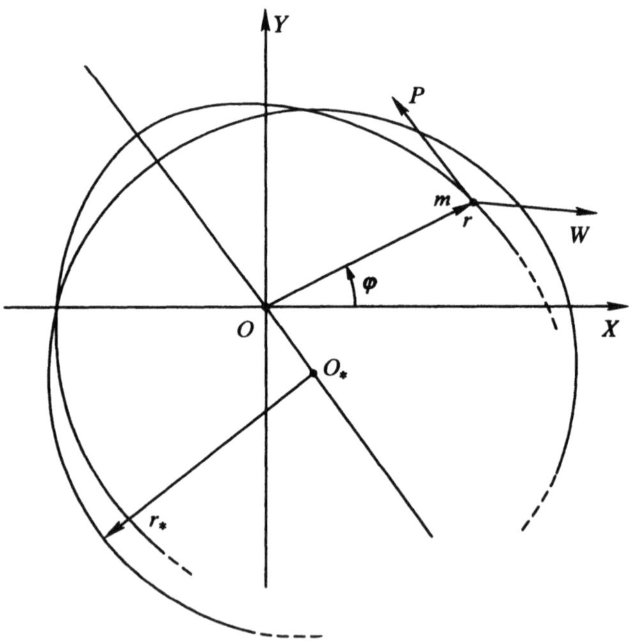

Fig. 1.4.

Here, r and φ are the polar coordinates, v_r and v_φ are the radial and transverse components of the velocity, respectively, m is the mass, k is the gravity coefficient, $W_{r,\varphi}$ are components of the perturbing force, and $P_{r,\varphi}$ are components of the thrust vector. Having introduced characteristic parameters of the motion r_* and $\nu^2 = kr_*^3$, we obtain the equations in dimensionless variables. Let the motion be close to a circular one ($\varepsilon \ll 1$):

$$r^0 = 1 + \varepsilon R^0, \qquad v_r^0 = \varepsilon V_r^0, \qquad W_r = \varepsilon^2 f_r, \qquad P_r = \varepsilon^2 u_r;$$
$$|\varphi^0| < \infty, \qquad v_\varphi^0 = 1 + \varepsilon V_\varphi, \qquad W_\varphi = \varepsilon^2 f_\varphi, \qquad P_\varphi = \varepsilon^2 u_\varphi.$$

Then, using the change

$$r = 1 + \varepsilon R, \qquad v_r = \varepsilon V_r, \qquad \varphi = \varphi, \qquad v_\varphi = 1 + \varepsilon V_\varphi$$

in (23) and cancelling $\varepsilon \neq 0$, we obtain a system of the type (1) with the generating solution of the form (3)

$$R = \alpha \sin(t - t_0) + \beta \cos(t - t_0) + 2c, \qquad \varphi = (t - t_0) + \gamma,$$
$$V_r = \dot{R}, \qquad V_\varphi = (1/2)(\ddot{R} + R); \qquad \alpha, \beta, c, \gamma = \text{const}.$$

The connection of the parameters α, β, c, and γ with the elements of the orbit, traditionally adopted in celestial mechanics, is simple; for instance, the eccentricity of the orbit is $e = \varepsilon(\alpha^2 + \beta^2)^{1/2} + \varepsilon^2 + \cdots$. The oscillating variables α, β, c, and γ are governed by equations of the type (14), provided $\varepsilon \neq 0$. For the obtained system, we can pose and solve various control problems; in particular, the time-optimal control problem (see Section 6.1).

(5) *Control of rotations of a rigid body with axial symmetry.* Consider controlled rotations of a rigid body with respect to a fixed point O (the centre of mass), see Fig. 1.5. They are governed by the vector equation [84, 91, 138, 154]

$$\dot{\mathbf{L}} + \omega \times \mathbf{L} = \mathbf{M}, \qquad \omega(t_0) = \omega^0 \quad (J_{1,2} = J \neq J_3). \tag{1.1.24}$$

Here, $\mathbf{L}$ is the angular momentum vector, J_i are principal moments of inertia relative to the centre of mass, and $\mathbf{M}$ is a control torque: $|\mathbf{M}| \leqslant M^0$. The control torque is

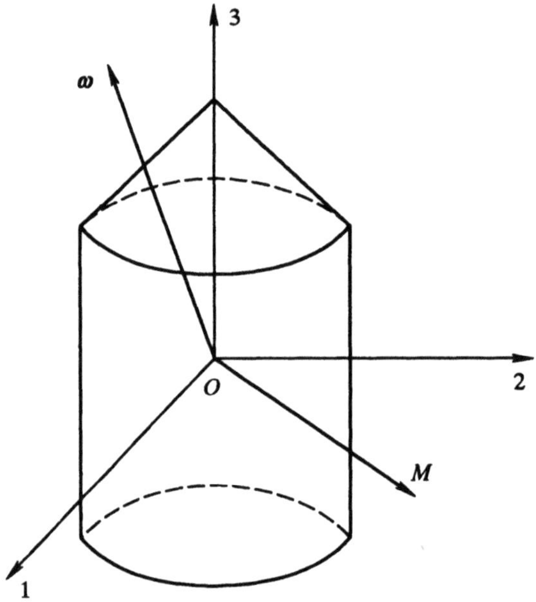

Fig. 1.5.

considered to be small, i.e. $\frac{1}{2}(\mathbf{L}^0 \cdot \omega^0) \gg M^0$. Then, in dimensionless variables, we have a system of the type (1)

$$\dot{\omega}_{1,2} \pm (d-1)\omega_3\omega_{2,1} = \varepsilon u_{1,2}, \qquad \omega = \varepsilon d^{-1} u_3$$
$$\left(t_* = \omega^0 t,\ \omega_* = \frac{\omega}{\omega^0},\ d = \frac{J_3}{J} \neq 1,\ \mathbf{u} = \frac{M}{M^0},\ \varepsilon = \frac{M^0}{J\omega^{02}} \ll 1\right). \tag{1.1.25}$$

Let the control u_3, governing an independent change of the axial component of the angular velocity ω_3, be given. In this case, $\omega_3(\tau)$ can be considered as a known function of τ. Then the change

$$\omega_1 = -\Omega \sin\psi, \qquad \omega_2 = \Omega\cos\psi, \quad \psi = \int_{t_0}^{t} \nu(\tau')\,\mathrm{d}t' + \psi^0,$$
$$\nu(\tau) = (d-1)\omega_3(\tau), \quad \tau = \varepsilon t,$$

leads to the equations of controlled motion in the standard form (6) for the equatorial component (ω_1, ω_2):

$$\begin{aligned}
&\dot{\Omega} = \varepsilon(-u_1 \sin\varphi + u_2 \cos\varphi), \qquad \Omega^0 = (\omega_1^{02} + \omega_2^{02})^{1/2},\\
&\dot{\varphi} = \nu(\tau) + \varepsilon\Omega^{-1}(u_1 \cos\psi + u_2 \sin\psi), \qquad\qquad (1.1.26)\\
&\sin\psi^0 = -\omega_1^0/\Omega^0, \qquad \cos\psi^0 = \omega_2^0/\Omega^0.
\end{aligned}$$

Having assumed

$$\begin{aligned}
&\omega_1 = a\cos\varphi - b\sin\varphi, \qquad \omega_2 = a\sin\varphi + b\cos\varphi,\\
&\varphi = \int_{t_0}^{t} \nu(\tau')\,dt',
\end{aligned}$$

we obtain a system of the form (12) or (14)

$$\begin{aligned}
&\dot{a} = \varepsilon(u_1 \cos\varphi + u_2 \sin\varphi) \qquad a^0 = \omega_1^0,\\
&\dot{b} = \varepsilon(-u_1 \sin\varphi + u_2 \cos\varphi) \qquad b^0 = \omega_2^0.
\end{aligned} \qquad (1.1.27)$$

For the system (26) or (27), the problem of quenching the equatorial component of the angular velocity is of practical interest, see Section 1.4. It is worth paying attention to the fact that Equation (26) for ψ has a singularity when $\Omega = 0$.

Thus, the examples above, which can be extended (see Chapters 5, 6), show the significance of the investigation of control problems for quasilinear oscillatory systems.

1.2. Method of Averaging in Optimal Control Problems for Motions of Quasilinear Oscillatory Systems for a Fixed Time Interval

1.2.1. STATEMENT OF THE PROBLEM

Here, we consider a controlled system of standard form (1.6). It is required to choose an admissible control $u \in U$ which brings the state point of the system (a, ψ) to the set defined by

$$M(a,\psi)\big|_T = 0, \qquad M = (M_1, \ldots, M_l), \quad 0 \leqslant l \leqslant n+1, \qquad (1.2.1)$$

at some specified time instant $t = T$, $T = \Theta\varepsilon^{-1}$, $\Theta = \text{const}$, and gives a minimum of the functional

$$J[u] = g(a, \psi)\big|_T + \varepsilon \int_0^T G(\tau, a, \psi, u)\, dt \longrightarrow \min_{u \in U}, \tag{1.2.2}$$

reflecting the control quality. The specified functions M, g, G, as well as the right-hand side of system (1.6), are assumed to be sufficiently smooth and 2π-periodic in ψ. The optimal control, the phase trajectory and the minimal value of the functional are to be found.

We suppose here that the optimal control problem (1.6), (1), (2) has a solution for $\varepsilon \in (0, \varepsilon_0]$. The existence of a solution can be proved under the rather general conditions of Filippov's lemma [62, 64, 65, 109, 155]. To find the solution, we use necessary optimality conditions in the form of Pontryagin's maximum principle: for all $t \in [t_0, T]$ the equality

$$\begin{aligned} &H(\tau, a, \psi, p, q, u^*, \varepsilon) = \max_{u \in U} H(\tau, a, \psi, p, q, u, \varepsilon), \\ &H \equiv \varepsilon(p \cdot f) + q(\nu + \varepsilon F) - \varepsilon G, \end{aligned} \tag{1.2.3}$$

holds. Here, H is the Hamiltonian function of the problem, p and q are adjoint variables corresponding to a and ψ, respectively. They are governed by the set of linear adjoint equations with transversality conditions at $t = T$:

$$\begin{aligned} \dot p &= -\frac{\partial H}{\partial a}, \qquad p(T) = \frac{\partial}{\partial a}[(\lambda \cdot M) - g]\Big|_T, \\ \dot q &= -\frac{\partial H}{\partial \psi}, \qquad q(T) = \frac{\partial}{\partial \psi}[(\lambda \cdot M) - g]\Big|_T. \end{aligned} \tag{1.2.4}$$

Here, λ is a constant l-vector of Lagrange's multipliers; the functions $a(t)$ and $\psi(t)$ are determined from (1.6), if $u = u^*(t)$; $a \in D_a \subset R^n$, $p \in R^n$, ψ, $q \in R^1$.

The optimal control to be found is assumed to be nonsingular, i.e. the function u^* is determined uniquely:

$$u^* = u(\tau, a, \psi, p, q), \qquad u^* = \arg\max_{u \in U} H. \tag{1.2.5}$$

We choose the function u^* to be 2π-periodic in ψ and consider it to be sufficiently smooth in all arguments within some admissible domain. Having substituted u^* from (5) into (3), we obtain an expression for the maximal value H^* of the Hamiltonian:

$$H^*(\tau, a, \psi, p, q, \varepsilon) = q\nu(\tau) + \varepsilon h^*(\tau, a, \psi, p, q). \tag{1.2.6}$$

Below, we use the differentiability of the function H^*. This function is differentiable if the set U does not depend on a and ψ, and the vector u^* is unique (the latter has been suggested). It results in (see [1, 28, 40, 43, 46, 110, 131])

$$\begin{aligned}
\partial H/\partial p\big|_{u^*} &= \partial H^*/\partial p = \varepsilon \partial h^*/\partial p = \varepsilon f^*,\\
\partial H/\partial q\big|_{u^*} &= \partial H^*/\partial q = \nu + \varepsilon \partial h^*/\partial q = \nu + \varepsilon F^*,\\
\partial H/\partial a\big|_{u^*} &= \partial H^*/\partial a = \varepsilon \partial h^*/\partial a,\\
\partial H/\partial \psi\big|_{u^*} &= \partial H^*/\partial \psi = \varepsilon \partial h^*/\partial \psi.
\end{aligned} \tag{1.2.7}$$

Here f^*, F^*, and G^* are known functions of τ, a, ψ, p, and q, obtained after substitution of u^* according to (5). As a result, we obtain the boundary-value problem of the maximum principle governed by the Hamiltonian standard system of the order $2(n+1)$ with rotating phase:

$$\begin{aligned}
&\dot{a} = \varepsilon f^*(\tau, a, \psi, p, q), && \dot{\psi} = \nu(\tau) + \varepsilon F^*(\tau, a, \psi, p, q),\\
&\dot{p} = -\varepsilon \frac{\partial}{\partial a}\left[(p \cdot f^*) + qF^* - G^*\right], && \dot{q} = -\varepsilon \frac{\partial}{\partial \psi}\left[(p \cdot f^*) + qF^* - G^*\right].
\end{aligned} \tag{1.2.8}$$

The initial and boundary conditions (the total number of which is $2(n+1)+l$) have the form

$$\begin{aligned}
&a(t_0) = a^0, \qquad \psi(t_0) = \psi^0, && M(a, \psi)\big|_T = 0,\\
&p(T) = \frac{\partial}{\partial a}\left[(\lambda \cdot M) - g\right]\Big|_T, && q(T) = \frac{\partial}{\partial \psi}\left[(\lambda \cdot M) - g\right]\Big|_T.
\end{aligned} \tag{1.2.9}$$

The solution a^*, ψ^*, p^*, and q^* of the boundary-value problem (8), (9) exists due to the assumptions made. If it is unique, then these functions determine the solution of the optimal control problem stated above. In the case of nonuniqueness, the solution is found from the minimality condition for the functional $J(2)$.

Let the functions a^*, ψ^*, p^*, and q^* be determined:

$$
\begin{aligned}
a^* &= a(t, t_0, a^0, \psi^0, \varepsilon), \qquad & \psi^* &= \psi(t, t_0, a^0, \psi^0, \varepsilon), \\
p^* &= p(t, t_0, a^0, \psi^0, \varepsilon), \qquad & q^* &= q(t, t_0, a^0, \psi^0, \varepsilon).
\end{aligned}
\tag{1.2.10}
$$

Then their substitution into (5) results in the open loop (programmed) optimal control u_p^*:

$$
u_p^*(t, \varepsilon) = u(\tau, a^*, \psi^*, p^*, q^*). \tag{1.2.11}
$$

If these functions are constructed for all initial values $t_0 \leqslant T$, $a^0 \in D_a$, and $\psi^0 \leqslant \psi_T < \infty$, then the optimal control can be also expressed in the feedback form

$$
u_s^*(t, a, \psi, \varepsilon) = u\big(\tau, a, \psi, p(t, t, a, \psi, \varepsilon), q(t, t, a, \psi, \varepsilon)\big). \tag{1.2.12}
$$

It is appropriate to mention here that one can successfully construct an exact analytical solution for the boundary-value problem (8), (9) only in a few particular cases. The numerical solution (see [43, 62, 88, 135]) causes significant difficulties, since the right-hand side of (8) is a rapidly oscillating function of ψ and since the time interval is large ($t - t_0 \sim \varepsilon^{-1}$). All this gives rise to a much of calculation. Hence, the development of asymptotic techniques which enables simplifying the procedure of a numerical solution, and which sometimes lead to the explicit solution of the boundary-value problem, seems to be important.

1.2.2. SEPARATION OF VARIABLES IN EQUATIONS OF THE BOUNDARY-VALUE PROBLEM

For this purpose, we apply the asymptotic method of averaging which enables eliminating the fast variable (the phase ψ) from the right-hand side of (8). As is known [35, 36, 38, 45, 69, 79, 100, 119, 120, 130, 137, 149, 150], the averaging method envisages a change of variables eliminating the phase from the right-hand side with a given accuracy in terms of ε; this accuracy being restricted only by the smoothness of the system with respect to slow variables.

A brief description of such a change is given below. To make the notation more compact, we combine the variables a, p, and q in one vector z of dimension $2n + 1$. Then Equations (8) are written as

$$
\dot{z} = \varepsilon Z(\tau, z, \psi), \qquad \dot{\psi} = \nu(\tau) + \varepsilon \Psi(\tau, z, \psi). \tag{1.2.13}
$$

We introduce the change of variables $(z, \psi) \to (\zeta, \varphi)$

$$
\begin{aligned}
z &= \zeta + \varepsilon v(\tau, \zeta, \varphi, \varepsilon) = \zeta + \varepsilon v_1(\tau, \zeta, \varphi) + \varepsilon^2 v_2 + \cdots + \varepsilon^j v_j + \cdots, \\
\psi &= \varphi + \varepsilon w(\tau, \zeta, \varphi, \varepsilon) = \varphi + \varepsilon w_1(\tau, \zeta, \varphi) + \varepsilon^2 w_2 + \cdots + \varepsilon^j w_j + \cdots,
\end{aligned}
\tag{1.2.14}
$$

so that the equations for ζ and φ do not contain the phase, i.e.

$$
\begin{aligned}
\dot{\zeta} &= \varepsilon Z(\tau, \zeta, \varepsilon) = \varepsilon Z_1(\tau, \zeta) + \varepsilon^2 Z_2 + \cdots + \varepsilon^j Z_j + \cdots, \\
\dot{\varphi} &= \nu(\tau) + \varepsilon \Phi(\tau, \zeta, \varepsilon) = \nu(\tau) + \varepsilon \Phi_1(\tau, \zeta) + \varepsilon^2 \Phi_2 + \cdots + \varepsilon^j \Phi_j + \cdots .
\end{aligned}
\tag{1.2.15}
$$

The functions v, w and Z, Φ in (14), (15) are to be determined, v and w being 2π-periodic in φ. For an exact determination of the change functions v and w (14), one has to know the general solution of the original system (8). The functions v and w are constructed approximately by means of the recursive calculation of the coefficients v_j and w_j, $j = 1, 2, \ldots$.

Differentiation of (14), according to system (13), and the substitution of the expansions (15) for $\dot{\zeta}$ and $\dot{\varphi}$ yield the sequence of sets of $2(n+1)$ partial differential equations. The recursive solution of these sets of equations, taking the periodicity in φ into account, results in the expressions for the coefficients v_j and w_j [1, 46, 38, 150]:

$$
\begin{aligned}
Z_1(\tau, \zeta) &= \langle Z(\tau, \zeta, \varphi) \rangle \equiv Z_0(\tau, \zeta), \\
\Phi_1(\tau, \zeta) &= \langle \Psi(\tau, \zeta, \varphi) \rangle \equiv \Psi_0(\tau, \zeta), \\
v_1(\tau, \zeta, \varphi) &= \frac{1}{\nu(\tau)} \int [Z(\tau, \zeta, \varphi) - Z_0(\tau, \zeta)] \, d\varphi + d_1(\tau, \zeta), \\
w_1(\tau, \zeta, \varphi) &= \frac{1}{\nu(\tau)} \int [\Psi(\tau, \zeta, \varphi) - \Psi_0(\tau, \zeta)] \, d\varphi + e_1(\tau, \zeta), \\
Z_2(\tau, \zeta) &= \left\langle \frac{\partial Z}{\partial \zeta} v_1 + \frac{\partial Z}{\partial \varphi} w_1 - \frac{\partial v_1}{\partial \tau} - \frac{\partial v_1}{\partial \zeta} Z_1 - \frac{\partial v_1}{\partial \varphi} \Phi_1 \right\rangle, \\
\Phi_2(\tau, \zeta) &= \left\langle \frac{\partial \Psi}{\partial \zeta} v_1 + \frac{\partial \Psi}{\partial \varphi} w_1 - \frac{\partial w_1}{\partial \tau} - \frac{\partial w_1}{\partial \zeta} Z_1 - \frac{\partial w_1}{\partial \varphi} \Phi_1 \right\rangle, \\
&\cdots\cdots\cdots\cdots\cdots\cdots\cdots\cdots
\end{aligned}
\tag{1.2.16}
$$

Here, broken brackets denote averaging with respect to φ, and the zero in the subscript indicates the zeroth Fourier harmonic. The functions Z_j and Φ_j are expressed through integrals, by using functions calculated from the previous steps, but v_j and w_j, in addition, require a knowledge of Z_j and Φ_j. It follows from (16) that the transformation is determined not uniquely but with an accuracy up to the arbitrary functions $d_j(\tau,\zeta)$ and $e_j(\tau,\zeta)$. This nonuniqueness disappears when returning to the original variables z and ψ [1, 38, 150].

The transformation of the j_*th order is determined by the j_*th term in expansions (14) and (15). As a result, the right-hand side of system (15) is constructed with the error $O(\varepsilon^{j_*+1})$; the integration is carried out in the slow time over a short interval: $\tau \in [\tau_0, \Theta]$. If the general solution of the j_*th approximation has been constructed, then its substitution into (14) gives the desired general solution with the same accuracy. It is worth noting that the calculation of the functions v_{j_*} and w_{j_*} is excessive, they do not increase the accuracy but give rise to the so-called "proved j_*th approximation" which is used for substantiating the method. As has been established in [150], the general solution of the j_*th approximation can be constructed by means of differentiation and integration on the basis of the solution of the first approximation (Sections 2.1, 2.2). After that, the integration constants are determined with the required accuracy from the initial and boundary conditions (9).

If we succeed in satisfying these conditions with the error $O(\varepsilon^{j_*})$, the construction of the solution of the optimal control problem of the j_*th approximation with respect to the control, the phase trajectory and the functional, will be completed. The detailed scheme of construction of the solution based on the canonical averaging technique, as well as its justification, is given in Chapter 2. The convergence, as $j_* \to \infty$, is not investigated [35, 38, 119, 149, 150].

1.2.3. THE FIRST APPROXIMATION

In applications, one can usually confine oneself to the first approximation of the solution of the optimal control problem. This approximation gives the qualitative behaviour of the system over an asymptotically large time interval $t - t_0 \sim \varepsilon^{-1}$, with the error $O(\varepsilon)$. Besides, this circumstance is often caused by the nonsmoothness of system (8).

The boundary-value problem for the first approximation has the form

$$\begin{aligned}
&\dot{\xi} = f_0(\tau,\xi,\eta,\beta), \qquad \dot{\varphi} = \varepsilon^{-1}\nu(\tau) + F_0(\tau,\xi,\eta,\beta),\\
&\dot{\eta} = -\frac{\partial}{\partial\xi}\left[(\eta\cdot f_0) + \beta F_0 - G_0\right], \qquad \beta = \text{const}, \quad (\dot{\ }) \equiv \left(\frac{\mathrm{d}}{\mathrm{d}\tau}\right),\\
&\xi(\tau_0) = a^0, \quad \varphi(\tau_0) = \psi^0, \quad M(\xi,\varphi)\big|_\Theta = 0, \quad \tau = \varepsilon t \in [\tau_0, \Theta], \qquad (1.2.17)\\
&\eta(\Theta) = \frac{\partial}{\partial\xi}\left[(\lambda\cdot M(\xi,\varphi)) - g(\xi,\varphi)\right]\Big|_\Theta,\\
&\beta = \frac{\partial}{\partial\varphi}\left[(\lambda\cdot M(\xi,\varphi)) - g(\xi,\varphi)\right]\Big|_\Theta.
\end{aligned}$$

System (17) is defined by the averaged Hamiltonian H_0 (6):

$$H_0\varepsilon^{-1} = \varepsilon^{-1}\beta\nu(\tau) + h_0(\tau, \xi, \eta, \beta).$$

Here only the solution of the set of $2n$ equations for ξ and η is to be constructed; the phase φ is found through simple integration. The system for ξ and η is also canonical, with the Hamiltonian $h_0(\tau, \xi, \eta, \beta)$. Integration must be carried out over a relatively small interval of the slow time $\tau \in [\tau_0, \Theta]$. Besides, $h_0 = \text{const}$ if h_0 does not depend on τ. The integration constants, and the unknown l-vector λ, are determined through the boundary conditions (17). If the root is unique and the corresponding Jacobian (see Section 2.2) does not vanish, then the obtained solution does determine the desired solution with the error $O(\varepsilon)$. In the case that the boundary-value problem admits a number of simple roots, the criterion for choosing the optimal solution is the functional $J(2)$ calculated with the error $O(\varepsilon)$:

$$J = J_0 + O(\varepsilon), \qquad J_0 = g(\xi, \varphi)\big|_\Theta + \int_{\tau_0}^{\Theta} G_0(\tau, \xi, \eta, \beta)\, d\tau. \tag{1.2.18}$$

1.2.4. EXAMPLES OF THE CONTROL OF THE AMPLITUDE OF OSCILLATIONS FOR A SYSTEM HAVING A SINGLE DEGREE OF FREEDOM

(1) *The statement of the problem for a quasilinear oscillator.* Consider controlled oscillations of the system (see examples (1), (2) from Section 1.1.3)

$$\ddot{x} + \nu^2(\tau)x = \varepsilon f(\tau, x, \dot{x}, u), \qquad f = f^0(\tau, x, \dot{x}) + D(\tau)u.$$

Using a change of variables of the form (1.3) we obtain a system of the type (1.6)

$$\begin{aligned}
&\dot{a} = \varepsilon\nu^{-1}(f - a\nu'\cos\psi)\cos\psi, \qquad a^0 = \left[x^{02} + \dot{x}^{02}\nu^{-2}(\tau_0)\right]^{1/2},\\
&\dot{\psi} = \nu - \varepsilon\nu^{-1}a^{-1}(f - a\nu'\cos\psi)\sin\psi,\\
&\sin\psi^0 = \frac{x^0}{a^0}, \qquad \cos\psi^0 = \frac{\dot{x}^0}{a^0\nu(\tau_0)}\\
&\big(x = a\sin\psi,\ \dot{x} = \nu(\tau)a\cos\psi,\ a > 0\big).
\end{aligned} \tag{1.2.19}$$

For system (19), we consider the following optimal control problems with the functionals and terminal conditions ($T = \Theta\varepsilon^{-1}$):

$$\begin{aligned}
&\text{A:}\quad J_A = \frac{1}{2}ka^2(T) + \varepsilon\int_{t_0}^{T} G(t)u^2\,dt, \qquad |u| < \infty, \quad G \geqslant G_0 > 0;\\
&\text{B:}\quad J_B = \varepsilon\int_{t_0}^{T} G(t)u^2\,dt, \qquad |u| < \infty, \quad a(T) = a^* \geqslant 0; \qquad (1.2.20)\\
&\text{C:}\quad J_C = \pm\frac{1}{2}a^2(T), \qquad |u| \leqslant u_0.
\end{aligned}$$

In the cases A and B, according to (3), (6), and (5), the Hamiltonian function H, H^*, and the corresponding control u^* are

$$\begin{aligned}
H &= q\nu + \varepsilon\nu^{-1}(v + Du)w - \varepsilon Gu^2,\\
v &= v(\tau, a, \psi) \equiv f^0 - a\nu'\cos\psi,\\
w &= w(a, \psi, p, q) \equiv p\cos\psi - qa^{-1}\sin\psi; \qquad (1.2.21)\\
H^* &= q\nu + \varepsilon\nu^{-1}vw + \frac{1}{4}\varepsilon\nu^{-2}D^2w^2G^{-1}, \qquad u^* = \frac{1}{2}\nu^{-1}DwG^{-1}.
\end{aligned}$$

(2) *Solution of the first approximation.* The averaging of the function H^* from (21) leads, according to (17), to the averaged boundary-value problems A and B

$$\begin{aligned}
&\dot{\xi} = \nu^{-1}f_c^0(\tau,\xi) - \frac{1}{2}\nu^{-1}\nu'\xi + \frac{1}{4}(D^2/(\nu^2 G))\eta, \qquad \xi(t_0) = a^0,\\
&\dot{\eta} = \left[\frac{1}{2}\nu^{-1}\nu' - \nu^{-1}\partial f_c^0/\partial\xi\right]\eta, \qquad \beta = 0, \qquad f_c^0 \equiv \langle f^0\cos\varphi\rangle, \qquad (1.2.22)\\
&\dot{\varphi} = \varepsilon^{-1}\nu(\tau) - \xi^{-1}f_s^0(\tau,\xi), \qquad \varphi(\tau_0) = \psi^0, \qquad f_s^0 \equiv \langle f^0\sin\varphi\rangle;\\
&\text{A:}\quad \eta(\Theta) = -k\xi(\Theta), \qquad \text{B:}\quad \xi(\Theta) = a^*.
\end{aligned}$$

In the general case, it is impossible to obtain the solution of Equations (22) for ξ and η in a closed form. They can be integrated, in particular, if h_0 does not depend on τ ($h_0 = \text{const}$) or f_c^0 is linear in ξ (the equation for η is linear and does not depend on ξ). The latter takes place for examples (1) and (2) from Section 1.1.3 (if $u_y \equiv 0$).

The solution of the boundary-value problems (22) in the special case

$$f^0 = -2\chi\dot{x} + \mu x^3, \qquad \nu, \chi, \mu = \text{const}, \qquad D = G = 1 \qquad (1.2.23)$$

leads to the expressions

$$\begin{aligned}&\xi(\tau)=\left[a^0-(\eta_*/(8\nu^2\chi))e^{-\chi\Theta}\right]e^{-\chi\tau}+(\eta_*/(8\nu^2\chi))e^{-\chi(\Theta-\tau)},\\&\eta(\tau)=\eta_* e^{-\chi(\Theta-\tau)},\qquad f_c^0=-\nu\chi\xi,\qquad f_s^0=(3/8)\mu\xi^3.\end{aligned}\tag{1.2.24}$$

The parameter η_* for the problem A is

$$\eta_*=-ka^0e^{-\chi\Theta}\left[1+(k/(8\nu^2\chi))\left(1-e^{-2\chi\Theta}\right)\right]^{-1}.\tag{1.2.25}$$

It stems from (25) that the quantity $|\eta_*|$ is bounded as $k\to\infty$ $(k>0)$. Hence, according to (22), $\xi(\Theta)\to 0$ as $k\to\infty$.

For the problem B, the value of the parameter η_* is given by

$$\eta_*=8\nu^2\chi\left(a^*-a^0e^{-\chi\Theta}\right)\left(1-e^{-2\chi\Theta}\right)^{-1}.\tag{1.2.26}$$

Choosing the quantity $k^*=k(a^*)$ in the problem A, one can achieve that $\xi(\Theta)=a^*$, if $k=k^*$.

The averaged phase φ of the first approximation is found by simple integration on the basis of (22) and (24):

$$\varphi(t,\varepsilon)=\psi^0+\int_0^t\Omega(\tau',\varepsilon)\,dt',\qquad \Omega=\nu-\varepsilon(3/8)\mu\nu^{-1}\xi^2(\tau).\tag{1.2.27}$$

Here, $\Omega(\tau,\varepsilon)$ has the sense of a perturbed frequency.

The open-loop optimal control for both problems A and B has the form (see (11) and (21))

$$u_p(t,\varepsilon)=(1/2)\nu^{-1}\eta_* e^{-\chi(\Theta-\tau)}\cos\varphi(t,\varepsilon).\tag{1.2.28}$$

According to (12), the feedback optimal control is

$$\begin{aligned}&u_s(\Theta-\tau,a,\psi)=(1/2)\eta_*\nu^{-1}e^{-\chi(\Theta-\tau)}\cos\psi,\\&\cos\psi=\dot{x}/(\nu a),\qquad a=\left(x^2+\dot{x}^2/\nu^2\right)^{1/2}.\end{aligned}\tag{1.2.29}$$

To obtain the optimal feedback control one has to make the changes $a_0 \to a$, $\Theta \to \Theta - \tau$ in expressions (25) and (26) for η_*. The minimal values of the functionals (20) in the first approximation are calculated by direct integration and are given by

$$
\begin{aligned}
J_A &= (1/2)\eta_*^2\left[k^{-1} + (1/(8\nu^2\chi))\left(1 - e^{-2\chi\Theta}\right)\right],\\
J_B &= \left(\eta_*^2/(16\nu^2\chi)\right)\left(1 - e^{-2\chi\Theta}\right), \qquad J_A\big|_{k^*} = J_B.
\end{aligned}
\tag{1.2.30}
$$

It is worth noting that for the fixed value of $a(T) = a^*$ (the condition B in (22)) the control u_s has a singularity as $\tau \to \Theta$, viz. $u_s \to \infty$ (see (26)). Proximity estimates for the controls become nonuniform. This irregular situation also takes place in the general case of a fixed terminal manifold $M(1)$.

For the problem C, the functions H, H^*, and u^* are found according to (3)–(6):

$$
\begin{aligned}
H &= q\nu + \varepsilon\nu^{-1}(v + Du)w, \qquad D \geqslant D_0 > 0,\\
H^* &= q\nu + \varepsilon\nu^{-1}\left(vw + Du_0|w|\right), \qquad u^* = u_0 \operatorname{sign} w, \quad w \neq 0.
\end{aligned}
\tag{1.2.31}
$$

If $\eta \not\equiv 0$ on some time interval, then the averaged boundary-value problem of the first approximation for ξ and η has the form

$$
\begin{aligned}
&\dot{\xi} = \nu^{-1}\left[f_c^0(\tau,\xi) - (1/2)\nu'\xi + (2/\pi)Du_0 \operatorname{sign}\eta\right],\\
&\xi(\tau_0) = a^0 > 0,\\
&\dot{\eta} = \nu\left((1/2)\nu' - \partial f_c^0/\partial\xi\right)\eta,\\
&\eta(\Theta) = \pm\xi(\Theta), \qquad \beta = 0.
\end{aligned}
\tag{1.2.32}
$$

As follows from Equation (32) for η, the function $\eta(\tau) \not\equiv 0$ does not change its sign, if $\xi(\Theta) \neq 0$. In this case,

$$
u_p(t,\varepsilon) = u_0 v(\tau) \operatorname{sign}\cos\varphi(t,\varepsilon), \qquad v = \operatorname{sign}\eta = \mp 1. \tag{1.2.33}
$$

To analyse the solution of the boundary-value problem (32) it is important to note that $f_{c,s}^0 = O(\xi)$ for all τ and sufficiently small ξ. This property follows from the requirement of smoothness in x and $\dot{x}$ for the function $f^0(\tau, x, \dot{x})$ and from the change (19). Thus, the sign of the right-hand side of Equation (32) coincides with the sign of η, provided ξ is sufficiently small. Consider, at first, the case of the sign "−" for J_C in (20), which corresponds to the maximization of the amplitude of the oscillations. Then, the boundary-value problem is satisfied for $\eta(\Theta) > 0$, since $\eta(\tau) > 0$ for $\tau \in [\tau_0, \Theta]$, and the right-hand side of the equation for ξ is positive, at least for small $\xi > 0$. It provides $\xi(\tau) > 0$ for all $\tau \in [\tau_o, \Theta]$. The boundary condition is satisfied by the normalization of $\eta(\tau)$. Thus, the averaged optimal amplitude $\xi(\tau)$ is determined as the solution of the initial-value problem for the first Equation (32) with $v = \operatorname{sign} \eta = +1$, i.e. $u_s = u_0 \operatorname{sign} \dot{x}$, $u_p = u_0 \operatorname{sign} \cos \varphi(t, \varepsilon)$.

The case of the sign "+" for J_C in (20) corresponds to the minimization of the amplitude. If $\xi(\tau) > 0$, $\tau \in [\tau_0, \Theta]$, then $u_p = -u_0 \operatorname{sign} \cos \varphi(t, \varepsilon)$, and $u_s = -u_0 \operatorname{sign} \dot{x}$ are open-loop and feedback optimal controls, respectively. If the solution $\xi(\tau^*) = 0$ at the time instant $\tau = \tau^* < \Theta$, then the singular optimal control takes place [28, 31, 43, 62, 65, 66, 105, 109, 155]. In this case, $\eta(\tau) \equiv 0$ on the part of the motion interval where optimal control is determined nonuniquely, the absolute minimum (zero) of the functional being achieved to a first approximation: $J_C = (1/2)\xi^2(\Theta) = 0$. For the function $v(\tau)$, $|v(\tau)| \leqslant 1$, in (33) one can take an arbitrary piecewise constant (bang-bang) function such that $\xi(\tau) \geqslant 0$, $\xi(\Theta) = 0$, for instance, $v(\tau) = -1$, $\tau \in [\tau_0, \tau^*]$; $v(\tau) \equiv 0$, $\tau \in (\tau^*, \Theta]$. The fulfillment of the above conditions for each current time instant determines the control of the first approximation in the feedback form.

The solution of optimal control problems for quasilinear oscillations in mechanical systems is given in Section 1.4 and in Chapters 5, 6.

1.3. The Asymptotic Solution of Optimal Control Problems with Non-Fixed Time

1.3.1. STATEMENT OF THE PROBLEM ANALOGOUS TO THE MINIMUM-TIME PROBLEM

As was done in Section 1.2, we consider here standard controlled system with a rotating phase (1.6). However, the time of the process termination is not specified but is to be chosen from the condition of achieving the set (τ is an argument)

$$N(\tau, a, \psi)\big|_T = 0, \qquad N = (N_1, \ldots, N_r), \qquad 0 \leqslant r \leqslant n+1, \tag{1.3.1}$$

by the state point (a, ψ).

For the sake of compactness we assume that the performance index (the functional to be minimized) of type (2.2) does not include the integral term [131]:

$$J[u] = g(\tau, a, \psi)_T \longrightarrow \min_u, \qquad u \in U. \tag{1.3.2}$$

This assumption does not restrict the statement of the problem, because the functional (2.2) can be reduced to the particular case (2), the dimension of the slow vector a increasing by unity. If $N = a - a^*$ and $g = \tau - \tau_0$, then relationships (1), (2) and (1.6) define a two-point time-optimal control problem in slow variables.

As in Section 1.2, it is natural to suppose that the functions N and g are 2π-periodic in ψ and sufficiently smooth in all arguments. Using the necessary optimality conditions for the control u^*, one can obtain the boundary-value problem of the maximum principle [131] similar to (2.9)

$$\begin{aligned}
&\dot{a} = \varepsilon f^*(\tau, a, \psi, p, q), \qquad \dot{p} = -\varepsilon \partial h^*/\partial a,\\
&\dot{\psi} = \nu(\tau) + \varepsilon F^*(\tau, a, \psi, p, q), \qquad \dot{q} = -\varepsilon \partial h^*/\partial \psi,\\
&a(t_0) = a^0, \qquad \psi(t_0) = \psi^0, \qquad N(\tau, a, \psi)\big|_T = 0,\\
&p(T) = \frac{\partial}{\partial a}[(\lambda \cdot N) - g]\Big|_T, \qquad q(T) = \frac{\partial}{\partial \psi}[(\lambda \cdot N) - g]\Big|_T.
\end{aligned} \tag{1.3.3}$$

Here, the functions H^*, h^*, and the corresponding control u^* are determined by the relationships

$$\begin{aligned}
&q\nu + \varepsilon \max_{u \in U}\left[(p \cdot f) + qF\right] = H^* \equiv q\nu + \varepsilon h^*,\\
&u^* = u(\tau, a, \psi, p, q), \qquad u^* = \arg\max_{u \in U} h,
\end{aligned} \tag{1.3.4}$$

the fulfillment of the transversality condition for H^*

$$H^*\big|_T = \varepsilon \frac{\partial}{\partial \tau}[g - (\lambda \cdot N)]\Big|_T, \qquad T = \Theta\varepsilon^{-1}, \tag{1.3.5}$$

being necessary.

Thus, the boundary-value problem (3)–(5) contains $2(n+1) + r + 1$ unknown parameters: $2(n+1)$ integration constants for the system (3), the r-vector λ, and the time T, which are to satisfy the same number of initial and boundary conditions given above.

1.3.2. CONSTRUCTION OF AN APPROXIMATE SOLUTION

Using the technique developed in Sections 1.2.2 and 1.2.3 for fixed T, we construct an approximate solution of the boundary-value problem (3) with a prescribed accuracy with respect to ε. Then we use relationship (5) to determine the optimal value for T, $T = \Theta\varepsilon^{-1}$. However, the approximate analysis of the boundary conditions shows (see the example from Section 1.3.4) that the solution is essentially nonunique, since (due to (3)) the left-hand side of (5) is a rapidly oscillating function of Θ with an amplitude of order unity and an average of order ε. Moreover, it turns out that the points of local extrema are determined by oscillations of the functions N and g. This circumstance, as a rule, makes the asymptotic analysis of the control problem uninteresting and rather difficult.

Having analysed a number of applied problems, we come to the conclusion that it is natural to suppose independence or weak dependence of the functions N and g on the phase ψ, when considering problems of the control of oscillatory systems with small control forces. In this case, important from the viewpoint of applications, the construction of optimal control is significantly simplified and one can obtain comparatively simple sufficient conditions of local and global optimality of the approximate solution, though, as is established below, the solution of the boundary-value problem is also essentially nonunique.

Thus, we consider the boundary-value problem (3)–(5) under the simplifying assumption of the independence of the functions N and g of ψ:

$$N(\tau,a)\big|_T = 0, \qquad J[u] = g(\tau,a)\big|_T \longrightarrow \min_{u\in U}. \tag{1.3.6}$$

The solution of the averaged system (2.15) of the j_*th approximation is considered to have been constructed in the form

$$\begin{aligned} &\xi = \xi(\tau,\Theta,\xi^0,\eta_*,\beta,\varepsilon), \qquad \eta = \eta(\tau,\Theta,\xi^0,\eta_*,\varepsilon),\\ &\xi\big|_{\tau_0} = \xi^0, \qquad \eta\big|_\Theta = \eta_*, \qquad \beta = \text{const},\\ &\varphi = \varphi^0 + \varepsilon^{-1}\int_{\tau_0}^{\tau} \nu(\tau')\,\delta\tau' + \int_{\tau_0}^{\tau} \Phi(\tau',\xi,\eta,\beta,\varepsilon)\,d\tau'. \end{aligned} \tag{1.3.7}$$

Here, ξ^0, η_*, Θ, and φ^0 are unknown constants to be determined from the boundary conditions. Using (2.14) we represent the unknown variables of the original boundary value problem in the form

$$\begin{aligned} &a = \xi + \varepsilon A(\tau,\xi,\varphi,\eta,\beta,\varepsilon), \qquad \psi = \varphi + \varepsilon\Psi(\tau,\xi,\varphi,\eta,\beta,\varepsilon),\\ &p = \eta + \varepsilon P(\tau,\xi,\varphi,\eta,\beta,\varepsilon), \qquad q = \beta + \varepsilon Q(\tau,\xi,\varphi,\eta,\beta,\varepsilon), \end{aligned} \tag{1.3.8}$$

where A, Ψ, P, and Q are known 2π-periodic functions of φ, calculated with the required accuracy, the error being $O(\varepsilon^{j_*})$. Then we write down (with the same accuracy) the conditions for determining the unknown parameters ξ^0, φ^0, η_*, β, and Θ

$$\begin{aligned}
&\xi^0 + \varepsilon A\big|_{\tau_0} = a^0, \qquad \varphi^0 + \varepsilon\Psi\big|_{\tau_0} = \psi^0, \qquad N(\tau, \xi + \varepsilon A)_\Theta = 0,\\
&\eta_* + \varepsilon P\big|_\Theta = \frac{\partial}{\partial a}[(\lambda \cdot N) - g]\Big|_\Theta, \qquad \beta + \varepsilon Q\big|_\Theta = 0, \\
&\varepsilon^{-1} H^*\big|_\Theta \equiv ((\eta + \varepsilon P)\cdot f^*)\big|_\Theta = \frac{\partial}{\partial \tau}[g - (\lambda \cdot N)]\Big|_\Theta.
\end{aligned} \tag{1.3.9}$$

Provided (6), transversality conditions for q imply that $\beta = O(\varepsilon)$. The left-hand side of the latter of Equations (9) rapidly oscillates in Θ with a frequency of order $\nu\varepsilon^{-1}$ and the amplitude of order unity. Similarly to that mentioned above, it gives rise, in general, to a large number (of the order of ε^{-1}) of roots Θ_α, the distance between adjacent roots being of the order $O(\varepsilon)$.

Consider, first, the solution of the optimal control problem to a first approximation with respect to ε. Based on this (see Chapter 2), arbitrary approximations are constructed easily. As is mentioned in Section 1.2, the first approximation often turns out to be sufficient for applications.

1.3.3. THE SOLUTION OF THE FIRST APPROXIMATION

(1) *The properties of the boundary-value problem of the first approximation*. The relationships (3) to a first approximation with respect to ε have the form

$$\begin{aligned}
&\dot\xi = f_0(\tau, \xi, \eta), \qquad \dot\eta = -\partial(\eta \cdot f_0)/\partial\xi, \qquad \beta_0 = 0,\\
&\xi(\tau_0) = a^0, \qquad N(\tau, \xi)\big|_\Theta = 0, \qquad \eta_* = \frac{\partial}{\partial \xi}[(\lambda \cdot N) - g]\Big|_\Theta,\\
&\dot\varphi = \varepsilon^{-1}\nu(\tau) + F_0(\tau, \xi, \eta), \qquad \varphi(\tau_0) = \varphi^0, \qquad (\dot{}) \equiv (\mathrm{d}/\mathrm{d}\tau).
\end{aligned} \tag{1.3.10}$$

The functions f_0 and F_0 are defined similarly to (2.17), taking into account the estimate $\beta = O(\varepsilon)$. We suppose that, for a given Θ, the solution of the boundary-value problem (10) has been constructed (the dependence on a^0 and τ_0 is not indicated for conciseness):

$$\begin{aligned}
&\xi = \xi(\tau, \Theta), \qquad \eta = \eta(\tau, \Theta),\\
&\varphi = \psi^0 + \varepsilon^{-1}\int_{\tau_0}^{\tau} \nu(\tau')\,\mathrm{d}\tau' + \int_{\tau_0}^{\tau} F_0\,\mathrm{d}\tau', \qquad \lambda = \lambda(\Theta).
\end{aligned} \tag{1.3.11}$$

The unknown variables a, p, ψ, and q based on (2.14) (see Section 2.3) differ from the expressions (11) by amounts $O(\varepsilon)$ for $\tau \in [\tau_0, \Theta]$:

$$a = \xi + O(\varepsilon), \quad p = \eta + O(\varepsilon), \quad q = O(\varepsilon), \quad \psi = \varphi + O(\varepsilon). \tag{1.3.12}$$

The parameter Θ, yet unknown, is determined from the latter of relationships (9) after substitution of the expression (12):

$$(\eta \cdot f^*(\tau, \xi, \varphi, \eta, O))\big|_\Theta + \frac{\partial}{\partial \tau}[(\lambda(\Theta) \cdot N(\tau, \xi)) - g(\tau, \xi)]\Big|_\Theta = O(\varepsilon). \tag{1.3.13}$$

Having omitted the terms $O(\varepsilon)$, Equation (13) takes the form

$$(\eta \cdot f_0)\big|_\Theta + \frac{\partial}{\partial \tau}[(\lambda(\Theta) \cdot N) - g]\Big|_\Theta = -(\eta \cdot (f^* - f_0))\big|_\Theta. \tag{1.3.14}$$

Relationship (14), as a rule, admits an asymptotically large number of roots $\{\Theta_\alpha\}$, the distance between adjacent roots being of the order ε. The sufficient condition for this assertion is the requirement that the averaged equation (14), i.e. with a zero right-hand side, has a simple real root $\Theta_0 > \tau_0$. Indeed, (11) and (14) imply that the right-hand side rapidly oscillates in Θ with a frequency of the order of ε^{-1} and a finite amplitude relative to the average value $O(\varepsilon)$ as $\varepsilon \to 0$. Due to this, the roots $\{\Theta_\alpha\}$, in some neighbourhood of the value $\Theta = \Theta_0$, develop the behaviour qualitatively shown in Fig. 1.6. The requirement for the root Θ_0 to be simple (nonmultiple) is essential for further considerations.

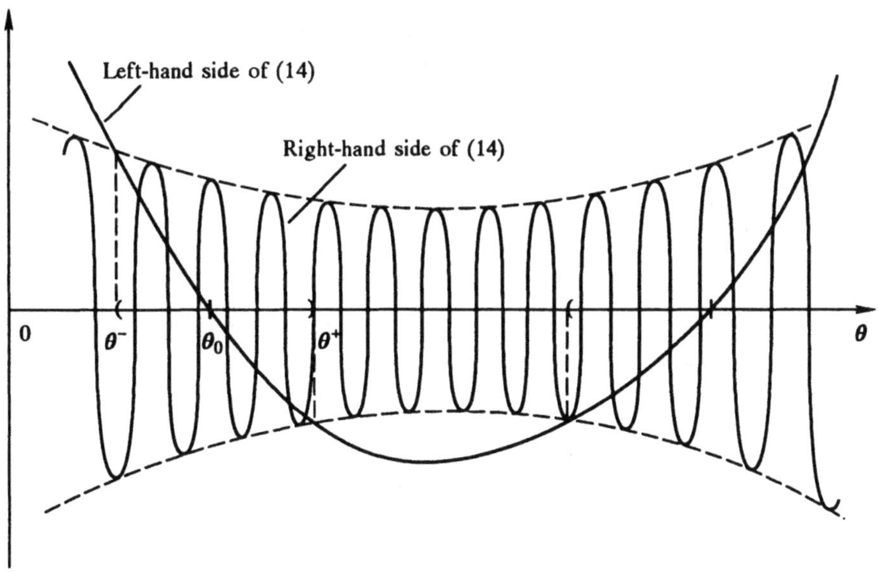

Fig. 1.6.

Since the root Θ_0 of the averaged equation to a first approximation is assumed to be simple, consideration of terms of order ε in the original equation (13) will lead to its shift by an amount $O(\varepsilon)$, and the roots $\{\Theta_\alpha\}$ will shift by amounts $O(\varepsilon^2)$. As a result, the established property of the roots $\{\Theta_\alpha\}$ leads to an important conclusion, namely concerning the essential nonuniqueness of the solution of the maximum principle boundary-value problem, corresponding to the optimal control problem analogous to the minimum-time one, and to the accumulation of the roots $\{\Theta_\alpha\}$ as $\varepsilon \to 0$. The latter is caused by the degeneration when $\varepsilon = 0$.

Thus, the roots $\{\Theta_\alpha\}$ are determined with an accuracy sufficient for the construction to the first approximation, the error being $O(\varepsilon^2)$ if the terms $O(\varepsilon)$ are omitted in (13), i.e. if Equation (14) is considered. The optimal value $\Theta^* \in \{\Theta_\alpha\}$ and the optimal solution of the boundary-value problem can be found from the minimality condition for the functional to a first approximation

$$J_0(\Theta) = g\big(\Theta, \xi(\Theta, \Theta)\big) \longrightarrow \min_{\Theta}, \qquad \Theta \in \{\Theta_\alpha\}. \tag{1.3.15}$$

(2) *The determination of the parameter* Θ. It is sufficient to determine the parameter Θ with the error $O(\varepsilon)$. This does not lead to a loss of accuracy of the solution of the boundary-value problem with respect to slow variables. In this case, we may consider the admissible set for Θ as being continuous, since there are exact roots in the ε-neighbourhood of any value of Θ. Due to this, problem (15) reduces to a determination of the boundaries Θ^- and Θ^+ for the corresponding interval $I_\Theta = (\Theta^-, \Theta^+)$ and to the minimization of the function $J_0(\Theta)$, $\Theta \in I_\Theta$. This minimization can be achieved by using well-known methods, both analytical and numerical [43, 63]. It is worth noting that the set I_Θ may consist of several intervals (Θ_s^-, Θ_s^+), and

$$I_\Theta = \bigcup_{s=1}^{S} (\Theta_s^-, \Theta_s^+).$$

For a wide class of control problems, the optimal value of Θ is determined analytically. Equation (14) is rewritten as

$$(\eta \cdot f_0)\big|_\Theta + \frac{\partial}{\partial \tau}\big[(\lambda \cdot N) - g\big]\Big|_\Theta + \nu(\Theta)\varkappa = 0. \tag{1.3.16}$$

Here, $\varkappa = \varepsilon^{-1}\beta$ is a parameter whose values $\varkappa \in I_\varkappa$ when $\Theta \in I_\Theta$.

Since the left-hand side of (14) vanishes when $\Theta = \Theta_0 \in I_\Theta$, the following inclusion holds: $\varkappa = 0 \in I_\varkappa$. Then, due to the simplicity of each root and the inequality $\nu(\Theta_0) \geqslant 0$, the one-to-one correspondence between Θ and $\varkappa$ follows from the implicit function theorem: $\Theta = \Theta(\varkappa)$ and $\varkappa = \varkappa(\Theta)$.

Provided the dependence $\Theta = \Theta(\varkappa)$ has been established, the problem of finding the minimum for the function $J_0[\varkappa]$ may be stated as:

$$J_0[\varkappa] = g(\tau, \xi)\big|_{\Theta(\varkappa)} \longrightarrow \min_{\varkappa}, \qquad \varkappa \in I_\varkappa. \tag{1.3.17}$$

LEMMA 1.1. *If the smoothness conditions are fulfilled for the function* $J_0[\varkappa]$, $\varkappa \in I_\varkappa$, *then the point* $\varkappa = 0$ *is suspected to be extremal. If* $\Theta'(0) > 0$, *then* $\varkappa = 0$ *and* $\Theta(0) = \Theta_0$ *are the points of local minimum; if* $\Theta'(\varkappa) > 0$ *for all* $\varkappa \in I_\varkappa$, *then* $\varkappa = 0$ *defines the global minimum for* J_0.

The function $J_0[\varkappa]$ is considered to be continuously differentiable due to proper smoothness assumptions for the functions f, N, and g. Hence, the necessary condition of the minimum is represented in the form

$$J_0'[\varkappa] = \left[\frac{\partial g}{\partial \tau} + \left(\frac{\partial g}{\partial \xi} \cdot \left(f_0 + \frac{\partial \xi}{\partial \theta}\right)\right)\right]\Bigg|_{\Theta(\varkappa)} \Theta'(\varkappa) = 0. \tag{1.3.18}$$

Here, $\dot{\xi}(\tau, \Theta)$ is replaced with $f_0(\tau, \xi, \eta)$, according to (10).

Substituting the expression $(\partial/\partial\xi)(\lambda \cdot N)|_\Theta - \eta_*$ into (18) instead of $(\partial g/\partial \xi)|_\Theta$, according to (10), and then instead of $(\eta \cdot f_0)|_\Theta$ its expression from (16), we find the derivative $\mathrm{d}g_*/\mathrm{d}\Theta$ in the form

$$\frac{\mathrm{d}g_*}{\mathrm{d}\Theta} = \left(\lambda \cdot \frac{\mathrm{d}N_*}{\mathrm{d}\Theta}\right) + \nu(\Theta)\varkappa \qquad \left(g_* \equiv g\big|_\Theta,\ \eta_* = \eta\big|_\Theta,\ N_* = N\big|_\Theta\right).$$

The vector function $N_* \equiv 0$ with respect to Θ, and, hence, $N_*' = 0$. As a result, for the derivative to be found, the following expression is valid

$$J_0[\varkappa] = \Theta'(\varkappa)\nu\big(\Theta(\varkappa)\big)\varkappa = 0. \tag{1.3.19}$$

It stems from (19) that the point $\varkappa = 0$ gives the local minimum if $\Theta'(0) > 0$, and the quantity $\Theta(0) = \Theta_0$ defined above is the optimal value for Θ. If $\Theta'(\varkappa) > 0$ for all $\varkappa \in I_\varkappa$, then Θ_0 is the point of global minimum for $J_0(\Theta)$.

Using (19), one can also write down the condition of the global minimum for the function $J_0[\varkappa]$ at the point $\varkappa = 0$ as

$$J_0[\varkappa] - J[0] = \int_0^\varkappa \nu\big(\Theta(\chi)\big)\Theta'(\chi)\chi \,\mathrm{d}\chi \geqslant 0, \qquad \varkappa \in I_\varkappa. \tag{1.3.20}$$

Having used the expression for $\nu(\Theta(\varkappa))\varkappa$ from (16), it is possible to obtain the optimality condition for $\Theta = \Theta_0$ in the alternative form:

$$J_0(\Theta) - J_0(\Theta_0) = \int_{\Theta_0}^{\Theta} \rho(\theta)\,\mathrm{d}\theta \geqslant 0, \qquad \Theta \in I_\Theta,$$
$$\rho(\Theta) = \frac{\partial}{\partial\tau}[g - (\lambda \cdot N)]\Big|_\Theta - (\eta \cdot f_0)|_\Theta. \tag{1.3.21}$$

Verification of these inequalities virtually completes the construction of the solution for the optimal control problem (1.6), (6), in the first approximation with respect to the parameter $\varepsilon > 0$. The accuracy is understood here in terms of the functional and boundary conditions. A detailed investigation of the boundary-value problem of the maximum principle, the construction of higher approximations and a discussion of the issues concerning the substantiation of proximity are given in Chapter 2.

(3) *Remarks and conclusions.* Thus, the algorithm of the asymptotic solution of the optimal control problem reduces to the solution of the boundary-value problem for a Hamiltonian system with respect to $2n$ variables ξ and η on a bounded interval of the slow variable $\tau \in [\tau_0, \Theta]$. The minimal value of the functional J (6) with an error $O(\varepsilon)$ equals J_0 from (15), if $\Theta = \Theta_0$, or (16), if $\varkappa = 0$, provided the conditions of Lemma 1.1 are fulfilled. The optimal control is determined by substitution of the obtained expressions for ξ, φ, and η into (4):

$$u_p = u(\tau, \xi, \varphi, \eta, 0), \qquad u_s = u\big(\tau, a, \psi, \eta(\tau, a), 0\big). \tag{1.3.22}$$

As has been mentioned in Section 1.2.3, the Hamiltonian $h_0 = (\eta \cdot f_0) = \text{const}$ is conserved, if f_0 does not depend on τ. The boundary-value problem for the system with a single degree of freedom in this case reduces to quadratures. If, in addition, the functions g and N do not depend on τ, then Θ_0 is a root of the equation $h_0 = 0$.

In the construction of optimal controls in Sections 1.2 and 1.3, we have supposed that u^* from (2.5) or (4) is uniquely determined from condition (2.3) or (4) and is a sufficiently smooth function with respect to slow variables. This enables us to apply the well-known scheme of averaging with respect to the phase ψ, which gives rise to a significant simplification of the solution of the maximum principle boundary-value problem. Optimal control problems with constraints on the control function $u \in U$, where U is a bounded set with a nonsmooth boundary, e.g.,

$$U = \big\{u\colon\ |u_i| \leqslant u_i^0,\ u_i^0 > 0,\ i = 1, \ldots, m\big\},$$

give, as a rule, discontinuous (bang-bang in particular) functions of time, state and adjoint variables. The values of these functions at discontinuity points are then determined nonuniquely. The presence of discontinuities in the right-hand sides of the equations causes difficulties for using the averaging method, see [1, 35, 38, 60, 104, 128, 149, 150]. This significantly complicates the construction of the approximate solution for the boundary-value problem of the maximum principle and the optimal control, as well as the proximity estimation for the formal constructions. The technique given in Sections 1.2 and 1.3 can be used formally to construct the first approximation for problems with discontinuous controls also. The basic difficulty arising in such cases is connected with the possibility of singular or sliding control modes, which needs special consideration in each particular case.

One example has been considered in Section 1.2.4, item (2). Asymptotic solutions for a number of quasilinear problems with bang-bang controls are given below in Section 1.3.5 and in Chapter 5. The features of using the averaging method in irregular cases are discussed in Sections 2.5 and 2.6.

1.3.4. AN ILLUSTRATIVE EXAMPLE

Consider a simple model problem which admits exact integration and the analysis of typical situations described in Sections 1.3.2 and 1.3.3:

$$\begin{aligned}
&\dot{x} = \varepsilon u \sin\psi, \qquad \dot{z} = \varepsilon k^2 + (1/2)\varepsilon u^2, \qquad \dot{\psi} = 1, \qquad t \in [0, T],\\
&x(0) = z(0) = \psi(0) = 0, \qquad N(x, \psi) = x - \alpha(\psi),\\
&\alpha(\psi + 2\pi) \equiv \alpha(\psi); \qquad J[u] = z(T), \qquad |u| < \infty.
\end{aligned}$$

Here, εk^2 is the weighting coefficient, reflecting "the penalty for the time expenditure", and the second term in the expression for $\dot{z}$ describes the consumption of control resources. According to Section 1.3.1, we write down the boundary-value problem of the maximum principle. Its solution is

$$\begin{aligned}
&x = \frac{1}{2} p_x \left(\tau - \frac{\varepsilon}{2} \sin 2\frac{\tau}{\varepsilon}\right), \qquad z = k^2 \tau + \frac{1}{2} p_x x(\tau, \varepsilon),\\
&\psi = t = \frac{\tau}{\varepsilon}, \qquad p_x = 2\frac{\alpha(\Theta/\varepsilon)}{x(\Theta, \varepsilon)}, \qquad p_z = -1,\\
&q(\tau, \varepsilon) = \frac{\varepsilon}{2} p_x^2 \left(\sin^2 \frac{\Theta}{\varepsilon} - \sin^2 \frac{\tau}{\varepsilon}\right) + q(\Theta, \varepsilon), \qquad u = p_x \sin\psi,\\
&q(\Theta, \varepsilon) = -\frac{\alpha^{2\prime}(\Theta/\varepsilon)}{2x(\Theta, \varepsilon)} p_x, \qquad J(\Theta, \varepsilon) = k^2 \Theta + \frac{1}{2} p_x x(\Theta, \varepsilon).
\end{aligned}$$

Here, p_x, p_z, and q are the adjoint variables corresponding to x, z, and ψ, respectively.

The transversality condition (5) defining Θ has the form

$$H^* = H^*\big|_\Theta = -\frac{\alpha^{2'}(\theta/\varepsilon)}{2x(\Theta,\varepsilon)}p_x + \frac{\varepsilon}{2}\frac{\alpha^2(\Theta/\varepsilon)}{x^2(\Theta,\varepsilon)}\sin^2\frac{\Theta}{\varepsilon}p_x^2 - \varepsilon k^2 = 0.$$

It follows from here that, to a first approximation with respect to ε, the roots Θ_α correspond to the zero points of the derivative of the periodic function $\alpha^2(\Theta/\varepsilon)$, i.e. $\{\Theta_\alpha\} = \operatorname{Arg}\alpha^{2'}(\Theta/\varepsilon)$, and form a sequence which becomes dense and covers the whole number axis, as $\varepsilon \to 0$. Minimization of $J\{\Theta,\varepsilon\}$ over $\{\Theta_\alpha\}$ is an uninteresting and difficult problem. It is caused by fast oscillation of p_x^2 in Θ due to a periodic dependence of α on ψ.

Corresponding to the remarks of Section 1.3.2, we investigate below the case $\alpha = \text{const}$ (without loss of generality it is assumed that $\alpha = 1$). The condition (5) defining $\{\Theta_\alpha\}$ is rewritten in the form ($\alpha^{2'} \equiv 0$), see (13):

$$\frac{1}{\Theta^2} - k^2 = \frac{1}{\Theta^2}\cos 2\frac{\Theta}{\varepsilon} - 2\varepsilon\frac{\sin^2(\Theta/\varepsilon)\sin(2\Theta/\varepsilon)(\Theta - (\varepsilon/4)\sin 2\Theta/\varepsilon)}{\Theta^2(\Theta - (\varepsilon/2)\sin 2\Theta/\varepsilon)^2}.$$

The left- and right-hand sides of this relationship are shown in Fig. 1.7(a) for $\Theta > 0$ ignoring small terms $O(\varepsilon)$. This picture qualitatively coincides with that given in

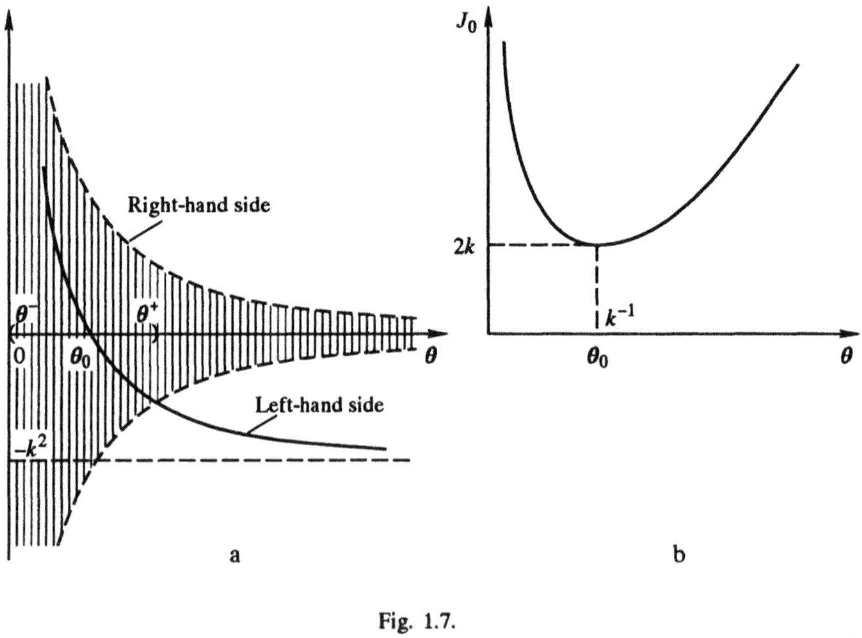

Fig. 1.7.

Fig. 1.6. Here, $\Theta_0 = k^{-1}$, $\Theta^- = 0$, $\Theta^+ = \sqrt{2}k^{-1} > \Theta_0$; hence, $\Theta, \Theta_0 \in (\Theta^-, \Theta^+)$. The functional J is given by the expression

$$J(\Theta,\varepsilon) = z\left(\frac{\Theta}{\varepsilon}\right) = k^2\Theta + \frac{1}{\Theta} + \frac{\varepsilon}{2}\,\frac{\sin 2\Theta/\varepsilon}{\Theta(\Theta - (\varepsilon/2)\sin 2\Theta/\varepsilon)} \equiv J_0(\Theta) + \varepsilon J_1(\Theta, \Theta/\varepsilon),$$

which implies that $\Theta = \Theta_0$ is a minimum point for the functional of the first approximation: $J_{0\,\min} = J_0(\Theta_0) = 2k$. It is appropriate to mention here that the addition of the quantity $O(\varepsilon^d)$, $d \geqslant \frac{1}{2}$, to Θ_0 does not change the value of J_0 in the first approximation. Indeed, $J_0(\Theta) = J_0(\Theta_0) + O(\varepsilon^{2d})$, provided $\Theta = k^{-1} + O(\varepsilon^d)$. Hence, for an arbitrary nonsmooth (with respect to Θ) function $J_1(\Theta, \Theta/\varepsilon)$, one cannot guarantee that the value Θ found from the minimality condition of $J_0(\Theta)$ lies within an ε-neighbourhood of the exact optimal value. But this value is ε-close in terms of the functional J. However, in the case examined, the obtained expression for $J_1(\Theta, \Theta/\varepsilon)$ implies that the minimum point $\Theta^*(\varepsilon)$ for the exact functional $J(\Theta, \varepsilon)$ lies within an ε-neighbourhood of Θ_0: $\Theta^* = \Theta_0 + \varepsilon\Theta_1$. This assertion stems from the geometrical fact, that the envelope of the function $J_1(\Theta, y)$ with respect to y is smooth in $\Theta, \Theta \in (\Theta^-, \Theta^+)$. The qualitative behaviour of $J(\Theta, \varepsilon)$ and $J_0(\Theta)$ is shown in Fig. 1.7(b).

Analogously, one can establish the ε-proximity of the point $\Theta = \Theta_0$ to the exact value $\Theta^*(\varepsilon)$ for the general system (1.6), (6) in the case that it possesses sufficient smoothness.

1.3.5. CONTROL OF OSCILLATIONS OF THE SYSTEM WITH A SINGLE DEGREE OF FREEDOM

(1) *Statement of the problem of the time-optimal type.* For the system (2.19), the function N and the functional J from (6) are taken to have the form

$$\begin{aligned} &N = a - \alpha(\tau), \qquad b = \varepsilon\left[\gamma(\tau) + G(\tau)u^2\right], \\ &b(t_0) = 0, \qquad J[u] = b(T) \longrightarrow \min_u, \qquad |u| < \infty. \end{aligned} \tag{1.3.23}$$

Here, the time instant T is not fixed, γ takes account of the time expenditure, and Gu^2 characterizes the expenditure of control resources. Then the Hamiltonian function is written down, taking into account that the variable b is cyclic. The averaged system of the type (10) for a, ψ, p, and q coincides with (2.22); the initial and boundary conditions, according to (10), (16), are reduced to the form (ζ is the average of b):

$$\begin{aligned} &\xi(\tau_0) = a^0, \quad \xi(\Theta) = \alpha(\Theta), \quad \varphi(\tau_0) = \psi^0, \quad \zeta(\tau_0) = 0, \\ &\left[\frac{\eta}{\nu}\left(f_c^0 - \frac{\nu'}{2}\varepsilon\right) + \frac{D^2\eta^2}{4\nu^2 G}\right]\bigg|_\Theta - \left(\gamma + \frac{D^2\eta^2}{8\nu^2 G}\right)\bigg|_\Theta - \eta_*\alpha'(\Theta) + \nu(\Theta)\varkappa = 0. \end{aligned} \tag{1.3.24}$$

The last relationship (24) connects Θ and $\varkappa$; their optimal values are determined from the conditions

$$J_0(\Theta) = \zeta(\Theta) = \int_{\tau_0}^{\Theta} \left(\gamma + \frac{D^2\eta^2}{8\nu^2 G}\right) d\tau \longrightarrow \min_{\Theta}, \qquad \Theta \in I_\Theta;$$
$$J_0(\varkappa) = \zeta(\Theta(\varkappa)) = \int_{\tau_0}^{\Theta(\varkappa)} \left(\gamma + \frac{D^2\eta^2}{8\nu^2 G}\right) d\tau \longrightarrow \min_{\varkappa}, \qquad \varkappa \in I_\varkappa. \tag{1.3.25}$$

Then one has to integrate the system (2.22) with the specified Θ and, after that, to find the optimal value of Θ or $\varkappa$ from the conditions (25), taking (24) into account. To establish the local optimality of the point $\varkappa = 0$, it is sufficient to verify that the corresponding inequality is valid. This can be done by calculating the implicit derivative $\Theta'(\varkappa)$.

(2) *Construction of the optimal solution.* We consider a specific case for the system (2.19) of the form (2.23), provided $\alpha = a^* = \text{const} \geqslant 0$, taking into account the results of integration (see Section 1.2.4, item (2)) and, in particular, the expression (2.26) for η_*. Then we find the function $\eta_*(\varkappa)$ by solving Equation (24), which is quadratic in η_*. After that, from relationship (2.26), which can be reduced to the quadratic equation relative to the variable $z = e^{-\chi\Theta}$, the positive root is found for both branches of $\eta_*(\varkappa)$. Finally, we obtain two expressions for $z = z_{1,2} \leqslant 1$:

$$z_{1,2} = e^{-\chi\Theta_{1,2}} = a^0/(2c_{1,2}) + \left[\left(a^0/(2c_{1,2})\right)^2 + 1 - a^*/c_{1,2}\right]^{1/2},$$
$$c_{1,2}(\varkappa) = \frac{1}{2}\left[a^* \pm \left(a^{*2} + \frac{\gamma - \nu\varkappa}{2\nu^2\chi^2}\right)^{1/2}\right]. \tag{1.3.26}$$

Here, the sign "+" in $c_{1,2}$ means that $\eta_* > 0$, i.e. the problem of increasing the amplitude is solved ($a^* > a^0$), and the sign "−" corresponds to a decrease in the amplitude ($a^* < a^0$). Evidently, $z_{1,2} = 1$ if $a^* = a^0$, i.e. $\Theta_{1,2} = 0$. Formulae (26) define the dependence $\Theta(\varkappa)$ for admissible $\varkappa$ corresponding to non-negative expressions under the square root sign: according to (24), $\gamma - \nu\varkappa > 0$.

The value $\varkappa = 0$ is proved to correspond to the optimal solution. To verify this it is sufficient, according to (19) (or (20)) and (26), to verify the inequalities $dz_i/d\varkappa < 0$, $i = 1, 2$. Since $dc_1/d\varkappa < 0$ and $dc_2/d\varkappa > 0$, then $z_i' < 0$ if $dz_1/dc_1 > 0$ and $dz_2/dc_2 < 0$ for all admissible values of $\varkappa$ within a neighbourhood of $\varkappa = 0$. According to (26), the derivatives dz_i/dc_i are

$$\frac{dz_i}{dc_i} = \frac{B_i - \frac{1}{2}A_i^2 - A_i\left(\frac{1}{4}A_i^2 + 1 - B_i\right)^{1/2}}{2c_i\left(\frac{1}{4}A_i^2 + 1 - B_i\right)^{1/2}},$$
$$A_i = a^0/c_i, \qquad B_i = a^*/c_i, \qquad i = 1, 2. \tag{1.3.27}$$

Consider, at first, the case $a^* > a^0 > 0$; then $c_1 > a^*$, $0 < A_1 < B_1 \leqslant 1$. Hence, the derivative $\mathrm{d}z_1/\mathrm{d}c_1$ exists and is continuous with respect to A_1 and B_1, since the expression under the square root sign in (27) is positive. This derivative is positive, if $B_1 \leqslant 1$ and $A_1 \geqslant 0$. We can prove, using the rule of contraries, that the derivative does not vanish, if $B_1 > A_1$. Hence, due to the continuity, the inequality $\mathrm{d}z_1/\mathrm{d}c_1 \leqslant 0$ is also impossible. For the problem of the amplitude decrease ($a^* < a^0$), we have $c_2 < 0$, $-\infty < A_2 < B_2 < 0$. Analogously, it can be proved that the inequality $\mathrm{d}z_2/\mathrm{d}c_2 \geqslant 0$ does not hold in the domain $\gamma - \nu\varkappa > 0$. Thus, $\varkappa = 0$, $\Theta_0 = \Theta(0)$ are the optimal values of the parameters of the problem:

$$\Theta_{1,2}(0) = \frac{1}{\chi}\ln\frac{1}{z_{1,2}(0)},$$
$$c_{1,2}(0) = \frac{1}{2}\left[a^* \pm \left(a^* + \frac{\gamma}{2\nu^2\chi^2}\right)^{1/2}\right]. \tag{1.3.28}$$

Optimal controls u_p (2.28) or u_s (2.29) and the value of the functional J_B (2.30) are obtained after the substitution $\eta_* = \pm\sqrt{8\gamma}\,\nu$ is made.

(3) *Comparison with the bang-bang control.* Consider the time-optimal control problem for the system (2.19) and (2.23), provided the modulus of the control function u is restricted: $|u| \leqslant u_0$ (see the case C in (2.20)). Then, the application of the averaging method under the assumption $\chi = 0$ results in the boundary-value problem of the first approximation:

$$\dot{\xi} = \left(2u_0/(\pi\nu)\right)v(\tau), \qquad \xi(0) = a_0, \qquad \xi(\Theta) = a^*,$$
$$\left(2u_0/(\pi\nu)\right)|\eta| - 1 + \nu\varkappa = 0, \tag{1.3.29}$$
$$v = \operatorname{sign}\eta, \qquad \eta = \text{const} \neq 0.$$

The solution of problem (29) depends on the sign of η which is obviously $\operatorname{sign}(a^* - a^0)$. The last condition in (29) can always be satisfied by the choice of $\varkappa$. A similar situation also exists for many other problems with bang-bang controls, see Chapter 5. To a first approximation, the control, according to (22), is

$$\bar{u}_p = u_0 \operatorname{sign}\left[\left(a^* - a^0\right)\cos\varphi(t,\varepsilon)\right],$$
$$\bar{u}_s = u_0 \operatorname{sign}\left[\left(a^* - a\right)\cos\psi\right]. \tag{1.3.30}$$

Here, $\varphi(t,\varepsilon)$ is defined according to (27) and ξ varies linearly:

$$\xi(\tau) = a^0 + (a^* - a^0)\tau/\Theta_*,$$

where Θ_* is the optimal time: $\Theta_* = (\pi\nu/(2u_0))|a^* - a^0|$.

For comparison, we present here the solution of the optimal control problem which is obtained from the solution constructed in Section 1.3.5, items (1) and (2), as a limiting case under $\chi \to 0$:

$$\begin{aligned}\xi(\tau) &= \frac{1}{4}\nu^{-2}\eta_*\tau + a^0,\\ \eta(\tau) &= \eta_* = -4\nu^2(a^0 - a^*)\Theta_0,\\ \Theta_0 &= \nu|a^* - a^0|(2/\gamma)^{1/2}.\end{aligned} \tag{1.3.31}$$

We require that $\Theta_0 = \Theta_*$, which implies the relationship $u_0 = (\pi/2)(\gamma/2)^{1/2}$.

Then we compare the integral functionals characterizing the expenditure of control resources for both types of the obtained controls (2.28) and (30). In the case of the quadratic functional reflecting the energy consumption for control, the following relationships are obtained:

$$\begin{aligned}J_0^{(2)} &= \int_0^{\Theta_0} u_p^2\,d\tau = \nu(2\gamma)^{1/2}|a^* - a^0|,\\ \overline{J}_0^{(2)} &= \int_0^{\Theta_*} \bar{u}_p^2\,d\tau = \frac{\pi^2}{8}\nu(2\gamma)^{1/2}|a^* - a^0| = \frac{\pi^2}{8}J_0^{(2)} > J_0^{(2)}.\end{aligned} \tag{1.3.32}$$

It stems from (32) that the control, optimal with respect to the performance index of type (23), expends the integral quadratic control resource more economically, which seems to be rather evident. However, it is interesting to note that just the same relationship $\overline{J}_0^{(1)} = (\pi^2/8)J_0^{(1)}$ is obtained for functionals of the type

$$J_0^{(1)} = \int_0^{\Theta} |u_p|\,d\tau,$$

describing the expenditure of the impulse (working medium) of the control system. Thus, the smooth control corresponding to the quadratic performance index expends integral resources of control $J_0^{(1)}$, $J_0^{(2)}$ more economically. This apparently is true for all $J_0^{(\mu)}$, $0 < \mu < \infty$.

1.4. Examples of Control of Quasilinear Oscillations in Mechanical Systems

1.4.1. CONTROL OF A PENDULUM WITH VARIABLE LENGTH

Consider the time-optimal control problem for the pendulum-type system. The control is implemented by moving the suspension point along the horizontal guide, see Section 1.1.4, item (2) and Fig. 1.2. By the change (2.19), Equation (1.19), and relationships (1.20, b) with $u_y \equiv 0$, are reduced to the standard form:

$$\begin{aligned}
\dot a &= -\varepsilon\left(u - \frac{1}{6}a^3\sin^3\psi + \frac{3}{2}\frac{\sigma' a}{\sigma^{1/2}}\cos\psi\right)\frac{1}{\sigma^{1/2}}\cos\psi,\\
\dot\psi &= \frac{1}{\sigma^{1/2}} + \varepsilon\left(u - \frac{1}{6}a^3\sin^3\psi + \frac{3}{2}\frac{\sigma' a}{\sigma^{1/2}}\cos\psi\right)\frac{1}{a\sigma^{1/2}}\sin\psi,\\
N\big|_T &= (a-\alpha)\big|_T = 0, \quad J[u] = \Theta, \quad u_1 \leqslant u \leqslant u_2 \qquad (u \equiv u_x).
\end{aligned} \tag{1.4.1}$$

Based on Section 1.3.5, (3), we obtain the optimal control of the first approximation in the form of the bang-bang function

$$\begin{aligned}
u &= \frac{1}{2}(u_1+u_2) + \frac{1}{2}(u_2-u_1)v, \qquad |v| \leqslant 1,\\
v_p &= -\operatorname{sign}\left[(z_0 - z^*)\cos\varphi(t,\varepsilon)\right], \qquad v_s = -\operatorname{sign}\left[(z-z^*)\dot\varphi\right],\\
z &= a\sigma^{3/4}, \qquad z^0 = a^0\sigma(0)^{3/4}, \qquad z^* = \alpha\sigma(\Theta)^{3/4}.
\end{aligned} \tag{1.4.2}$$

The equation of the first approximation for the amplitude ξ after substitution of v from (2) into (1) and averaging is integrated in closed form:

$$\xi(\tau) = a^0\sigma^{-3/4} - \frac{u_2-u_1}{\pi\sigma^{3/4}}\operatorname{sign}(z^0 - z^*)\int_0^T \sigma^{1/4}(\theta)\,d\theta. \tag{1.4.3}$$

The optimal (minimal) time Θ is determined from the condition $\xi(\Theta) = \alpha$ and is the positive root of the equation

$$\alpha\sigma^{3/4}(\Theta) - a^0 = -\frac{u_2-u_1}{\pi}\operatorname{sign}\left[z^0 - \alpha\sigma(\Theta)^{3/4}\right]\int_0^\Theta \sigma^{1/4}(\tau)\,d\tau. \tag{1.4.4}$$

It should be noted that, due to natural constraints on the relative length of the pendulum σ , e.g., $0 < \sigma_1 \leqslant \sigma \leqslant \sigma_2 < \infty$, the solution does not exist for all laws of length variation, provided the parameters of the problem a^0, α, and $u_{1,2}$ are given.

For instance, for quenching oscillations ($\alpha \simeq 0$) when the length is permanently decreasing ($\sigma' = -V,\ V > 0$), the solvability condition for Equation (4) is

$$a^0 \leqslant a^0_{\max} = \frac{4}{5\pi}\frac{u_2 - u_1}{V}, \qquad a^0 = \frac{4}{5\pi}\frac{u_2 - u_1}{V}\left[1 - (1 - V\Theta)^{5/4}\right]. \tag{1.4.5}$$

If a^0 is greater than $a^0_{\max}$, then the statement of the problem can be extended by admitting the motion with $\sigma \equiv \sigma_1$. Estimates and assertions similar to (5) may also be applied for $\sigma' = V$, $V = \text{const} > 0$.

1.4.2. CONTROL OF ROTATIONS OF A RIGID BODY WITH DYNAMICAL SYMMETRY

The minimum-time problem of quenching the equatorial component of the angular velocity (ω_1, ω_2), provided $\omega_3(\tau)$ is a known function ($|\omega_3| \geqslant \omega_{30} > 0$), is described by the set of equations (1.27) in osculating variables a and b and the relationships

$$a(T) = b(T) = 0, \qquad \Theta \longrightarrow \min_{u_{1,2}}, \qquad (u_1, u_2) \in U^{(2)}. \tag{1.4.6}$$

Below, we construct the solution for some specific control systems encountered in applications, see Fig. 1.8.

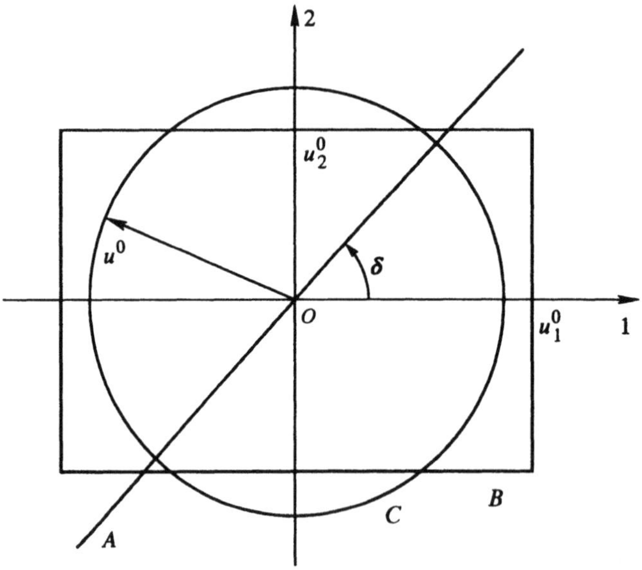

Fig. 1.8.

(A) Braking by means of a pair of engines rigidly attached to the body. In this case, the set $U^{(2)}$ degenerates in the segment, see Fig. 1.8, A:

$$U^{(2)} = \{u_1, u_2 \colon \ u_1 = u\cos\delta, \ u_2 = u\sin\delta, \ |u| \leqslant u^0, \ \delta = \text{const}\}. \tag{1.4.7}$$

To solve the problem (1.27), (6), and (7) we use the technique given in Section 1.3. Since the variables a and b are cyclic, the momenta (adjoint variables) p_a and p_b are constants ($p_a, p_b = \text{const}$). According to (3.4),

$$\begin{gathered} u = u^0 \operatorname{sign}\cos(\varphi - \delta + \gamma), \\ \cos\gamma = p_a/p, \qquad \sin\gamma = p_b/p, \qquad p = \left(p_a^2 + p_b^2\right)^{1/2}. \end{gathered} \tag{1.4.8}$$

Substitution of (8) into (1.27) and averaging with respect to φ leads to the equations of the first approximation for the averaged variables α and β:

$$\dot\alpha = \frac{2}{\pi}u^0\cos\gamma, \qquad \dot\beta = \frac{2}{\pi}u^0\sin\gamma, \qquad \gamma = \text{const}. \tag{1.4.9}$$

The solution of the boundary-value problem (9), (1.27), and (6) yields

$$\begin{gathered} \cos\gamma = -\frac{\omega_1^0}{\omega_\perp^0}, \quad \sin\gamma = -\frac{\omega_2^0}{\omega_\perp^0}, \quad \omega_\perp = \left(\omega_1^2 + \omega_2^2\right)^{1/2}, \\ \alpha(\tau) = \omega_1^0\left(1 - \frac{\tau}{\Theta}\right), \quad \beta(\tau) = \omega_2^0\left(1 - \frac{\tau}{\Theta}\right), \quad \Theta = \frac{\pi}{2}\frac{\omega_\perp^0}{u^0}. \end{gathered} \tag{1.4.10}$$

After substitution of the current values of variables into (10), the optimal control is expressed in the feedback form:

$$\begin{gathered} u_{1s} = -u^0\cos\delta \operatorname{sign}\omega_0, \qquad u_{2s} = -u^0\sin\delta\operatorname{sign}\omega_\delta, \\ \omega_\delta = \omega_1\cos\delta + \omega_2\sin\delta. \end{gathered} \tag{1.4.11}$$

It is interesting to note that, in the first approximation, both the phase trajectory and the minimal time do not depend on the parameter δ. It is the result of averaging over the axial rotation of the body.

(B) Quenching by means of torques directed along the axes fixed within the body, see Fig. 1.8, *B*:

$$U^{(2)} = \{u_1, u_2\colon\ |u_1| \leqslant u_1^0,\ |u_2| \leqslant u_2^0\}. \tag{1.4.12}$$

Similar to that explained above, we find the open-loop controls:

$$u_1 = u_1^0 \operatorname{sign}\cos(\varphi + \gamma), \qquad u_2 = u_2^0 \operatorname{sign}\sin(\varphi + \gamma). \tag{1.4.13}$$

After substitution of (13) into (1.27) and averaging with respect to φ, the boundary-value problem of the maximum principle takes the form considered above, provided the change $u^0 = u_1^0 + u_2^0$ has been made. The expressions for α, β, γ, and Θ also coincide with (10), and feedback controls $u_{1,2}$ have the form

$$u_{1s} = -u_1^0 \operatorname{sign}\omega_1, \qquad u_{2s} = -u_2^0 \operatorname{sign}\omega_2. \tag{1.4.14}$$

(C) Consider now a "rotatory pair" of engines, see Fig. 1.8, *C*:

$$U^{(2)} = \{u_1, u_2\colon\ u_1^2 + u_2^2 \leqslant u^{02}\}. \tag{1.4.15}$$

Then the expressions to be found are presented as

$$\begin{gathered} u_1 = -\frac{u^0\omega_1}{\omega_\perp}, \quad u_2 = -\frac{u^0\omega_2}{\omega_\perp}, \quad \Theta = \frac{\omega_\perp^0}{u^0}, \\ \omega_\perp = \omega_\perp^0\left(1 - \frac{T}{\Theta}\right), \quad \omega_1 = \frac{\omega_1^0}{\omega_\perp^0}\omega_\perp \sin\varphi, \quad \omega_2 = \frac{\omega_2^0}{\omega_\perp^0}\omega_\perp \cos\varphi. \end{gathered} \tag{1.4.16}$$

It is worth noting that the optimal solution (16) coincides with the exact one, i.e. it is valid for any $\varepsilon > 0$. A similar solution is obtained for the variable $u^0(t) \geqslant 0$; the control remaining the same and the optimal time being determined from the condition

$$\omega(T) = 0, \quad T > t_0; \qquad \omega_\perp(t) = \omega_\perp^0 - \varepsilon\int_{t_0}^{t} u^0(s)\,ds. \tag{1.4.17}$$

A more detailed consideration of such problems is given in Chapter 7.

1.4.3. CONTROL OF THE ENERGY OF QUASILINEAR OSCILLATIONS OF A PLANE OSCILLATOR

Here, we investigate the optimal control problem for a system of the type (1.22) from Section 1.1.3, item (3), for an asymptotically large time interval $T = \Theta\varepsilon^{-1}$. Parameters of the system are assumed to be constant. The equations of motion in the polar coordinates have the form

$$\begin{aligned} &\dot{r} = v_r, \qquad \dot{v}_r = v_\varphi^2 r^{-1} - r + \varepsilon u_r + \varepsilon f_r, \\ &\dot{\varphi} = v_\varphi r^{-1}, \qquad \dot{v}_\varphi = -v_r v_\varphi r^{-1} + \varepsilon u_\varphi + \varepsilon f_\varphi. \end{aligned} \tag{1.4.18}$$

Perturbing functions f_r and f_φ may depend on r, v_r, and v_φ and are 2π-periodic in φ. System (18) relates to the form (1.1) and has the single-frequency generating solution of the type (1.3). The full set of integrals for the unperturbed system is

$$\begin{aligned} &\frac{1}{2}(v_r^2 + v_\varphi^2) + \frac{1}{2}r^2 = E, \qquad r v_\varphi = K, \\ &r^2 = Ez, \qquad \varphi = \delta + \tan^{-1}\left[(\tan\psi + d)(1-d^2)^{-1/2}\right]. \end{aligned} \tag{1.4.19}$$

Here, E is the total energy, K is the angular momentum, δ is an angular constant, and ψ is the phase,

$$z = 1 + d\sin 2\psi, \qquad d = \left(1 - K^2E^{-2}\right)^{1/2}, \qquad \psi = t + \psi^0. \tag{1.4.20}$$

The parameters E, K (or d), δ, and ψ^0 are constants of integration. It follows from (19) and (20), that

$$E(1-d) \leqslant r^2 \leqslant E(1+d), \quad v_r = d\left(\frac{E}{Z}\right)^{1/2}\cos 2\psi, \quad v_\varphi = \frac{K}{(Ez)^{1/2}}. \tag{1.4.21}$$

Trajectories of the unperturbed system are ellipses with a common centre at the origin, see Fig. 1.3. Their dimensions and orientations are determined by the parameters E, $K(d)$, and δ.

For system (18), it is convenient to take the integrals E, K, and δ as slow variables and ψ as the fast variable. Then, based on (18) and (1.5), we obtain

$$\begin{aligned} &\dot{E} = \varepsilon v_r(u_r + f_r) + \varepsilon v_\varphi(u_\varphi + f_\varphi), \qquad \dot{K} = \varepsilon r(u_\varphi + f_\varphi), \\ &\dot{\delta} = -\left(\frac{\dot{E}}{E} - \frac{\dot{K}}{K}\right)\frac{(1-d^2)^{1/2}}{dz}(\cos\psi + d\sin\psi)\cos\psi, \\ &\dot{\psi} = 1 - \varepsilon\frac{d + \sin 2\psi}{2d(Ez)^{1/2}}(u_r + f_r) - \varepsilon\frac{K\cos 2\psi}{2d(E^3z)^{1/2}}(u_\varphi + f_\varphi). \end{aligned} \tag{1.4.22}$$

The initial conditions for E, K, δ, and ψ are determined from the formulae of the change (19)–(21), δ^0 and ψ^0 being determined to an accuracy up to π.

Instead of the variable E or K one can take d which is assumed to be constrained by the inequalities $0 < d_1 \leqslant d \leqslant d_2 < 1$ (if $d = 0$, then the corresponding trajectory is a circle, and if $d = 1$ it is a segment).

We consider the terminal conditions and the functional in the form

$$E(T) = E^* \geqslant 0, \qquad J[u] = \int_0^\Theta (u_r^2 + u_\varphi^2)\,d\tau, \qquad |u| < \infty. \tag{1.4.23}$$

The integral in (23) is interpreted as the energy consumption for control. Here, we consider the perturbation to have the form $f_r = \sigma r^3$, $f_\varphi \equiv 0$, which corresponds to the addition of a nonlinear restoring force.

Based on Section 1.1.2, we find the adjoint variables corresponding to δ and ψ: $p_\delta \equiv 0$, $p_\psi = O(\varepsilon)$. According to (2.17), the following expressions are obtained

$$\begin{aligned}
h_0 &= \frac{1}{4}\xi\left(\eta_E^2 + \eta_K^2\right) + \frac{1}{2}\mu\eta_E\eta_K = \text{const},\\
\xi(\tau) &= \left[\sqrt{E^0} - \frac{\tau}{\Theta}\left(\sqrt{E^0} - \sqrt{E^*}\right)\right]^2,\\
\mu(\tau) &= K^0\frac{E^*}{E^0}\chi^2(\tau),\\
\eta_E(\tau) &= \frac{4}{\chi(\tau)\Theta}\left(1 - \frac{\sqrt{E^0}}{\sqrt{E^*}}\right), \qquad \eta_K(\tau) \equiv 0,\\
\chi(\tau) &= 1 - \frac{\Theta - \tau}{\Theta}\left(1 - \frac{\sqrt{E^0}}{\sqrt{E^*}}\right).
\end{aligned} \tag{1.4.24}$$

Here, ξ, μ, η_E, and η_K are the averaged variables E, K, p_E, and p_K, respectively. The optimal control and the functional in the first approximation are

$$u_{r,\varphi} = \frac{1}{2}\eta_E v_{r,\varphi}, \qquad J_0 = h_0\Theta, \qquad h_0 = \frac{1}{4}\xi\eta_E^2. \tag{1.4.25}$$

It is worth paying attention once more to a singularity of the control in the feedback form, as $\tau \to \Theta$ ($\eta_E \to \infty$, $E \neq E^*$). The control (25) is tangential to the trajectory. The functions η_E and $v_{r,\varphi}$, as well as E and K (or d), are determined to the first approximation according to (24), (21), and the phase ψ is given by (see (2.17))

$$\psi(t,\varepsilon) = t - \frac{3}{4}\sigma\int_{\tau_0}^{\tau} \xi^2(\theta)\,d\theta + \psi^0. \tag{1.4.26}$$

Formulae (24)–(26) give the solution of the optimal control problem. They determine both the open-loop and the feedback control. It is worth noting that expression (26) is similar to the corresponding relationship (2.27); (26) coincides with (2.27) with the change $E = \frac{1}{2}a^2$, where a is the "amplitude".

1.4.4. STABILIZATION OF AN UNBALANCED ROTOR

Consider a mechanical system with three degrees of freedom as shown in Fig. 1.9. This model, in some practical cases, describes oscillations of the body of a mechanism, containing an unbalanced rotor, in the vertical plane OXY. The axis of the rotor is rigidly attached to the body. The mass of the body is M, the mass of the rotor is m, the distance between the axis of rotation of the rotor and its centre of mass is l. The body is attached to a fixed base by means of two single-axis vibration isolators. The axes of the undeformed isolators are directed along the X- and Y-axes of the reference frame. It is supposed that damping (k_x, k_y) and stiffness (c_x, c_y) coefficients of the isolators can be adjusted within some, not so wide, limits. The angular velocity $\dot{\varphi} = \omega$ of the rotor is also assumed to be controlled, the angular acceleration $\dot{\varphi} = \gamma$ being taken for the control variable.

It is of practical interest to solve control problems aimed at reducing the amplitudes of unwanted oscillations which can arise both during transient processes of the rotor acceleration and deceleration and under steady-state rotations, see [1, 19].

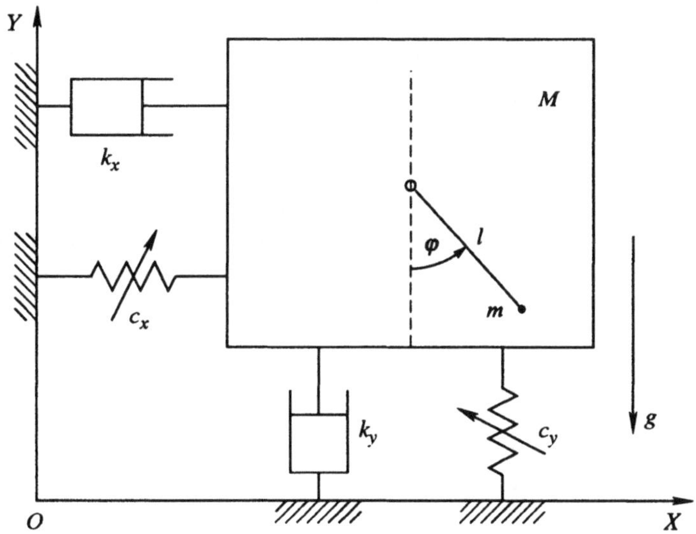

Fig. 1.9.

Having neglected small nonlinearities of the isolators, we obtain the set of equations of motion [19]

$$(M+m)\ddot{x}+k_x\dot{x}+c_x x = ml\big(\dot{\varphi}^2\sin\varphi-\ddot{\varphi}\cos\varphi\big),$$
$$(M+m)\ddot{y}+k_y\dot{y}+c_y y = -(M+m)g-ml\big(\dot{\varphi}^2\cos\varphi+\ddot{\varphi}\sin\varphi\big), \qquad (1.4.27)$$
$$\dot{\varphi}=\omega, \qquad \dot{\omega}=\gamma, \qquad \gamma(t)\in\Gamma.$$

We suppose that the stiffness coefficients c_x and c_y in (27) are weakly controllable, see [19], and the angular acceleration γ, the mass m and the damping coefficients are small, i.e.

$$c_{x,y}=c^0_{x,y}(1+\delta_{x,y}), \quad \delta_{x,y}=\varepsilon u_{x,y}, \quad \varepsilon\ll 1, \quad \gamma\nu^{-2}_{x,y}\sim\varepsilon,$$
$$\nu^2_{x,y}=\frac{c^0_{x,y}}{M+m}, \quad \frac{k_{x,y}}{(M+m)\nu_{x,y}}=\varepsilon k'_{x,y}, \quad \frac{m}{M+m}=\varepsilon\mu_*. \qquad (1.4.28)$$

Then we introduce new units of measurement $L\sim l$, $\nu\sim\nu_{x,y}$. If the function $\gamma=\varepsilon\gamma^*(\omega)$ is specified, then the function $\omega(\tau)$ can be considered to be known. As a result, we obtain two separate oscillatory systems acted on by a common periodic force (see (1.8))

$$\ddot{x}+\nu_x^2 x=-\varepsilon\nu_x^2 x u_x-\varepsilon k_x\dot{x}+\varepsilon\mu\omega^2\sin\varphi,$$
$$\ddot{y}+\nu_y^2 y=-g-\varepsilon\nu_y^2 y u_y-\varepsilon k_y\dot{y}-\varepsilon\mu\omega^2\cos\varphi, \qquad (1.4.29)$$
$$\dot{\varphi}=\omega(\tau), \qquad \tau=\varepsilon t\in[0,\Theta].$$

Here, $\nu_{x,y}\sim 1$, $|u_{x,y}|\leqslant u^0_{x,y}\sim 1$; $g\sim 1$ is the acceleration due to gravity divided by $\nu^2 L$; $\varepsilon\mu=ml/((M+m)L)$. We consider the minimization problem for the total energy of the oscillations corresponding to each degree of freedom, x and y:

$$J_{x,y}[u_{x,y}]=\int_0^\Theta \dot{E}_{x,y}\,d\tau\longrightarrow\min_{u_{x,y}}, \qquad |u_{x,y}|\leqslant u^0_{x,y},$$
$$E_x=\frac{1}{2}\big(\dot{x}^2+\nu_x^2x^2\big), \quad E_y=\frac{1}{2}\big[\dot{y}^2+\nu_y^2(y-y_0)^2\big], \quad y_0=-g\nu_y^{-2}. \qquad (1.4.30)$$

Below, the coefficients $\nu^2_{x,y}$, $k^2_{x,y}$, and μ are assumed to be constant. Using (1.9), the subsystems (29) are reduced to the form (1.10) with an external periodic force. In the "nonresonant" case $(\Omega_x - \omega) \sim 1$, $(\Omega_y - \omega) \sim 1$, $\tau \in [0, \Theta]$, where $\Omega_{x,y}(\tau, \varepsilon)$ are perturbed eigenfrequencies, the average value of the external force is small ($O(\varepsilon)$), and the control to be found has the form

$$u_x = u_x^0 \operatorname{sign}(x\dot{x}), \qquad u_y = u_y^0 \operatorname{sign}(y\dot{y}). \tag{1.4.31}$$

According to the method of equivalent linearization [38], the optimal laws (31) of oscillation damping to a first approximation give the same decrease of the amplitudes of oscillations as does viscous friction

$$\begin{gathered}
xu_x \sim \frac{2}{\pi}\frac{u_x^0}{\nu_x}\dot{x}, \quad yu_y \sim \frac{2}{\pi}\frac{u_y^0}{\nu_y}\mathrm{d}(\rho)\dot{y}, \\
\mathrm{d}(\rho) = 2/\rho, \quad 0 < \rho \leqslant 1; \qquad \mathrm{d}(\rho) = 1 + 1/\rho^2, \quad \rho \geqslant 1, \\
\rho = a_y/|y_0|, \quad a_y = \left[\dot{y}^2/\nu_y^2 + (y - y_0)^2\right]^{1/2} = (2E_y)^{1/2}/\nu_y.
\end{gathered} \tag{1.4.32}$$

Subsequent integration of the equations of motion for variable amplitudes a_x, a_y and phases ψ_x, ψ_y by using an averaging technique causes no difficulties, since a_x, $a_y > 0$ (singular controls do not appear, see Sections 1.2.4, and 2.6).

The analysis of the optimal control problem (29), (30) for amplitudes of the oscillatory subsystems is complicated if the resonant case takes place ($\nu_{x,y} - \omega = O(\varepsilon)$), and one has to consider the influence of the external force on the amplitude and the phase of free oscillations according to Section 2.5, see also [1, 19, 36, 38, 77, 123, 150].

CHAPTER 2

The Foundation of Asymptotic Methods for Controlled Quasilinear Systems and Some Generalizations

Using the method of the canonical separation of variables, we construct here the solutions for boundary-value problems arising from the maximum principle, with a given accuracy with respect to a small parameter. The method of averaging is extended to multi-frequency systems, as well as to more general systems. We also consider some critical cases: small frequencies, bang-bang controls, passing through resonances, etc.

2.1. The Method of Canonical Averaging for Standard Systems with Rotating Phase

2.1.1. CANONICAL CHANGE OF VARIABLES AND THE AVERAGING METHOD

(1) *Preliminary remarks.* It is well known [1, 4, 12, 35, 38, 46, 53, 69, 73, 100, 119, 120, 137, 149, 150] that the averaging method is connected with a change of variables which eliminates the fast variable (phase) from the right-hand side of a system, with a given accuracy with respect to ε (see Section 1.2). If the system is sufficiently smooth, then such a change is deduced from the solution of some sequence of sets of $2(n+1)$ partial differential equations [1, 38, 150]. Integration of these equations leads to arbitrary functions of slow variables (see [1, 38, 150] and (1.2.16)). These functions can be defined, for reasons of convenience, for investigating the new, averaged, system [1, 4, 69, 73, 150].

For instance, some authors have investigated the initial-value (Cauchy) problem for a canonical system with a small periodic (in time) Hamiltonian $H = \varepsilon h(t, x, p)$, where t, x, and p denote time, the coordinate, and the momentum, respectively, and where ε is a small parameter. They have chosen the arbitrary functions mentioned above so that the averaged (with respect to t) system is also canonical. In this case, the averaged Hamiltonian gives the energy integral, which is important in applications. In many cases, it allows qualitative conclusions and estimates to be made concerning the behaviour of the system over a large time interval. An additional methodical advantage of this approach is that one has to construct the solution of a partial differential equation for only one scalar generating function, instead of $2(n+1)$ functions: $v(\tau,\xi,\varphi,\varepsilon)$ and $w(\tau,\xi,\varphi,\varepsilon)$, see (1.2.14).

In [44, 69, 78–81, 94, 100, 118, 119, 130, 133, 141, 143], the method based on Lie transformations is developed for general Hamiltonian systems. We consider below the method of the canonical change of variables for a standard Hamiltonian system with rotating phase and slowly varying parameters, i.e. taking into account the slow time τ.

(2) *Construction of the generating function and averaged Hamiltonian.* For the system (1.2.8) (or (1.3.3)) with the Hamiltonian H^* (1.2.6) (or (1.3.4)), we construct a univalent canonical change of the variables a, ψ, p, and q for new, averaged, variables ξ, φ, η, and β, using the generating function $S = S(\tau, a, \psi, \eta, \beta, \varepsilon)$:

$$p = \frac{\partial S}{\partial a}, \qquad q = \frac{\partial S}{\partial \psi}, \qquad \xi = \frac{\partial S}{\partial \eta}, \qquad \varphi = \frac{\partial S}{\partial \beta}. \tag{2.1.1}$$

It is required that the new Hamiltonian K does not depend on φ: $K = K(\tau,\xi,\eta,\beta,\varepsilon)$, and, besides, the original and new variables coincide when $\varepsilon = 0$. The generating function S, and the Hamiltonian K, are connected by the nonlinear differential relationship [127]

$$\frac{\partial S}{\partial t} + H^*\left(\tau, \alpha, \psi, \frac{\partial S}{\partial a}, \frac{\partial S}{\partial \psi}, \varepsilon\right) = K(\tau,\xi,\eta,\beta,\varepsilon). \tag{2.1.2}$$

Within the framework of the assumptions made above, the unknown functions S, K and the change (1) have the form

$$\begin{aligned}
&S = (a\cdot\eta) + \psi\beta + \varepsilon\sigma(\tau, a, \psi, \eta, \beta, \varepsilon),\\
&p = \eta + \varepsilon\frac{\partial\sigma}{\partial a}, \qquad q = \beta + \varepsilon\frac{\partial\sigma}{\partial\psi}, \qquad \xi = a + \varepsilon\frac{\partial\sigma}{\partial\eta},\\
&\varphi = \psi + \varepsilon\frac{\partial\sigma}{\partial\beta}, \qquad K = \nu(\tau)\beta + \varepsilon k(\tau,\xi,\eta,\beta,\varepsilon).
\end{aligned} \tag{2.1.3}$$

Here, σ and k are unknown functions determined by the relationship (2) after substituting the expressions (3)

$$\begin{aligned}
&\nu(\tau)\frac{\partial\sigma}{\partial\psi} + h^*\left(\tau, a, \psi, \eta + \varepsilon\frac{\partial\sigma}{\partial a}, \beta + \varepsilon\frac{\partial\sigma}{\partial\psi}\right) + \varepsilon\frac{\partial\sigma}{\partial\tau}\\
&\quad = k\left(\tau, a + \varepsilon\frac{\partial\sigma}{\partial\eta}, \eta, \beta, \varepsilon\right).
\end{aligned} \tag{2.1.4}$$

The unknown functions σ and k can be found from Equation (4) as a series expansions in ε:

$$\begin{aligned} \sigma(\tau, a, \psi, \eta, \beta, \varepsilon) &= \sigma_0(\tau, a, \psi, \eta, \beta) + \varepsilon\sigma_1 + \cdots + \varepsilon^j \sigma_j + \cdots, \\ k(\tau, \xi, \eta, \beta, \varepsilon) &= k_0(\tau, \xi, \eta, \beta) + \varepsilon k_1 + \cdots + \varepsilon^j k_j + \cdots. \end{aligned} \tag{2.1.5}$$

Having compared the coefficients of like powers of ε (assuming that the function h^* is sufficiently smooth in the slow variables), we obtain from (3)–(5) a sequence of linking scalar differential equations. In particular:

$$\nu(\tau)\frac{\partial \sigma_0}{\partial \psi} + h^*(\tau, a, \psi, \eta, \beta) = k_0(\tau, a, \eta, \beta). \tag{2.1.6}$$

One can require the function σ (i.e. all coefficients $\sigma_0, \sigma_1, \ldots, \sigma_j, \ldots$) to be 2π-periodic in ψ. This results in the following relationships for the unknown functions k_0 and σ_0:

$$\begin{aligned} k_0(\tau, \xi, \eta, \beta) &= \langle h^*(\tau, \xi, \psi, \eta, \beta)\rangle \equiv h_0^*(\tau, \xi, \eta, \beta), \\ \sigma_0(\tau, a, \psi, \eta, \beta) &= -\frac{1}{\nu(\tau)} \int [h^*(\tau, a, \psi, \eta, \beta) - h_0^*(\tau, a, \eta, \beta)]\,\mathrm{d}\psi. \end{aligned} \tag{2.1.7}$$

Similar expressions are also valid for other unknown coefficients k_i and σ_i $(i \geqslant 1)$

$$\begin{aligned} k_i(t, \xi, \eta, \beta) &= h_{i0}(\tau, \xi, \eta, \beta), \qquad i = 1, \ldots, j, \ldots, \\ \sigma_i(\tau, a, \psi, \eta, \beta) &= -\frac{1}{\nu(\tau)} \int [h_i(\tau, a, \psi, \eta, \beta) - h_{i0}(\tau, a, \eta, \beta)]\,\mathrm{d}\psi. \end{aligned} \tag{2.1.8}$$

The functions h_i in (8) are determined in terms of coefficients found in previous steps. For instance,

$$\begin{aligned} h_1(t, a, \psi, \eta, \beta) &= \frac{\partial \sigma_0}{\partial \tau} + \left(\frac{\partial h^*}{\partial \eta} \cdot \frac{\partial \sigma_0}{\partial a}\right) + \frac{\partial h^*}{\partial \beta}\frac{\partial \sigma_0}{\partial \psi} - \left(\frac{\partial k_0}{\partial a} \cdot \frac{\partial \sigma_0}{\partial \eta}\right), \\ h_2(\tau, a, \psi, \eta, \beta) &= \frac{\partial \sigma_1}{\partial \tau} - \left(\frac{\partial h^*}{\partial \eta} \cdot \frac{\partial \sigma_1}{\partial a}\right) + \frac{1}{2}\left(\frac{\partial^2 h^*}{\partial \eta^2}\frac{\partial \sigma_0}{\partial a} \cdot \frac{\partial \sigma_0}{\partial a}\right) + \\ &\quad + \frac{\partial h^*}{\partial \beta} \cdot \frac{\partial \sigma_1}{\partial \psi} + \frac{1}{2}\frac{\partial^2 h^*}{\partial \beta^2}\left(\frac{\partial \sigma_0}{\partial \psi}\right)^2 + \end{aligned} \tag{2.1.9}$$

$$+\left(\frac{\partial^2 h^*}{\partial\eta\partial\beta}\cdot\frac{\partial\sigma_0}{\partial a}\right)\frac{\partial\sigma_0}{\partial\psi}-\left(\frac{\partial k_0}{\partial a}\cdot\frac{\partial\sigma_1}{\partial\eta}\right)-$$
$$-\frac{1}{2}\left(\frac{\partial^2 k_0}{\partial a^2}\frac{\partial\sigma_0}{\partial\eta}\cdot\frac{\partial\sigma_0}{\partial\eta}\right)-\left(\frac{\partial k_1}{\partial a}\cdot\frac{\partial\sigma_0}{\partial\eta}\right),\ldots.$$

It is worth noting that, analogously to (1.2.16), the coefficients σ_i are determined not uniquely but with an accuracy up to an arbitrary function of the slow variables. This arbitrariness disappears when returning to the original variables.

To construct a canonical averaged system of the $(j+1)$th approximation, one has to calculate the first $j+1$ coefficients in the expansion (5) for k_i $(0 \leqslant i \leqslant j)$ and j coefficients for σ_i $(i \leqslant j-1)$. Since, in the averaged system for ξ, φ, η, and β, the terms $O(\varepsilon^{j+2})$ are omitted, integration of this system, after introducing the slow time $\tau = \varepsilon t$ $(\tau \sim 1)$, generally leads to an error $O(\varepsilon^{j+1})$. The original variables a, ψ, p, and q are also determined with the same accuracy from (3). It is well known [1, 46, 38, 150] that taking into account the coefficient σ_j gives the so-called improved approximation of the $(j+1)$th order, which is used when proving the averaging method.

In the slow time $\tau = \varepsilon t$, the averaged system of the $(j+1)$th approximation has the form:

$$\dot\xi = \frac{\partial k}{\partial\eta}, \qquad \dot\eta = -\frac{\partial k}{\partial\xi}, \qquad \dot\varphi = \frac{\nu(\tau)}{\varepsilon} + \frac{\partial k}{\partial\beta}, \qquad \beta = \text{const},$$
$$k(\tau,\xi,\eta,\beta,\varepsilon) = \sum_{i=0}^{j} \varepsilon^i k_i(\tau,\xi,\eta,\beta), \qquad \tau \in [\tau_0, \Theta]. \tag{2.1.10}$$

The index j, which indicates the order of approximation with respect to ε is omitted for compactness. In (10), one has to integrate simultaneously only the Hamiltonian equations for ξ and η, considering β as a parameter. Knowing ξ and η, the phase φ is then determined by means of simple integration with the same accuracy. If k does not depend on τ, then the system for ξ and η has the integral $k = k(\xi,\eta,\beta,\varepsilon) = \text{const}$. As has been noted, the general solution of Equations (10), required to obtain the solution for the boundary-value problem of the maximum principle, can be constructed on the basis of the solution of the first approximation.

2.1.2. CONSTRUCTION OF THE FIRST APPROXIMATION

The system for the first approximation for ξ and η (see Sections 1.2.3 or 1.3.3) is obtained from (10) when $j = 0$:

$$\dot{\xi} = \frac{\partial k_0}{\partial \eta} = f_0(\tau, \xi, \eta, \beta), \qquad \beta = \text{const},$$
$$\dot{\eta} = -\frac{\partial k_0}{\partial \eta} = -\frac{\partial}{\partial \xi}\left[(\eta_0 \cdot f_0) + \beta F_0 - G_0\right]. \tag{2.1.11}$$

In what follows, we assume that the general solution for the system (11) is constructed in the form

$$\xi = \xi_0\big(\tau, \tau_0, \Theta, \xi^0, \eta_*, \beta\big), \qquad \xi_0\big|_{\tau_0} = \xi^0,$$
$$\eta = \eta_0\big(\tau, \tau_0, \Theta, \xi^0, \eta_*, \beta\big), \qquad \eta_0\big|_\Theta = \eta_*. \tag{2.1.12}$$

The phase φ is obtained from (12) and (11) by means of integration

$$\varphi = \varphi^0 + \varepsilon^{-1}\varphi_{-1}(\tau, \tau_0) + \varphi_0\big(\tau, \tau_0, \Theta, \xi^0, \eta_*, \beta\big),$$
$$\varphi_{-1} = \int_{\tau_0}^{\tau} \nu(\theta)\, d\theta, \tag{2.1.13}$$
$$\varphi_0 = \int_{\tau_0}^{\tau} F_0(\theta, \xi_0(\theta), \eta_0(\theta), \beta)\, d\theta.$$

When Θ is fixed, Equations (1.2.17) or (1.3.10) imply the boundary conditions of the first approximation which the free parameters ξ^0, φ^0, η_*, β, and λ must satisfy:

$$\xi^0 = a^0, \quad \varphi^0 = \psi^0, \quad M(\xi, \varphi)\big|_\Theta = 0,$$
$$\eta_* = \frac{\partial}{\partial \xi}\left[(\lambda \cdot M) - g\right]\Big|_\Theta, \quad \beta = \frac{\partial}{\partial \varphi}\left[(\lambda \cdot M) - g\right]\Big|_\Theta \quad (M \to N). \tag{2.1.14}$$

Here, expressions similar to (1.3.12) are substituted into the functions M (or N), g, and then a series expansion is produced in the neighbourhood of the values (12), (13); with $\xi^0 = a^0$ and $\varphi^0 = \psi^0$ being known. With the assumptions made above, there exists a root (λ, η_*, β) which depends on other parameters of the problem. If this root is simple and unique, then, as follows from the theory of implicit functions, the expressions derived define the optimal solution, i.e. the trajectory (a, ψ), the functional J, and the control u, with the error $O(\varepsilon)$. If there is a number of simple roots, then the optimal value is determined from the minimality condition for J_0 (1.2.18). If the optimal (in the sense of J_0) root is multiple, then proximity estimates require additional investigations, since, in this case, one cannot guarantee ε-proximity under perturbations of the order of ε. The error of the first approximation, as a rule, has an order of a fractional power of ε.

2.1.3. THE CANONICAL TRANSFORMATION OF ADJOINT VARIABLES INDUCED BY THE CHANGE OF STATE VARIABLES

Suppose, in a Hamiltionian system with the Hamiltonian $H(t,x,p)$, where x is the coordinate vector (state variable), and p is the momentum vector (adjoint variable), that a sufficiently smooth nonsingular point change is produced:

$$x = \varphi(t,y), \qquad y = \psi(t,x). \tag{2.1.15}$$

The change (15) often takes place when investigating optimal control problems using the maximum principle and the averaging method. For instance, such a change is produced for transformations from the original (Lagrange's) coordinates and velocities to osculating variables of the "amplitude-phase" type (see Chapter 1) or from one set of osculating variables to another (see Section 6.2).

We require finding the transformation of momenta, such that the Hamiltonian structure of the system of equations will be conserved. This transformation is defined by the generating function S [127]:

$$S = S(t,x,q), \qquad \frac{\partial S}{\partial x} = cp, \qquad \frac{\partial S}{\partial q} = y, \tag{2.1.16}$$

where q is the new momentum, and c is the valency of the canonical transformation ($c \neq 0$). It follows from the last equality that

$$S(t,x,q) = \big(q \cdot \psi(t,x)\big) + \alpha(t,x). \tag{2.1.17}$$

Here, α is a sufficiently smooth arbitrary function of t and x which can be chosen for reasons of convenience. Differentiation of S (17) with respect to x gives the expression sought for the new momentum q and the new Hamiltonian K:

$$\begin{gathered} q = c\left(\frac{\partial \psi}{\partial x}\right)^{-1} p - \left(\frac{\partial \psi}{\partial x}\right)^{-1} \frac{\partial \alpha}{\partial x} \equiv \eta(t,x,p), \\ K = K(t,y,q) = cH\left(t, x, \frac{1}{c}\frac{\partial S}{\partial x}\right) + \frac{\partial S}{\partial t}, \qquad x = \varphi(t,y). \end{gathered} \tag{2.1.18}$$

For the sake of simplicity, we consider, in what follows, univalent canonical transformations ($c = 1$). It is worth noting that the coordinate transformation (15) leads to the linear (in p) transformation η for momenta, see (18). This property is immediately established if the Hamiltonian system (t,x,p,H) can be represented in the

Lagrangian form ($q = p\partial\varphi/\partial y$). If the problem is time-invariant, i.e. if the functions H, φ, and ψ do not depend on t, then the transformations mentioned above, as well as the resultant new system, can be chosen to be independent of t. If some components of the vector x are transformed identically, then the respective momenta can also be transformed identically.

EXAMPLE. Consider a perturbed Hamiltonian system with the Hamiltonian H:

$$H = p_x v - \nu^2(\tau)p_\nu x + \varepsilon h(t, \tau, x, v, p_x, p_v), \qquad \tau = \varepsilon t. \tag{2.1.19}$$

Here, x and v are state variables, p_x and p_v are the respective momenta, ε is a small parameter, and h is a sufficiently smooth perturbing function. Let the "point" transformation (15) to the variables of Van der Pohl's type [1, 35, 38, 46, 150] be produced:

$$\begin{aligned} &x = a\cos\varphi + b\sin\varphi, \qquad v = \nu(\tau)(-a\sin\varphi + b\cos\varphi), \\ &\varphi = \int \nu(\tau)\,\mathrm{d}t = \frac{1}{\varepsilon}\int \nu(\tau)\,\mathrm{d}\tau, \qquad \tau = \varepsilon t. \end{aligned} \tag{2.1.20}$$

Then, using the generating function S and the change $p_x = \partial S/\partial x$, $p_v = \partial S/\partial v$, one can obtain for q_a and q_b:

$$\begin{aligned} &S(\tau, \varphi, x, v, q_a, q_b) \\ &\quad = q_a\big(x\cos\varphi - \nu^{-1}v\sin\varphi\big) + q_b\big(x\sin\varphi + \nu^{-1}v\cos\varphi\big), \\ &q_a = p_x\cos\varphi - \nu p_v\sin\varphi, \qquad q_b = p_x\sin\varphi + \nu p_v\cos\varphi, \\ &K = H + \frac{\partial S}{\partial t} = \varepsilon h^* + \varepsilon\frac{\nu'}{\nu^2}(q_a\sin\varphi + q_b\cos\varphi)(-a\sin\varphi + b\cos\varphi). \end{aligned} \tag{2.1.21}$$

Here, h^* is the perturbing function (19), where x, v, p_x, and p_v are replaced by their expressions through a, b, q_a, and q_b using (20) and (21). By means of differentiation with respect to q_a, q_b and a, b we obtain the equations for new Hamiltonian variables which appear to be osculating.

Let the change (15) correspond to the transformation to the "amplitude (A) – phase (ψ)" variables:

$$x = A\cos\psi, \qquad v = -A\nu\sin\varphi. \tag{2.1.22}$$

Then, using the generating function S

$$S = q_A\left(x^2 + \nu^{-2}v^2\right)^{1/2} + q_\psi \tan^{-1}(v/(x\nu)),$$

one can obtain the expressions for the transformed momenta q_A, q_ψ, and the Hamiltonian K:

$$\begin{aligned} q_A &= p_x \cos\psi - \nu p_\nu \sin\psi, \qquad q_\psi = -A(p_x \sin\psi + \nu p_v \cos\psi), \\ K &= \nu q_\psi + \varepsilon h^* - \varepsilon \nu^{-1}\nu'\left(q_A A \sin^2\psi + q_\psi \sin\psi\cos\psi\right). \end{aligned} \tag{2.1.23}$$

Here, h^* is the function of new variables obtained from the expression for h in (19) after the substitutions x, v, p_x, and p_v according to (22) and (23). Formulae (20), (21) and (22), (23) enable making the transition from the variables a, b, p_a, and p_b to the variables A, ψ, p_A, and p_ψ.

2.2. Construction of Higher Approximations to Solutions for Boundary-Value Problems of the Maximum Principle

2.2.1. GENERAL SOLUTION OF THE AVERAGED SYSTEM

For the system (1.10), we construct a solution, satisfying conditions $\xi(\tau_0) = \xi^0$ and $\eta(\Theta) = \eta_*$ with a required accuracy in terms of the small parameter (ξ^0 and η_* are unknown parameters belonging to the corresponding domain). The solution is constructed as a series expansion

$$\xi = \xi_0 + \varepsilon\xi_1 + \cdots + \varepsilon^j\xi_j, \qquad \eta = \eta_0 + \varepsilon\eta_1 + \cdots + \varepsilon^j\eta_j. \tag{2.2.1}$$

Here, ξ_0 and η_0 are known functions of the argument τ and the parameters of problem (1.12). The functions ξ_i and η_i ($i = 1, \ldots, j$) in (1) are unknown and can be determined by the successive solution of boundary-value problems of the following type:

$$\begin{aligned} \dot{\xi}_i &= \frac{\partial^2 k_0}{\partial\eta\partial\xi}\xi_i + \frac{\partial^2 k_0}{\partial\eta^2}\eta_i + f_i^\xi(\tau), \qquad \xi_i(\tau_0) = 0, \\ \dot{\eta}_i &= -\frac{\partial^2 k_0}{\partial\xi^2}\xi_i - \frac{\partial^2 k_0}{\partial\xi\partial\eta}\eta_i + f_i^\eta(\tau), \qquad \eta_i(\Theta) = 0. \end{aligned} \tag{2.2.2}$$

The n-dimensional vector-valued functions f_i^ξ and f_i^η are known at each step. They are determined through $k_0, k_1, \dots, k_{i-1}$ and the coefficients $\xi_0, \eta_0, \dots, \xi_{i-1}, \eta_{i-1}$ found on previous steps. For instance,

$$\begin{aligned} f_1^\xi &= \partial k_1/\partial\eta, \qquad f_1^\eta = -\partial k_1/\partial\xi, \\ f_2^\xi &= \frac{\partial k_2}{\partial\eta} + \frac{\partial}{\partial\eta}\left[\left(\frac{\partial k_1}{\partial\xi}\cdot\xi_1\right) + \left(\frac{\partial k_1}{\partial\eta}\cdot\eta_1\right) + \frac{1}{2}\left(\frac{\partial^2 k_0}{\partial\xi^2}\xi_1\cdot\xi_1\right) + \right. \\ &\quad \left. + \left(\frac{\partial^2 k_0}{\partial\xi\partial\eta}\xi_1\cdot\eta_1\right) + \frac{1}{2}\left(\frac{\partial^2 k_0}{\partial\eta^2}\eta_1\cdot\eta_1\right)\right], \\ f_2^\eta &= -\frac{\partial k_2}{\partial\xi} - \frac{\partial}{\partial\xi}\left[\left(\frac{\partial k_1}{\partial\xi}\cdot\xi_1\right) + \left(\frac{\partial k_1}{\partial\eta}\cdot\eta_1\right) + \frac{1}{2}\left(\frac{\partial^2 k_0}{\partial\xi^2}\xi_1\cdot\xi_1\right) + \right. \\ &\quad \left. + \left(\frac{\partial^2 k_0}{\partial\eta\partial\xi}\eta_1\cdot\xi_1\right) + \frac{1}{2}\left(\frac{\partial^2 k_0}{\partial\eta^2}\eta_1\cdot\eta_1\right)\right], \dots . \end{aligned} \tag{2.2.3}$$

The derivatives of the functions $k_0, k_1, k_2, \dots$ in (2), (3) are taken for the known generating solution ξ_0 and η_0 (1.12). If $f_i^\xi = f_i^\eta \equiv 0$, the homogeneous system (2) is the system for variations corresponding to equations of the first approximation (1.11). Hence, the fundamental matrix Z for system (2) can be obtained by means of differentiation of the general solution (1.12) with respect to the parameters ξ^0 and η_*:

$$Z(\tau) = \begin{bmatrix} \partial\xi_0/\partial\xi^0 & \partial\xi_0/\partial\eta_* \\ \partial\eta_0/\partial\xi^0 & \partial\eta_0/\partial\eta_* \end{bmatrix}. \tag{2.2.4}$$

In what follows, we assume that the $(2n \times 2n)$-matrix $Z(\tau)$ is nonsingular for $\tau \in [\tau_0, \Theta]$. Using the method of variation of integration constants based on (4), we can write the general solution of system (2) in the form

$$\begin{aligned} &\begin{pmatrix} \xi_i(\tau) \\ \eta_i(\tau) \end{pmatrix} \\ &= Z(\tau)\int_{\tau_0}^{\tau} Z^{-1}(\theta)\begin{pmatrix} f_i^\xi(\theta) \\ f_i^\eta(\theta) \end{pmatrix} d\theta + Z(\tau)\begin{pmatrix} c_i^\xi \\ c_i^\eta \end{pmatrix}, \qquad i = 1, \dots, j. \end{aligned} \tag{2.2.5}$$

The solution (5) is represented as a sum of a partial solution for the nonhomogeneous system $(\xi_i^*(\tau), \eta_i^*(\tau))$ and the general solution of the homogeneous system. Then, according to (1.12), we obtain the expressions:

$$\left.\frac{\partial\xi_0}{\partial\xi^0}\right|_{\tau_0} = \left.\frac{\partial\eta_0}{\partial\eta_*}\right|_{\Theta} = I, \qquad \left.\frac{\partial\xi_0}{\partial\eta_*}\right|_{\tau_0} = \left.\frac{\partial\eta_0}{\partial\xi^0}\right|_{\Theta} = 0,$$

where I and 0 are unit and zero $(n \times n)$-matrices, respectively. Using these equalities, the solution (5) and the boundary conditions (2), one can find the unknown constants c_i^ξ and c_i^η:

$$c_i^\xi = 0, \qquad c_i^\eta = -\eta_i^*(\theta), \qquad i = 1, \ldots j. \tag{2.2.6}$$

By means of the procedure described, we construct the solution of the $(j+1)$th approximation as a function of τ and the parameters of the boundary-value problem, part of which (ξ^0, η_*, and $\beta(\Theta)$) is unknown and has to be determined from the boundary conditions. Having determined ξ and η, we find φ from (1.10):

$$\varphi = \varphi^0 + \varepsilon^{-1}\varphi_{-1} + \varphi_0 + \varepsilon\varphi_1 + \cdots + \varepsilon^j \varphi_j. \tag{2.2.7}$$

Here, φ^0 is an unknown parameter, and φ_i are known functions.

2.2.2. CONSTRUCTION OF A GENERAL SOLUTION FOR THE ORIGINAL SYSTEM

Now we calculate (with required accuracy with respect to ε) the original variables a, ψ, p, and q related to the averaged variables ξ, φ, η, and β by equalities (1.3). The coefficients σ_i in the expansion (1.5) are known from (1.7)–(1.9). The obtained functions ξ, η (1) and φ (7) (β = const) are substituted into the transformation formulae (1.3), the terms $O(\varepsilon^{j+1})$ being omitted:

$$\begin{gathered} a = \xi - \varepsilon\frac{\partial\sigma}{\partial\eta}, \quad \psi = \varphi - \varepsilon\frac{\partial\sigma}{\partial\beta}, \quad p = \eta + \varepsilon\frac{\partial\sigma}{\partial a}, \quad q = \beta + \varepsilon\frac{\partial\sigma}{\partial\psi}, \\ \sigma = \sigma_0 + \varepsilon\sigma_1 + \cdots + \varepsilon^{j-1}\sigma_{j-1} \qquad (j \geqslant 1). \end{gathered} \tag{2.2.8}$$

We seek the unknown original variables in the form

$$\begin{aligned} a(t,\varepsilon) &= a_0 + \varepsilon a_1 + \cdots + \varepsilon^j a_j, \\ \psi(t,\varepsilon) &= \psi_0 + \varepsilon\psi_1 + \cdots + \varepsilon^j \psi_j, \\ p(t,\varepsilon) &= p_0 + \varepsilon p_1 + \cdots + \varepsilon^j p_j, \\ q(t,\varepsilon) &= q_0 + \varepsilon q_1 + \cdots + \varepsilon^i q_j. \end{aligned} \tag{2.2.9}$$

The expressions (9) are substituted into (8), and series expansions in ε are carried out. Then we compare the coefficients of like powers of ε^i. For the sake of convenience, one can use expansions in a neighbourhood of the averaged variables ξ, φ, η, and β or a double expansion in a neighbourhood of their first approximations ξ_0, $\varphi^0 + \varepsilon^{-1}\varphi_{-1} + \varphi_0$, η_0, and β_0. In particular,

$$\begin{aligned} &a_0 = \xi_0, \quad \psi_0 = \varphi^0 + \varepsilon^{-1}\varphi_{-1} + \varphi_0, \quad p_0 = \eta_0, \quad q_0 = \beta = \text{const}, \\ &a_1 = \xi_1 - \frac{\partial\sigma_0}{\partial\eta}, \quad \psi_1 = \varphi_1 - \frac{\partial\sigma_0}{\partial\beta}, \quad p_1 = \eta_1 + \frac{\partial\sigma_0}{\partial\xi}, \quad q_1 = \frac{\partial\sigma_0}{\partial\varphi}, \\ &\sigma_0 = \sigma_0\big(\tau, \xi_0, \varphi^0 + \varepsilon^{-1}\varphi_{-1} + \varphi_0, \eta_0, \beta\big). \end{aligned} \tag{2.2.10}$$

The solution of the $(j+1)$th approximation is represented in the form

$$\begin{aligned} &a = \xi_0 + \varepsilon A, \qquad \psi = \varphi^0 + \varepsilon^{-1}\varphi_{-1} + \varphi_0 + \varepsilon\Psi, \\ &p = \eta_0 + \varepsilon P, \qquad q = \beta + \varepsilon Q. \end{aligned} \tag{2.2.11}$$

Here, the "slow" functions ξ_0, φ_{-1}, φ_0, and η_0 are defined according to (1.12), (1.13), and the functions A, Ψ, P, and Q contain fast oscillations due to their periodic dependence on $\varepsilon^{-1}\varphi_{-1}(\tau, \tau_0)$. All these functions also depend on τ and the parameters τ_0, Θ, ξ^0, φ^0, η_*, and β.

2.2.3. SOLUTION OF THE BOUNDARY-VALUE PROBLEM OF THE MAXIMUM PRINCIPLE

An asymptotic solution and proximity estimates with respect to ε for the initial-value problem governed by a standard system with a rotating phase have been considered in many publications [12, 15, 38, 69, 74, 120, 150]. The asymptotic analysis of two-point boundary-value problems requires special consideration. Some results proving the proximity (ε-proximity in particular) of the first approximation can be found in [1–5, 8, 10, 20, 23, 24, 46, 71, 73, 86, 88, 107, 121, 128, 134], see also Section 2.3.

We consider the initial and terminal conditions for the boundary-value problem (1.2.9), and we consider two cases, $l = n + 1$ and $l < n + 1$.

In the first case, the transversality conditions are not taken into account, if the system with respect to ξ^0, φ^0, η_*, and β admits a finite number of roots. When considering the case $l < n + 1$, we assume that the rank of the linear system of algebraic equations with respect to the l-vector λ (Lagrange's multipliers), obtained from the transversality conditions for p and q after substitution of a, ψ, p, and q in the first approximation, provided $\xi^0 = a^0$ and $\varphi^0 = \psi^0$, equals l in some range of the parameters of the problem. Then this rank will be maximal in the $(j+1)$th approximation after substitution of a, ψ, p, and q from (11) for sufficiently small ε.

The linear system is solved with respect to λ, depending on the elements of the matrix $(\partial M/\partial a, \partial M/\partial \psi)|_\Theta$ and the vector $(\partial g/\partial a, \partial g/\partial \psi)|_\Theta$. Having substituted the obtained λ into the remaining $2(n+1)$ relationships, we have the following system of $2(n+1)$ equations with respect to the unknown parameters ξ^0, φ^0, η_*, and β:

$$\begin{aligned} &(\xi+\varepsilon A)\big|_{\tau_0} = a^0, \qquad (\varphi+\varepsilon\Psi)\big|_{\tau_0} = \psi^0, \qquad M\big|_\Theta = 0, \\ &R(a,\psi,p,q)\big|_\Theta = 0, \qquad R = (R_1,\dots,R_{n+1-l}). \end{aligned} \tag{2.2.12}$$

The root of system (12) is constructed as a function of ε, a^0, ψ^0, τ_0, and Θ by using series expansions in ε, analogously to (8)–(10):

$$\begin{aligned} &\xi^0 = a^0 + \varepsilon\xi_1^0 + \dots + \varepsilon^j\xi_j^0, \qquad \varphi^0 = \psi^0 + \varepsilon\varphi_1^0 + \dots + \varepsilon^j\varphi_j^0, \\ &\eta_* = \eta_*^0 + \varepsilon\eta_*^{(1)} + \dots + \varepsilon^j\eta_*^{(j)}, \qquad \beta = \beta^0 + \varepsilon\beta^{(1)} + \varepsilon^j\beta^{(j)}. \end{aligned} \tag{2.2.13}$$

After substitution of (13) into (12) and expansion of the functions A, Ψ, M, and R in a neighbourhood of the values a^0, ψ^0, η_*^0, and β^0, we obtain asymptotic representations for the unknown parameters with a required accuracy in terms of ε. It is worth noting that at $\tau = \Theta$, the expansion is produced in a neighbourhood of the value $\psi_0 \sim \varepsilon^{-1}$. The parameters η_*^0 and β^0 are given by the equations

$$\begin{aligned} &M\big(\xi_0(\Theta,a^0,\eta_*^0,\beta^0),\ \psi^0 + \varepsilon^{-1}\varphi_{-1}(\Theta) + \varphi_0(\Theta,a^0,\eta_*^0,\beta^0)\big) = 0, \\ &R\big(\xi_0(\Theta,a^0,\eta_*^0,\beta^0),\ \psi^0 + \varepsilon^{-1}\varphi_{-1}(\Theta) + \varphi_0(\Theta,a^0,\eta_*^0,\beta^0),\eta_*^0,\beta^0\big) = 0. \end{aligned} \tag{2.2.14}$$

The functions M and R in (14) rapidly oscillate when changing Θ, provided ε is small. Therefore, a root must exist for any ψ, for averages of these functions over ψ, i.e.

$$M_0\big(\xi_0(\Theta,a^0,\eta_*^0,\beta^0)\big) = 0, \qquad R_0\big(\xi_0(\Theta,a^0,\eta_*^0,\beta^0),\eta_*^0,\beta^0\big) = 0,$$

and also for maxima and minima over ψ. Moreover, the root η_*^0 and β^0 must be simple for all $\psi^0 \in [0,2\pi]$. In particular, if M and g do not depend on ψ, these difficulties do not occur.

Let the coefficients ξ_i^0, ψ_i^0, $\eta_*^{(i)}$, and $\beta^{(i)}$ in (13) be determined and substituted into a, ψ, p, and q (11). They define both the solution of the boundary-value problem and the value of the functional J (1.2.2) with the error $O(\varepsilon^{j+1})$. In the case of smooth functions u^* (1.2.5), a similar estimate takes place with respect to the control as well. If the controls are piecewise continuous, then the proximity is considered in terms of the mean-square (over the period) norm or the functional J.

Based on the foregoing, one can formulate the following statement.

THEOREM 2.1. *Suppose that necessary requirements for the smoothness and the conditions of Lemma* 1.1 *are fulfilled and, besides,*

(1) *problem* (1.11) *admits a solution in the form* (1.12);

(2) *system* (1.14) *has a real simple root* (η_*, β, λ);

(3) *the rank of the matrix* $Z(\tau)$ (4) *equals* $2n$.

Then the optimal control problem has the solution (1.2.10)–(1.2.12) which differs from the exact one by $O(\varepsilon^{j+1})$ in the functional and in the fulfillment of the boundary conditions. In the case of fixed Θ, this proximity also takes place with respect to the state trajectory.

In the case of fixed T and a specified terminal condition, the synthesis of optimal control and the proximity estimate have singularities as $t \to T$, and the uniform proximity estimate is not guaranteed.

2.3. Asymptotic Solution for Two-Point Boundary-Value Problems

2.3.1. BASIC ASSUMPTIONS AND STATEMENT OF THE PROBLEM

Consider a boundary-value problem for the system represented in the standard form (1.1.12). The combination of the slow variables (the state and adjoint ones) in one vector x and the elimination of the vector of Lagrange's multipliers from the transversality conditions (at the left, $t = 0$, and right, $t = T$, endpoints) enable obtaining the boundary-value problem in the form

$$\dot{x} = \varepsilon X(t, x), \qquad M\big(x(0), x(T)\big) = 0. \tag{2.3.1}$$

Here, x is the unknown slow vector, t is the independent variable, $t \in [0, T]$, $T = \Theta\varepsilon^{-1}$, $\Theta = \text{const} > 0$; ε is a small numerical parameter, $\varepsilon \in [0, \varepsilon_0]$; $X(t, x)$ and $M(y, z)$ are vector-valued functions of the dimension $n \geqslant 2$ defined in the domain $(t, x) \in [0, T] \times D_x$, $(y, z) \in D_x \times D_x$. Note that in boundary-value problems of the maximum principle the boundary condition (1) usually separates into two vector relationships for the beginning ($t = 0$) and for the end ($t = T$) of the control process:

$$M_0\big(x(0)\big) = 0, \qquad M_T\big(x(T)\big) = 0, \qquad n_0 + n_T = n. \tag{2.3.2}$$

Here, n_0 and n_T are dimensions of the vector-functions M_0 and M_T, respectively.

If one formally puts either $n_0 = 0$ or $n_T = 0$ (for boundary-value problems of the maximum principle the inequalities $n_0, n_T \geqslant 1$ always hold), then (1) and (2) become initial-value problems for the investigation of which the method of averaging is widely used [35, 38, 74, 81, 150]. Below, we establish comparatively general conditions imposed on the functions X and M, sufficient for the proximity of solutions of the original (1) and the averaged boundary-value problems.

We would like to recall that some more general systems

$$\begin{aligned} &\dot{a} = \varepsilon A(\tau, a, \psi), \qquad a = (a_1, \ldots, a_n), \qquad \tau = \varepsilon t, \\ &\dot{\psi} = \nu(\tau) + \varepsilon \Psi(\tau, a, \psi), \qquad \psi = (\psi_1, \ldots, \psi_r), \qquad \nu_j(\tau) \geqslant \nu_0 > 0 \end{aligned} \tag{2.3.3}$$

reduce to system (1) with the required properties.

For instance, if $r = 1$ and the functions A and Ψ are quasiperiodic in ψ, one can reduce system (3) to the form (1), introducing a new independent variable θ and the slow variable δ ($\psi = \theta + \delta$):

$$\theta = \int_0^t \nu(\varepsilon s)\, \mathrm{d}s \qquad \left(\varepsilon\theta = \sigma = \int_0^\tau \nu(\tau')\, \mathrm{d}\tau'\right). \tag{2.3.4}$$

If $r > 1$ and the frequencies $\nu_j(\tau) \equiv \nu_*(\tau)$, see Section 1.1, then the vector change of type (4) for ψ leads to problem (1). In the case of different or partly coinciding frequencies, the change (4) leads to system (1) if the functions A and Ψ can be represented as finite trigonometric sums in ψ_j.

Below, the requirements imposed on $X(t, x)$ and $M(y, z)$ are stated.

1. The function $X(t, x)$ is defined for all $t \geqslant 0$ and measurable in t for fixed values $x \in D_x$, where D_x is an open connected set (an open domain).

2. The well-known condition of the existence of the uniform (with respect to $x \in D_x$) average of the function $X(t, x)$ over t is replaced by a more restrictive one. It is assumed that $X(t, x)$ is uniformly quasiperiodic in t, i.e. it is a finite sum of periodic (in t) functions $X^{(j)}$ ($j = 1, \ldots, k$, $k \geqslant 1$) with arbitrary constant periods Π_j.

3. The functions $X(t, x)$ and $M(y, z)$ are defined for all $x, y, z \in D_x$ and uniformly continuous in x, y, and z.

4. There exist constants $C_X, C_M > 0$ depending on D_x such that

$$|X(t, x)| \leqslant C_X, \qquad |M(y, z)| \leqslant C_M, \qquad t \geqslant 0, \quad x, y, z \in D_x. \tag{2.3.5}$$

5. The functions X and M satisfy the Lipschitz conditions in $x, y, z \in D_x$, i.e. there exist constants λ_X and λ_M, depending on D_x in general case, such that

$$\begin{aligned} &\left|X(t,x') - X(t,x'')\right| \leqslant \lambda_X \left|x' - x''\right|, \qquad t \geqslant 0, \quad x', x'' \in D_x, \\ &\left|M(y',z') - M(y'',z'')\right| \leqslant \lambda_M \left(\left|y' - y''\right| + \left|z' - z''\right|\right); \\ &\quad y', y'', z', z'' \in D_x. \end{aligned} \tag{2.3.6}$$

To simplify the proofs, we strengthen the requirements (6) assuming the existence of continuous and bounded partial derivatives

$$\left|\frac{\partial X}{\partial x}\right| \leqslant \lambda_X, \quad \left|\frac{\partial M}{\partial y}\right| \leqslant \lambda_M, \quad \left|\frac{\partial M}{\partial z}\right| \leqslant \lambda_M, \quad t \geqslant 0, \quad x, y, z \in D_x. \tag{2.3.7}$$

6. Provided the requirements mentioned above are satisfied, there exists the average $X_0(x)$ over t for the function $X(t,x)$, uniform with respect to $x \in D_x$:

$$\begin{aligned} &X_0(x) = \lim_{S \to \infty} \int_t^{t+S} X(s,x)\,ds = \sum_{i=1}^{k} X_0^{(i)}(x), \\ &X_0^{(i)}(x) = \frac{1}{\Pi_i} \int_t^{t+\Pi_i} X(s,x)\,ds, \qquad i = 1, \dots, k, \quad t \geqslant 0. \end{aligned} \tag{2.3.8}$$

The function $X_0(x)$ satisfies the conditions 3–5.

It is usually assumed that the boundary-value problem (1) has a solution $x = x(t,\varepsilon) \in D_x$, $t \in [0, \Theta\varepsilon^{-1}]$, $\varepsilon \in (0, \varepsilon_0]$. The proof of the existence of this solution is a special problem, which we consider in Section 2.4, see also [12].

Parallel with the original boundary-value problem, we consider the simpler one, averaged according to (8):

$$\dot{\xi} = X_0(\xi), \qquad M\left(\xi(0), \xi(\Theta)\right) = 0, \qquad \tau \in [0, \Theta]. \tag{2.3.9}$$

We suppose that there exists a solution $\xi(\tau)$ of the boundary-value problem (9), belonging to the domain D_x for all $\tau \in [0, \Theta]$, and this solution is unique. It is known that provided conditions 1–5 are satisfied the solutions of initial-value problems for the original and averaged systems are ε-proximate for some range of common or sufficiently close initial values [1, 12, 35, 38, 150]. Below, we investigate, what additional conditions one has to impose on the functions $X(t,x)$ and

$M(y, z)$ to guarantee the existence of the solution $x(t, \varepsilon)$ for the original boundary-value problem (1) and how close is $x(t, \varepsilon)$ to the solution $\xi(\varepsilon t)$, $t \in [0, \Theta\varepsilon^{-1}]$, $\varepsilon > 0$. As has been mentioned, the construction of a solution by means of analytical or numerical methods is considerably easier for the averaged problem of the maximum principle (9) than for the original problem (1), because the averaged system does not depend on t and has a canonical structure.

2.3.2. ESTIMATE OF THE PROXIMITY BETWEEN SOLUTIONS OF THE ORIGINAL AND AVERAGED BOUNDARY-VALUE PROBLEMS

To investigate the existence of the solution $x(t, \varepsilon)$ for the original two-point problem (1) and its proximity to the solution $\xi(\tau)$, $\tau = \varepsilon t$, of the averaged problem (9), we apply the constructive approach given below.

Assuming Conditions 1–5 (see Section 2.3.1), the initial-value problem

$$\dot{x} = \varepsilon X(t, x), \qquad x(0) = a \in D_a \subseteq D_x, \qquad t \in [0, T], \tag{2.3.10}$$

has the solution of the form

$$x(t, a, \varepsilon) = \varphi(\tau, a) + v(t, a, \varepsilon), \qquad a \in D_a, \quad x \in D_x. \tag{2.3.11}$$

Here, D_a is a nonvoid open domain, $\varphi(\tau, a)$ is a function continuously differentiable with respect to $\tau \in [0, \Theta]$; $a \in D_a$. The function $\varphi(\tau, a)$ is the general solution of the averaged initial-value problem

$$\dot{\xi} = X_0(\xi), \qquad \xi(0) = a \in D_a, \qquad \xi = \varphi(\tau, a) \in D_x. \tag{2.3.12}$$

According to [1, 10, 12, 35, 38, 47, 75, 150], the function v in (11) is uniformly bounded with respect to t and ε and continuous in a, i.e.

$$\begin{gathered} |v| \leqslant \varepsilon C_v, \qquad t \in [0, T], \quad T = \Theta\varepsilon^{-1}, \quad C_v = \text{const}, \\ \lim_{k \to \infty} v(t, a_k, \varepsilon) = v(t, a_*, \varepsilon), \qquad a_k \to a_*; \quad a_k, a_* \in D. \end{gathered} \tag{2.3.13}$$

The solution $\xi(\tau)$ of the averaged boundary-value problem (9) corresponds to some value of the parameter $a = a_0 \in D_a$, in the function $\varphi(\tau, a)$, satisfying the following nonlinear system

$$M\big(a, \varphi(\Theta, a)\big) \equiv M^*(a) = 0, \qquad \xi(\tau) = \varphi(\tau, a_0). \tag{2.3.14}$$

Analogously, the solution $x(t,\varepsilon)$ for the original boundary-value problem (1) is obtained when substituting the value of the parameter a, defined by the relationship

$$M^*(a) + N(a,\varepsilon) = 0, \qquad N(a,\varepsilon) \equiv M\big(a, x(T,a,\varepsilon)\big) - M^*(a), \tag{2.3.15}$$

into (11). Here, the function N is continuous in $a \in D_a$, due to (7) and (13), and due to (5) and (13) it is uniformly bounded: $|N| \leqslant \varepsilon C_N$, $\varepsilon \in [0,\varepsilon_0]$. While investigating the implicit function $a = a(\varepsilon)$, defined by Equation (15), the root a_0 of system (14) is supposed to be simple:

$$\det\big(\partial M^*/\partial a\big)\big|_{a_0} \neq 0, \qquad a_0 \in D_a. \tag{2.3.16}$$

The root $a(\varepsilon)$ is sought in the form $a = a_0 + \alpha$, where the unknown α continuously depends on ε, $\varepsilon \in [0,\varepsilon_0]$, and vanishes when $\varepsilon = 0$. This root is defined by the relationship

$$\alpha = -\left(\frac{\partial M^*}{\partial a_0}\right)^{-1} \frac{\partial M^*(a_0,\xi(\Theta))}{\partial \xi}\, v\left(\frac{\Theta}{\varepsilon}, a_0, \varepsilon\right) + A(\alpha,\varepsilon). \tag{2.3.17}$$

Here, A is a continuous function of α and ε (in the domain $a_0 + \alpha \in D_a$, $\varepsilon \in [0,\varepsilon_0]$) vanishing at $\varepsilon = 0$, $\alpha = 0$, the estimate $A = o(\varepsilon + |\alpha|)$, $\varepsilon \to 0$, $|\alpha| \to 0$, taking place. To determine α from (17) we use the method of successive approximations considering A as a perturbation. It results in the sequence

$$\begin{aligned} &\alpha_{j+1} = \alpha_1 + A(\alpha_j,\varepsilon), \qquad j = 1,2,\dots; \\ &\alpha_1 = -\left(\frac{\partial M^*}{\partial a_0}\right)^{-1} \frac{\partial M^*}{\partial \xi}\, v\bigg|_{\Theta} = O(\varepsilon), \\ &M^* = M(a_0,\xi_*), \qquad \xi_* = \varphi(\Theta, a_0), \end{aligned} \tag{2.3.18}$$

which is uniformly bounded and equicontinuous, if $\varepsilon_0 > 0$ is sufficiently small.

Based on Arzela's theorem [49, 75], one can choose from (18) a sequence $\alpha_{j_l} \to \alpha_*$, converging to the solution (17), such that

$$a_{j_l} = a_0 + \alpha_{j_l} \longrightarrow a^* \in D_a, \qquad a^* = a_0 + \alpha^*, \qquad \big|\alpha^*(\varepsilon)\big| \leqslant \varepsilon C_\alpha.$$

Therefore, the following statement takes place.

THEOREM 2.2. *Provided the conditions* 1–6, (14) *and* (16), *the original boundary-value problem* (1) *has a solution* $x(t, \varepsilon)$ *in the considered domain* $t \in [0, \Theta\varepsilon^{-1}]$, $\varepsilon \in [0, \varepsilon_0]$, $x \in D_x$, *for sufficiently small* $\varepsilon_0 > 0$. *This solution belongs to an* ε-*neighbourhood of the generating solution* $\xi(\tau)$, $\tau = \varepsilon t$, *of the averaged problem* (9), *i.e.*

$$|x(t, \varepsilon) - \xi(\tau)| \leqslant \varepsilon C, \qquad \tau = \varepsilon t \in [0, \Theta], \quad C = \text{const.} \tag{2.3.19}$$

Moreover, problem (1) *does not admit solutions violating the estimate* (19) *in the considered domain. The value of the constant* C *in this estimate can be determined in terms of the coefficients of the problem and the dimensions of the domains* D_x *and* D_a.

2.3.3. REMARKS

1) The uniqueness of the solution of Equation (17) and the boundary-value problem (1) is not guaranteed. For the existence and uniqueness of the solution for the original boundary-value problem, turning into the generating solution $\xi(\tau)$ at $\varepsilon = 0$, it is sufficient to assume that, apart from the requirement (16), the function A in (17) satisfies the Lipschitz condition or will have uniformly bounded derivative with respect to α. These assumptions evidently imply similar restrictions on the smoothness (in a) of the function v from (11). It is worth noting that the uniform estimate of the derivative $\partial v/\partial a = O(\varepsilon)$ for all $t \in [0, \Theta\varepsilon^{-1}]$, $a \in D_a$, and $\varepsilon \in [0, \varepsilon_0]$ can be obtained by using a standard procedure based on the Gronwall–Bellman lemma [35, 38, 47, 75, 150]. In this case, however, more restrictive conditions have to be imposed on the smoothness of the functions X and M in $x, y, z \in D_x$. Namely, for the function X, the derivative $\partial X/\partial x$ must exist, which is uniformly bounded and satisfies the Lipschitz condition in x with constants independent of t and ε. Analogously, the smoothness properties of the function M in $y, z \in D_x$ must also be strengthened. For construction of the first approximation for the solution of the boundary-value problem these additional requirements of smoothness are not necessary.

2) The fulfillment of inequality (16) is essential because, otherwise, branching of the root a_0 by amounts of the order of a fractional power of ε is possible. It may turn out that the difference between the generating and exact solutions is of the order of the lowest nonzero power of the small parameter ε. For this critical case, special consideration is needed.

3) In the particular case of the terminal control problem (the performance index is $\Phi(z(T)) \to \min_u$), the boundary conditions for the vector $x = (z, p)$, where z and p are the state and adjoint vectors, respectively, have the form $z(0) = z^0$, $p(T) = -\Phi'(z(T))$. Inequality (16) is equivalent to the requirement of the existence of the general solution $\xi = (\zeta, \eta)$ for the averaged system, such that $\zeta(0, a, b) = a$

and $\eta(\Theta, a, b) = b$. Here, the parameters a and b belong to some neighbourhoods of the points z^0 and $b^* = b(z^0)$, respectively, b^* being a simple root of the equation $\Phi'(\zeta(\Theta, z^0, b)) + b = 0$.

4) Theorem 2.2 and the remarks just made are valid for any admissible root $a_0 \in D_a$. The proper value a^* for the optimal control problem is chosen from the additional requirement of minimality for the functional calculated with sufficient accuracy in terms of ε.

Thus, Theorem 2.2 justifies the application of the averaging method for the asymptotic solution of optimal control problems to a first approximation with respect to the small parameter ε. Justification of recursive expansions from Section 2.1 and error estimates when constructing higher approximations requires either stronger smoothness of the controlled system or another technique for constructing the general solution with sufficient accuracy in terms of ε, e.g., combination of the averaging method and the method of successive approximations (see Section 2.4 and [1, 12]).

2.4. General Solution of the Standard System and Its Sensitivity

Here, we consider the issue of the existence of a solution of the initial-value problem for the standard (in the sense of Bogolyubov) set of ordinary differential equations [1, 12, 35, 38, 150], which govern a wide variety of nonlinear oscillation and rotation processes. Constructive sufficient conditions for the existence and uniqueness of this solution on an asymptotically large time interval are established by using the method of successive approximations [1, 12, 45, 47, 49, 61, 69, 74, 75, 115, 118, 151]. The smoothness and sensitivity properties of the non-steady-state process with respect to the parameters of the problem, and the initial data in particular, are investigated. These results are used when constructing solutions for boundary-value problems described in Sections 2.2.

2.4.1. STATEMENT OF THE PROBLEM AND INITIAL ASSUMPTIONS

Consider a standard set of ordinary differential equations on an asymptotically large time interval with specified initial conditions

$$\dot{x} = \varepsilon X(t, x), \qquad x(t_0) = x^0, \qquad t \in [t_0, T], \quad T = \Theta\varepsilon^{-1}. \tag{2.4.1}$$

Here, x and X are vectors of an arbitrary dimension $n \geqslant 1$, t_0, x^0, and Θ are given constants, and $\varepsilon > 0$ is a small numerical parameter, $\varepsilon \in [0, \varepsilon_0]$. We suppose the following requirements to be fulfilled for the vector-valued function X:

1) this function is defined for all $t \geqslant t_0$ and $x \in D_x$ ($D_x \subset R^n$ is an open connected set), 2π-periodic and piece-wise continuous in t;

2) the function X has a continuous derivative with respect to x, $x \in D_x$, satisfying the Lipschitz condition in the mentioned domain;

3) for all $\varepsilon \in (0, \varepsilon_0]$ and $x^0 \in D_{x^0} \subset D_x$, there exists a unique solution of the initial-value problem (1)

$$x = x(t, t_0, x^0, \varepsilon), \qquad t \in [t_0, \Theta\varepsilon^{-1}], \tag{2.4.2}$$

belonging to the open set D_x (this will be proved below);

4) the initial-value problem for the averaged system (1.1)

$$\dot{\xi} = X_0(\xi), \qquad \xi(0) = x^0, \qquad \tau = \varepsilon t, \quad \tau \in [\tau_0, \Theta], \tag{2.4.3}$$

where X_0 is the average of the function X with respect to t over the period 2π and τ is the slow time, has the general solution

$$\xi = \xi(\tau - \tau_0, x^0), \qquad \xi \in D_x, \quad x^0 \in D_{x^0}. \tag{2.4.4}$$

in the mentioned domain. This solution is assumed to be known.

Remark. In many investigations (see [12, 35, 38, 150], as well as in the references of [1, 38, 150]) the proximity property, the ε-proximity in particular, is established between the unknown solution (2) of the original initial-value problem (1) and the solution (4) of the averaged system (3) with the same initial conditions. The averaged system is time-invariant and enables eliminating the parameter ε. Usually, the investigation and construction of the solution for the averaged system is substantially simpler than for the original complete system. The determination of the unknown general solution (2) to a higher accuracy with respect to ε on the asymptotically large time interval $t \in [t_0, \Theta\varepsilon^{-1}]$ is complicated, since the procedure of the averaging method connected with the change of variables leads to solving the perturbed set of partial differential equations. The last circumstance requires a corresponding high degree of smoothness of the function X in $x \in D_x$.

Below, we consider the existence and uniqueness of the general solution (2) for system (1), under the assumptions 1), 2), and 4). The algorithm for constructing this solution is developed and its properties are investigated with respect to the variation of the parameters of the problem.

2.4.2. SIMPLIFICATION OF THE ORIGINAL INITIAL-VALUE PROBLEM

In system (1), a series of changes of the unknown variable x is carried out. Analogously to [1, 12, 35, 38, 150], we introduce the error of the averaging method

$$\begin{gathered} \delta = x - \xi_*, \qquad \xi_*(t, \varepsilon) = \xi(\tau) + \varepsilon u(t, \varepsilon), \\ u(t, \varepsilon) \equiv \int_{t_0}^{t} \big[X(s, \xi(\sigma)) - X_0(\xi(\sigma))\big] ds, \qquad \sigma = \varepsilon s. \end{gathered} \tag{2.4.5}$$

Here, the function u is uniformly bounded with respect to t and ε; the dependence of u and ξ on t_0 and x^0 in (5) is not indicated for the sake of compactness. For the unknown δ, we obtain the initial-value problem

$$\begin{aligned}&\dot{\delta} = \varepsilon X'(t,\xi)\delta + \varepsilon\Phi(t,\delta,\varepsilon), \qquad \delta(t_0) = 0, \qquad t \in \left[t_0, \Theta\varepsilon^{-1}\right],\\ &\Phi(t,\delta,\varepsilon) = \left[X(t,\xi_*) - X(t,\xi)\right] + X(t,\xi_* + \delta) - \\ &\qquad - \left[X(t,\xi_*) - X'(t,\xi)\delta\right], \qquad X'(\tau,x) \equiv \partial X/\partial x,\end{aligned} \tag{2.4.6}$$

where Φ is a known function of t, δ, and ε. In the domain $(\xi_* + \delta) \in D_x$, as it arises from (6), the following estimate for Φ is valid

$$\begin{aligned}&\left|\Phi(t,\delta,\varepsilon)\right| \leqslant C_\Phi\left(\varepsilon + \varepsilon|\delta| + \delta^2\right), \qquad C_\Phi, \lambda_\Phi = \text{const},\\ &\left|\Phi(t,\delta_1,\varepsilon) - \Phi(t,\delta_2,\varepsilon)\right| \leqslant \lambda_\Phi|\delta_1 - \delta_2|\left(|\delta_1| + |\delta_2| + \varepsilon\right).\end{aligned} \tag{2.4.7}$$

The linear transformation of the vector δ, close to the identity transformation, can be carried out [1, 12, 69]

$$\delta = (I + \varepsilon U)z, \qquad U(t,\varepsilon) \equiv \int_{t_0}^{t} \left[X'(s,\xi(\sigma)) - X_0'(\xi(\sigma))\right] ds. \tag{2.4.8}$$

Here, I is the unit matrix, U is a uniformly bounded matrix-valued function similar to u from (5); the prime denotes the derivative with respect to the vector x. Having differentiated expression (8) according to system (6), we obtain the initial-value problem for the unknown z

$$\begin{aligned}&\dot{z} = \varepsilon X_0'(\xi(\tau))z + \varepsilon F(t,z,\varepsilon), \qquad z(t_0) = 0,\\ &F(t,z,\varepsilon) \equiv (I + \varepsilon U)^{-1}\left\{\left[X(t,\xi_*) - X(t,\xi)\right] - \varepsilon U X_0'(\xi)z + \right.\\ &\qquad + \varepsilon X'(t,\xi)Uz + \left[X(t,\xi_* + (I + \varepsilon U)z) - X(t,\xi_*) - \right.\\ &\qquad \left.\left. - X'(t,\xi_* - \varepsilon u)z\right]\right\}.\end{aligned} \tag{2.4.9}$$

Here, F is a known function of t, z, and ε. According to (9) and assumption 2) from Section 2.4.1, an estimate of type (7) occurs, and the Lipschitz condition

$$\begin{aligned}&\left|F(t,z,\varepsilon)\right| \leqslant C_F\left(\varepsilon + \varepsilon|z| + z^2\right), \qquad C_F = \text{const},\\ &\left|F(t,z_1,\varepsilon) - F(t,z_2,\varepsilon)\right| \leqslant \lambda_F|z_1 - z_2|\left(z_1 + |z_2| + \varepsilon\right), \qquad \lambda_F = \text{const},\end{aligned} \tag{2.4.10}$$

is valid for $[\xi_* + (I + \varepsilon U)z] \in D_x$.

Using the estimate (10), one can arrange that the perturbation is of the order of ε^2. Indeed, the function F can be represented as

$$F(t,z,\varepsilon) = \varepsilon F_\varepsilon(t,\varepsilon) + \varepsilon F_{\varepsilon z}(t,z,\varepsilon)z + zF_{z^2}(t,z,\varepsilon)z, \qquad (2.4.11)$$

where F_ε, $F_{\varepsilon z}$, F_{z^2} are uniformly bounded functions. Note that for $F \equiv 0$, (9) is the system for variations corresponding to (3). Taking the described properties into account, we make the change of the unknown variable z in system (9):

$$\begin{aligned} &z = \varepsilon Z(r + \varphi), \qquad Z(\tau) \equiv \frac{\partial \xi}{\partial x^0}, \\ &\varphi(t,\varepsilon) \equiv \int_{t_0}^{t} Z^{-1}(\sigma) F(s,0,\varepsilon)\, ds. \end{aligned} \qquad (2.4.12)$$

Here, Z is the fundamental matrix for the system for variations mentioned above, φ is a known uniformly bounded function and r is an unknown variable. The integral in (12) is uniformly bounded according to (10) and (11). As a result, we obtain the initial-value problem for the unknown vector r

$$\dot r = \varepsilon R(t,r,\varepsilon), \qquad r(t_0) = 0, \qquad t \in [t_0, \Theta\varepsilon^{-1}]. \qquad (2.4.13)$$

The function R is uniformly bounded and satisfies the Lipschitz condition:

$$\begin{aligned} &|R(t,r,\varepsilon)| \leqslant \varepsilon C_R, \qquad C_R = \text{const}, \\ &|R(t,r_1,\varepsilon) - R(t,r_2,\varepsilon)| \leqslant \varepsilon \lambda_R |r_1 - r_2|, \qquad \lambda_R = \text{const}. \end{aligned} \qquad (2.4.14)$$

To construct the solution for the initial-value problem (13) on an asymptotically large time interval $t \in [t_0, T]$, $T = \Theta\varepsilon^{-1}$, one can apply the recursive procedure of the method of successive approximations (Picard's method) in terms of powers of ε [1, 12, 47, 49, 75, 115].

2.4.3. CONSTRUCTION OF THE EXACT SOLUTION

The function r to be found is constructed by using the recursive scheme of Picard's method [47, 49, 75, 115]:

$$r_{j+1}(t,\varepsilon) = \varepsilon \int_{t_0}^{t} R(s, r_j(s,\varepsilon), \varepsilon)\, ds, \qquad r_0 \equiv 0, \quad j = 0, 1, \ldots. \qquad (2.4.15)$$

The unknown r also depends on the parameters t_0, x^0, but this dependence is not indicated for the sake of brevity. The following statement is valid.

THEOREM 2.3. *For sufficiently small* $\varepsilon > 0$, *successive approximations* (15) *uniformly converge to the unique solution of the initial-value problem* (13) *on an asymptotically large time interval* $t \in [t_0, \Theta\varepsilon^{-1}]$.

At first, we prove that any j_*th approximation is bounded by the value εM, where $M > 0$ is the constant defined below. Indeed, the quantity r_j can be represented as the sum

$$r_{j_*} = r_0 + (r_1 - r_0) + (r_2 - r_1) + \cdots + (r_{j_*} - r_{j_*-1}). \tag{2.4.16}$$

Then we use (14) and the estimate (for $t \in [t_0, \Theta\varepsilon^{-1}]$)

$$|r_1| \leqslant \varepsilon N, \qquad N = \max_t \left|Z^{-1}[F(\tau, \varepsilon Z\varphi, \varepsilon) - F(t, 0, \varepsilon)]\right|\Theta\varepsilon^{-2}.$$

Majorizing the sum (16) one can obtain

$$\begin{aligned} |r_{j_*}| &\leqslant \sum_{j=1}^{j_*} |r_j - r_{j-1}| \leqslant \sum_{j=1}^{\infty} |r_j - r_{j-1}| \\ &\leqslant \varepsilon N\left[1 + \sum_{j=1}^{\infty} (\varepsilon\lambda_R\Theta)^j\right] = \frac{\varepsilon N}{1 - \varepsilon\lambda_R\Theta} \leqslant \varepsilon M \end{aligned} \tag{2.4.17}$$

for $t \in [t_0, \Theta\varepsilon^{-1}]$, $\varepsilon \leqslant \varepsilon_0 < \lambda_R/\Theta$.

Moreover, relationships (15) and (17) imply that the successive approximations uniformly converge to a continuously differentiable function r_* when $\varepsilon < 1/\lambda_r\Theta$:

$$\lim_{j\to\infty} r_j = r_*(t, \varepsilon), \qquad \lim_{j\to\infty} \dot{r}_j = \dot{r}_*(t, \varepsilon). \tag{2.4.18}$$

After that we establish that the limit function $r_*(t, \varepsilon)$ from (18) is the solution of the initial-value problem (13). This assertion follows from the fact that the solution $\rho(t, \varepsilon)$ of the system

$$\dot{\rho} = \varepsilon R(t, r_*(t, \varepsilon), \varepsilon), \qquad \rho(t_0) = 0, \qquad t \in \left[t_0, \Theta\varepsilon^{-1}\right],$$

coincides with $r_*(t, \varepsilon)$.

Finally, by using the rule of contradiction one can prove that the constructed solution $r_*(t,\varepsilon)$ is unique. Let ρ_1 and ρ_2 be two different solutions of the initial-value problem (13). Then the identities

$$\rho_i(t,\varepsilon) \equiv \varepsilon \int_{t_0}^{t} R\big(s, \rho_i(s,\varepsilon), \varepsilon\big)\, \mathrm{d}s, \qquad i = 1, 2,$$

and (14) imply the inequality

$$\max_t |\rho_1 - \rho_2| \leqslant \varepsilon\lambda_R\Theta \max_t |\rho_1 - \rho_2|, \qquad \varepsilon\lambda_R\Theta < 1,$$

which holds only if $\rho_1 \equiv \rho_2$ for all $t \in [t_0, \Theta\varepsilon^{-}1]$. This completes the proof of the theorem.

Thus, having substituted the constructed function $r_*(t,\varepsilon)$ into (12), then the known $z(t,\varepsilon)$ into the expression (8) for δ and, finally, the known function $\delta(t,\varepsilon)$ into (5), we obtain the exact solution of the initial-value problem (1) in the form (η is a uniformly bounded function)

$$\begin{aligned} &x\big(t, t_0, x^0, \varepsilon\big) = \xi\big(\tau - \tau_0, x^0\big) + \varepsilon\eta\big(t, t_0, x^0, \varepsilon\big), \\ &\eta \equiv u + (I + \varepsilon U)Z(r_* + \varphi), \qquad t \in \big[t_0, \Theta\varepsilon^{-1}\big]. \end{aligned} \tag{2.4.19}$$

2.4.4. REMARKS

(1) By means of the change

$$\begin{aligned} &x = \xi^* + \delta, \qquad \xi^*(t,\tau,\varepsilon) + \varepsilon v\big(t, \xi(\tau)\big), \\ &v = v(t,\xi) = \int_{t_0}^{t} \big[X(s,\xi) - X_0(\xi)\big]\, \mathrm{d}s, \end{aligned} \tag{2.4.20}$$

where v and ξ^* are 2π-periodic functions of the explicit argument t, the original set of equations (1), is reduced to the form (6) with another function $\Phi = \Phi^*(t,\tau,\delta,\varepsilon)$. This function is 2π-periodic in t and satisfies the uniform estimate (7) with respect to ε and δ, since

$$\Phi^* = X\big(t, \xi^* + \delta\big) - X(t,\xi) - X'(t,\xi)\delta - \varepsilon^2 V' X_0(\xi). \tag{2.4.21}$$

Here, twice differentiation X with respect to x is assumed. Using the linear transformation of the unknown δ, close to the identity transformation, we obtain a system of the type (9)

$$\begin{aligned} &\dot{y} = \varepsilon X_0'(\xi)y + \varepsilon G(t,\tau,y,\varepsilon), \qquad y(t_0)=0, \\ &\delta = \big[I+\varepsilon V(t,\xi)\big]y, \qquad V(t,\xi) \equiv \frac{\partial v}{\partial \xi}, \qquad \xi = \xi(\tau), \\ &G \equiv (I+\varepsilon V)^{-1}\big[\Phi^* + \varepsilon X'(t,\xi)Vy - \varepsilon V' X_0 y - \varepsilon V X_0' y\big]. \end{aligned} \tag{2.4.22}$$

The function G in (22) is 2π-periodic in the explicit argument t. The estimate of the type (10) with respect to the variables ε and y is evidently valid for G, and this estimate can be represented in the form (11).

We apply, to the system (22), the method of successive approximations in terms of powers of ε, similar to that developed in Section 2.4.3. As a first approximation, vanishing for all $t \in [t_0, \Theta\varepsilon^{-1}]$ as $\varepsilon \to 0$, we consider the function

$$y_0 = \varepsilon\varphi(t,\varepsilon) \equiv \varepsilon Z(\tau)\int_{t_0}^{t} Z^{-1}(\sigma)G(s,\sigma,0,\varepsilon)\,ds, \qquad \sigma = \varepsilon s, \tag{2.4.23}$$

where φ is a function uniformly bounded on the above mentioned interval. The recursive scheme of Picard's method yields the solution $y(t,\varepsilon)$ with an arbitrary prescribed accuracy in terms of ε with the integrals:

$$y_{j+1}(t,\varepsilon) = \varepsilon Z(\tau)\int_{t_0}^{t} Z^{-1}(\sigma)G(s,\sigma,y_j(s,\varepsilon),\varepsilon)\,ds, \qquad j=0,1,\dots. \tag{2.4.24}$$

Successive approximations (23) and (24) uniformly converge to the unique solution of the initial-value problem (22), if $\varepsilon \in (0,\varepsilon_0]$ is sufficiently small. Thus, we again construct the exact solution of the form (20) for the initial-value problem (1). The approach given above may appear to be more convenient when constructing analytically the functions v and V, as well as Φ^* and G, for instance, in the form of a Fourier series in t. However, from (21), higher smoothness of the function X is required with respect to x, $x \in D_x$.

(2) Using the approach [1, 12, 75, 137] one can establish continuous differentiability of the solution (2) with respect to the initial data t_0 and x^0. At first, the uniform continuity with respect to some parameter μ, $\mu \in D_\mu$, is established for the limit function $r_*(t,\varepsilon,\mu)$ and, hence, for the solution x of the original system (1), provided the function X continuously depends on μ. This arises from the uniform convergence of the sequence (18) obtained by the method of Picard (15). After that,

based on the uniform continuity, the possibility of proceeding to the limit as $\Delta \to 0$ in the matrix differential equation

$$\frac{\mathrm{d}}{\mathrm{d}t}\left(\frac{x_\Delta - x}{\Delta}\right) = \varepsilon A(\tau, \Delta)\frac{x'_\Delta - x}{\Delta} \tag{2.4.25}$$

is proved. Here, x_Δ is the solution of system (1) with the initial condition $x_\Delta(t_0) = x^0 + \Delta$. The matrix A, continuously depending on the parameter Δ, is obtained, based on the mean-value theorem (Lagrange's). As a result of proceeding to the limit in (25), the linear matrix equation is obtained for the matrix P of partial derivatives of the solution x with respect to the initial data x^0 ($P = \partial x/\partial x^0$):

$$\dot{P} = \varepsilon X'\big(t, x(t,\varepsilon)\big)P, \qquad P(t_0) = I. \tag{2.4.26}$$

Here, I is the unit matrix. For the ith column p_i of the matrix P, the system for variations [47, 75] is obtained with corresponding initial conditions:

$$\dot{p}_i = \varepsilon X'(t,\varepsilon)p_i, \qquad p_{ii}(t_0) = 1, \qquad p_{ik}(t_0) = 0 \qquad (k \neq i). \tag{2.4.27}$$

An equation similar to (27) is obtained for the vector of the derivative of the solution with respect to t_0; the initial value being equal to $\partial x/\partial t_0|_{t_0} = -\varepsilon X(t_0, x^0)$ [75].

We construct the solution of the initial-value problem (26) in the form

$$P = (I + \varepsilon U_*)W, \qquad U_*(t,\varepsilon) = \int_{t_0}^{t} \left[X'(s, x(s,\varepsilon)) - X'_0(\xi(\sigma))\right]\mathrm{d}s. \tag{2.4.28}$$

Here, W is the unknown matrix, and the matrix U is uniformly bounded for $t \in [t_0, \Theta\varepsilon^{-1}]$. Differentiation of the expression (28) according to the system (26) gives

$$(I + \varepsilon U_*)\dot{W} = \varepsilon X'_0 W + \varepsilon^2 X' U_* W, \qquad W(t_0) = I. \tag{2.4.29}$$

The matrix W can be represented as a series with unknown matrix coefficients C_j ($j = 1, 2, \ldots$):

$$W = Z\big(I + \varepsilon C_1 + \varepsilon^2 C_2 + \cdots\big), \qquad C_j(t_0) = 0. \tag{2.4.30}$$

Substituting this expression into (29) one can obtain the recursive relationships for uniformly bounded coefficients $C_j(t,\varepsilon)$; $j \geqslant 1$:

$$C_{j+1} = \int_{t_0}^{t} Z^{-1}\big[\varepsilon\big(X'U_* - X_0'\big)ZC_j - U_*Z\dot{C}_j\big]\,ds,$$
$$C_1 = \varepsilon\int_{t_0}^{t} Z^{-1}\big[X'(s,\xi(\sigma))U_*(s,\varepsilon) - X_0'(\xi(\sigma))\big]Z\,ds,\ldots. \tag{2.4.31}$$

The series (30) uniformly converges to the unique solution of system (29) for all $t \in [t_0, \Theta\varepsilon^{-1}]$, provided $\varepsilon \in (0,\varepsilon_0]$ is sufficiently small. Thus, we have defined, according to (20), the matrix P characterizing the "sensitivity" of the solution $x(t,t_0,x^0,\varepsilon)$ (19) with respect to variations of the initial data [47, 75, 136].

(3) In the similar way, we can determine the sensitivity matrix K for the solution $x(t,t_0,x^0,\varepsilon,\mu)$ of the general system

$$\dot{x} = \varepsilon X(t,x,\mu), \qquad x(t_0) = x^0, \quad t_0 = t_0(\mu), \quad x^0 = x^0(\mu), \quad \mu \in D_\mu,$$

with respect to the parameter vector μ of an arbitrary dimension m, which may also include the initial data. To do this one can solve, using the approach expounded above, the initial-value problem for the matrix equation with initial conditions obtained by differentiation of the functions $x^0(\mu)$ and $t_0(\mu)$ with respect to μ:

$$\dot{K} = \varepsilon X'(t,x,\mu)K + \varepsilon\partial X(\tau,\xi,\mu)/\partial\mu,$$
$$K = \frac{Dx}{D\mu}, \qquad K(t_0) = \frac{\partial x^0}{\partial\mu} - \varepsilon X\big(t_0,x^0,\mu\big)\frac{\partial t_0}{\partial\mu}.$$

In particular, if the parameters x^0 and t_0 are fixed and do not depend on μ, a linear nonhomogeneous matrix equation with zero initial conditions will result.

(4) Consider now the initial-value problem for the standard system with rotating phase on an asymptotically large time interval $t \in [t_0, \Theta\varepsilon^{-1}]$ [38, 150]:

$$\dot{a} = \varepsilon A(a,\psi j), \qquad a(t_0) = a^0,$$
$$\dot{\psi} = \omega(a) + \varepsilon\Psi(a,\psi), \qquad \psi(t_0) = \psi^0. \tag{2.4.32}$$

Here, a is an n-vector of the slow variables, ψ is a scalar phase; A and Ψ are 2π-periodic functions of the fast variable ψ, the frequency $\omega(a) \geqslant \omega_0 > 0$. The right-hand sides of Equations (32) are supposed to be sufficiently smooth with respect to $a \in D_a$ and $\psi \in [\psi^0, \infty)$. The system (32) describes a wide class of perturbed, essentially nonlinear, oscillation and rotation processes.

The usual technique [150] (division of $\dot{a}$ by $\dot{\psi}$) allows obtaining the standard system of the form (1):

$$\frac{da}{d\psi} = \varepsilon \frac{A(a,\psi)}{\omega(a) + \varepsilon\Psi(a,\psi)}, \qquad a(\psi^0) = a^0. \tag{2.4.33}$$

For system (33), the assumptions of Section 2.4.1 are supposed to be fulfilled. Then, using the approach of Sections 2.4.2 and 2.4.3, we obtain a solution in a form similar to (19) on an asymptotically large interval of changing the fast variable ψ, $\psi \in [\psi^0, \psi_T]$, $\psi_T \sim \varepsilon^{-1}$:

$$\begin{aligned} &a = a(\psi, \psi^0, a^0, \varepsilon) = \alpha(\theta - \theta^0, a^0) + \varepsilon\gamma(\psi, \psi^0, a^0, \varepsilon), \\ &\theta = \varepsilon\psi, \quad \theta \in [\theta^0, \theta_T], \quad \theta^0 = \varepsilon\psi^0, \quad \theta_T = \varepsilon\psi_T \sim 1. \end{aligned}$$

This solution is smooth with respect to ψ^0 and $a^0 \in D_{a^0} \subseteq D_a$. Here, α is the solution of the averaged (with respect to ψ) system of the first approximation

$$d\alpha/d\theta = A_0(\alpha)/\omega(\alpha), \qquad \alpha(\theta^0) = a^0,$$

and γ is a function uniformly bounded for all $\psi \in [\psi^0, \psi_T]$.

Substitution of the expression for a into the equation for ψ (32) leads to the separable equation, uniquely connecting ψ and t:

$$t - t_0 = \int_{\psi_0}^{\psi} \frac{d\varphi}{\omega(a) + \varepsilon\Psi(a,\varphi)}, \qquad a = a(\psi, \varepsilon). \tag{2.4.34}$$

The dependence of the bounded function a on other arguments is not indicated for the sake of brevity. The unknown function $\psi(t,\varepsilon)$ and, hence, $a(t,\varepsilon)$ can be found from relationship (34) as successive approximations, based on the solution of the first approximation. The quantity $T = \Theta\varepsilon^{-1}$ is determined from (34) with $\psi = \psi_T$.

Thus, the successive approximation technique (the method of Picard) expounded above proves to be an efficient mode of analysis of non-steady-state oscillation processes on an asymptotically large time interval. The results obtained by this method are used when solving boundary-value problems of Pontryagin's maximum principle (Section 2.2).

2.5. Averaging in Optimal Control Problems for More General Systems

2.5.1. ASYMPTOTIC INVESTIGATION OF CONTROLLED QUASI-LINEAR SYSTEMS WITH AN ARBITRARY NUMBER OF PHASES

(1) *Statement of the problem.* For definiteness, we consider here the case of a fixed time interval. Some results obtained in Chapters 1 and 2 for single-frequency oscillations are also valid for multi-frequency controlled systems of the standard form (1.1.6), where $\psi = (\psi_1, \dots, \psi_{n_\psi})$, $n_\psi \geqslant 2$. Let the optimal control problem be reduced to the boundary-value problem of the maximum principle (1.2.8) and (1.2.9), where the phase ψ and the respective adjoint variable q are vectors of the dimension n_ψ. We also suppose that the right-hand sides of system (1.2.8) are sufficiently smooth functions of slow variables and phases and can be represented as absolutely and uniformly convergent Fourier series in ψ [38, 150]. Then the variable change (1.1.11) reduces Equation (1.2.8) to the form (1.1.12) where the number of slow variables is $2(n_a + n_\psi)$. In general, the right-hand side of the obtained system of the standard form will be a complicated nonperiodic function of t. The formal application of the methods expounded in Sections 1.2, 2.1, and 2.2, the calculation and investigation of averages over t for this function, and different combinations of its derivatives depending on slow variables is a rather difficult problem made more difficult by the variability of frequencies. Therefore, such an approach to the investigation of multi-frequency systems does not seem to be sufficiently constructive.

(2) *Difficulties of separating motions in multi-frequency time-varying systems.* It turns out that the most suitable way to investigate applicability conditions for different schemes of averaging is to use canonical transformations. The Hamiltonian function H^*, according to (1.2.6), has the form:

$$H^* = (\nu(\tau) \cdot q) + \varepsilon h^*(\tau, a, \psi, p, q), \qquad \tau = \varepsilon t. \tag{2.5.1}$$

In the general case of arbitrary behaviour of the vector $\nu(\tau)$, $\tau \in [\tau_0, \Theta]$, nobody succeeded in investigating this problem, since the formal application of the method of canonical separation of motions led to "small denominators" and divergent Fourier series [38, 69, 150]. Indeed, Equations (1.2)–(1.4) for the generating function S are reduced to

$$\begin{aligned} &S = (\psi \cdot \beta) + (a \cdot \eta) + \varepsilon\sigma(\tau, a, \psi, \eta, \beta, \varepsilon), \\ &\left(\nu(\tau) \cdot \frac{\partial \sigma}{\partial \psi}\right) + h^*\left(\tau, a, \psi, \eta + \varepsilon \frac{\partial \sigma}{\partial a}, \ \beta + \varepsilon \frac{\partial \sigma}{\partial \psi}\right) + \varepsilon \frac{\partial \sigma}{\partial \tau} \\ &\quad = k\left(\tau, a + \varepsilon \frac{\partial \sigma}{\partial \eta}, \eta, \beta, \varepsilon\right). \end{aligned} \tag{2.5.2}$$

The solution σ of Equation (2) and the averaged Hamiltonian k are constructed as series expansions in powers of ε in the form (1.5). For the so-called nonresonant case [1, 38, 69, 150],

$$|(\nu \cdot s)| \geqslant \delta(s) > 0, \qquad |s| \neq 0, \quad \tau \in [\tau_0, \Theta],$$
$$s = (s_1, s_2, \ldots, s_{n_\psi}), \qquad s_j = 0, \pm 1, \pm 2, \ldots, \quad j = 1, \ldots, n_\psi,$$

the unknown functions k_0 and σ_0 have the form

$$\begin{aligned} k_0(\tau, \xi, \eta, \beta) &= \frac{1}{(2\pi)^{n_\psi}} \int_0^{2\pi} \cdots \int_0^{2\pi} h^*(\tau, \xi, \psi, \eta, \beta)\, d\psi_1 \cdots d\psi_{n_\psi} \\ &= h_0(\tau, \xi, \eta, \beta), \qquad h^* = \sum_{|s| \geqslant 0} h_s^* e^{i(\psi \cdot s)}, \\ \sigma_0(\tau, a, \psi, \eta, \beta) &= \sum_{|s|>0} \frac{h_s^*(\tau, a, \eta, \beta)}{(\nu(\tau) \cdot s)} e^{i(\psi \cdot s)}, \qquad i = \sqrt{-1}. \end{aligned} \tag{2.5.3}$$

Since the set of rational numbers is everywhere dense, the nonresonance property can be fulfilled only under appropriate restrictive conditions on the behaviour of the vector function $\nu(\tau)$, $\tau \in [\tau_0, \Theta]$. Expression (3) will converge to a differentiable (with respect to τ, a, η, and β) function, if $h_s = O(\varepsilon)$ for $|s| \geqslant s^*$, where s^* is fixed, or a multiple Fourier series is transformed under $|s| \geqslant s^*$ into the sum of n_ψ one-dimensional series for each phase ψ_i, $i = 1, \ldots, n_\psi$. It is worth noting that the function σ_0 is not determined uniquely. It is not used to construct the solution of the first approximation but only to estimate the accuracy of this solution. Similar difficulties arise when calculating higher approximations, i.e. when constructing $\sigma_l(\tau, a, \psi, \eta, \beta)$ and $k_l(\tau, \xi, \eta, \beta)$, $l \geqslant 1$.

2.5.2. SPECIAL CASES

Those cases when there is a resonance between some frequencies (the average with respect to time differs from the average with respect to phases), a situation which can change in time, require a concrete investigation taking the specific features into account. Below, we consider some special cases encountered in applications which allow one to use constructive schemes of the averaging method. It is supposed that definite relationships between partial frequencies $\nu_i(\tau)$ are fulfilled uniformly for all $\tau \in [\tau_0, \Theta]$.

(1) *Commensurable frequencies*. Let all frequencies $\nu_i(\tau)$ be commensurable, i.e. there exist "not so large" positive integers r_i and r_j such that

$$\frac{\nu_i(\tau)}{r_i} = \frac{\nu_j(\tau)}{r_j} + \varepsilon\chi_{ij}(\tau), \qquad \chi_{ij} = -\chi_{ji}, \quad i, j = 1, 2, \ldots, n_\psi. \tag{2.5.4}$$

Then, according to Section 1.1, system (1.2.8) is reduced to a single-frequency system by a linear change of phases. Let us fix some i in (4), for instance, $i = n_\psi$; then the equations for the slow variables α_j are written as

$$\begin{aligned} &\dot{\alpha}_j = -\varepsilon r_j \chi_{ij} + \varepsilon\left(F_j - \frac{r_j}{r_i} F_i\right), \qquad \alpha_j(t_0) = \psi_j^0 - \frac{r_j}{r_i}\psi_i^0, \\ &\alpha_j \equiv \psi_j - (r_j/r_i)\psi_i, \qquad r_i \neq 0, \\ &\psi_i \equiv \psi_i \quad (a_i \equiv 0), \qquad i = n_\psi, \quad j = 1, \ldots, n_\psi - 1. \end{aligned} \tag{2.5.5}$$

The equation for the only phase $\psi = \psi_i$ conserves the form (1.2.8), taking account of the change (5) for the phases ψ_j. A linear time-invariant transformation of the state variables (phases ψ_i) with a nonsingular matrix A results in a linear transformation for the adjoint vector ρ corresponding to the vector α (Section 2.1.3):

$$\begin{aligned} &\begin{pmatrix} \alpha \\ \psi_i \end{pmatrix} = A\psi, \qquad \begin{pmatrix} \rho \\ p_i \end{pmatrix} = Bq, \\ &A = \begin{bmatrix} 1 & 0 & \ldots & -r_1/r_i \\ 0 & 1 & \ldots & -r_2/r_i \\ \ldots & \ldots & \ldots & \ldots \\ 0 & 0 & \ldots & 1 \end{bmatrix}, \qquad B^{-1} = A^T, \quad \det A = 1, \\ &\rho_j = q_j, \quad j = 1, \ldots, n_\psi - 1, \quad q_i = p_i - \sum_{j=1}^{i-1} \frac{r_j}{r_i}\rho_j, \quad i = n_\psi. \end{aligned} \tag{2.5.6}$$

The Hamiltonian function $H^*(1)$ is transformed as follows ($\partial S/\partial t \equiv 0$):

$$\begin{aligned} H^*(\tau, a, \psi, p, q, \varepsilon) &= K(\tau, a, \alpha, \psi_i, p, \rho, p_i, \varepsilon) \\ &= \nu_i p_i - \varepsilon \sum_{j=1}^{i-1} r_i \chi_{ij}\rho_j + \varepsilon k^*(\tau, a, \alpha, \psi_i, p, \rho, p_i, \varepsilon), \qquad i = n_\psi. \end{aligned} \tag{2.5.7}$$

Here, the function k^* is obtained from $h^*(1)$ after substitutions (6). Therefore, the boundary-value problem of the maximum principle for a multi-frequency system with the Hamiltonian H^* (1), where ψ is a vector of dimension $\eta_\psi > 1$, is reduced to the boundary-value problem for a single-frequency system with the Hamiltonian K (7), where ψ_i is a scalar phase, provided that full commensurability of the frequencies (4) takes place. However, this results in increasing the number of slow variables: the $2(n_\psi - 1)$-vector (α, ρ) is added. The averaging technique described in Chapters 1 and 2 is applicable to the new Hamiltonian system. Initial and terminal conditions and transversality conditions (1.2.9), as well as functional (1.2.2) and optimal control (1.2.5), are rewritten in new variables according to (6).

It is worth noting that, due to the periodicity of the function k^* in α, the averaging with respect to ψ_i may sometimes result in the situation when the dependence on some components of α disappears (false resonances [38, 150]). For example, for a two-frequency system with the function

$$h^* = h_1^*(\tau, a, \psi_1, p, q) + h_2^*(\tau, a, \psi_2, p, q) + O(\varepsilon),$$

the average $h_0 = \langle h^* \rangle$ is given by the expression

$$\begin{aligned} &h_0 = h_{10}(\tau, a, p, q) + h_{20}(\tau, a, p, q), \qquad q = (q_1, q_2), \\ &q(t, \varepsilon) = \text{const} + O(\varepsilon), \qquad t \in \left[t_0, \Theta\varepsilon^{-1}\right], \end{aligned}$$

where the relationship between $\nu_1(\tau)$ and $\nu_2(\tau)$ is arbitrary. A similar situation can arise when the function h^* in the first approximation is represented as a finite trigonometric sum.

(2) *Strictly incommensurable frequencies*. Suppose that, for $\tau \in [\tau_0, \Theta]$ under consideration, the relationship (4) is not satisfied for all $i, j = 1, 2, \ldots, n_\psi$, $i \neq j$, i.e. the rough incommensurability of frequencies takes place

$$\left| \frac{\nu_i(\tau)}{r_i} - \frac{\nu_j(\tau)}{r_j} \right| \geqslant \delta > 0, \qquad i \neq j. \tag{2.5.8}$$

Here, r_i and r_j are "not so large" relatively prime integers:

$$|r| = |r_1| + |r_2| + \cdots + |r_{n_\psi}| \leqslant r^*,$$

δ is fixed. For large r_i violating condition (8), Fourier coefficients of the function $h^*(\tau, a, \psi, p, q)$ decrease sufficiently rapidly:

$$h^* = \sum_{|r| \leqslant r^*} h_r^*(\tau, a, p, q) e^{i(r \cdot \psi)} + O(\varepsilon) \qquad \left(i = \sqrt{-1}\right). \tag{2.5.9}$$

Then the averaged system in the first approximation is described by the Hamiltonian function (see (1.3)–(1.7)):

$$K = \big(\nu(\tau)\cdot\beta\big) + \varepsilon h_0^*(\tau,\xi,\eta,\beta) + O(\varepsilon^2), \qquad \beta = \text{const.} \tag{2.5.10}$$

Therefore, in the case (8) and (9) of rough incommensurability of frequencies and for so-called "false resonances", the average system is obtained by independent averaging with respect to ψ_j, $j = 1,2,\ldots,n_\psi$. The expression for the accuracy estimate will contain (in the denominator) the quantity δ from (8). It is worth noting that the number of slow variables does not increase compared with Section 2.5.1, (1), and one has to integrate the set of $2n_a$ equations with respect to ξ, η. The constant n_ψ-vector β is to be determined from the transversality conditions (and the optimality of J_0 when T is not fixed, see Section 1.3). In particular, if the functional and terminal conditions do not depend on $\psi(T,\varepsilon)$, then $\beta = O(\varepsilon)$. The case when the commensurability relationships (4) leading to "false resonances" take place for some frequencies together with conditions (8) is considered analogously. The investigation is substantially simplified if the frequencies are nearly constant:

$$\nu_j(\tau) = \nu_j^0 + O(\varepsilon), \qquad \nu_j^0 = \text{const}, \quad j = 1,2,\ldots,n_\psi.$$

More general techniques for the investigation of multi-frequency systems, based on the method of successive changes (the averaging method and its varieties) and using metric and arithmetic properties of incommensurable frequencies, are given in [69, 123, 124, 150] and other publications. Significant results are obtained for two-frequency Hamiltonian systems [38, 69, 120, 123, 124, 130, 150].

(3) *General "rough" case*. Consider the situation when the frequencies $\nu_j(\tau)$, $j = 1,\ldots,n_\psi$, separate into l groups of the form

$$\begin{gathered}
\{\nu_j\} = \{\{\nu_{j_1}\},\ldots,\{\nu_{j_l}\}\}; \qquad \{\nu_{j_k}\} = \nu_{j_k}^{(k)}, \quad 1 \leqslant j_k \leqslant m_k, \\
k = 1,\ldots,l, \quad 1 \leqslant l \leqslant n_\psi, \quad m_1 + \cdots + m_l = n_\psi, \\
\nu_{j_k}^{(k)}(\tau)/r_{j_k}^{(k)} = \nu_{m_k}^{(k)}(\tau)/r_{m_k}^{(k)} + \varepsilon\chi_{m_k j_k}^{(k)}(\tau), \\
\big|\nu_{j_k}^{(k)}(\tau)/r_{j_k}^{(k)} - \nu_{j_s}^{(s)}(\tau)/r_{j_s}^{(s)}\big| \geqslant \delta > 0, \qquad k \neq s.
\end{gathered} \tag{2.5.11}$$

Here, r_{j_k} are "not so large" integers. Frequencies from different groups, i.e. corresponding to different k, are considered to be strongly incommensurable, while frequencies within each group are commensurable. Then, for $l = 1$, we have $k = 1$,

$m_k = m_l = n_\psi$, and the special case of commensurable frequencies takes place, similar to (4). If $l = n_\psi$, then all frequencies are roughly incommensurable.

Within each group, we make linear transformations of phases and respective adjoint variables of the type (5) and (6); $(m_k - 1)$-vectors of phase amounts off-resonance for each group are

$$\alpha_{j_k}^{(k)} = \psi_{j_k}^{(k)} - \left(r_{j_k}^{(k)}/r_{m_k}^{(k)}\right)\psi_{m_k}^{(k)}, \qquad \psi_{(k)} = \psi_{m_k}^{(k)}, \quad j_k = 1,\dots,m_k-1,$$
$$\rho_{j_k}^{(k)} = q_{j_k}^{(k)}, \qquad \rho_{m_k}^{(k)} = q_{m_k}^{(k)} + \sum_{j_k=1}^{m_k-1} \frac{r_{j_k}^{(k)}}{r_{m_k}^{(k)}}\rho_{j_k}^{(k)}. \tag{2.5.12}$$

The matrices $A^{(k)}$ and $B^{(k)}$ and adjoint vectors ($\rho^{(k)}$ and $q_{m_k}^{(k)}$) are written, according to (6), independently for different k; the general transformation has a block-diagonal structure. So, for the transformed Hamiltonian function K we obtain an expression similar to (7) ($\partial S/\partial t \equiv 0$):

$$\begin{aligned} H^* &= K \\ &= \sum_{k=1}^{l} \nu_{m_k}^{(k)}(\tau)\rho_{m_k}^{(k)} + \varepsilon\sum_{k=1}^{l}\sum_{j_k=1}^{m_k-1} \chi_{m_k}^{(k)}(\tau) r_{j_k}^{(k)}\rho_{j_k}^{(k)} + \\ &+ \varepsilon k^*\left(\tau, a, \alpha^{(1)}, \psi_{m_1}^{(1)}, \dots, \alpha^{(l)}, \psi_{m_l}^{(l)}, p, \rho^{(1)}, \rho_{m_1}^{(1)}, \dots, \rho^{(l)}, \rho_{m_l}^{(l)}, \varepsilon\right). \end{aligned} \tag{2.5.13}$$

The function k^* in (13) is obtained from h^* (1) by changes (12).

As a result of averaging with respect to the phases $(\psi_{(1)}, \dots \psi_{(l)})$ (the frequencies corresponding to them are strongly incommensurable), the order of the system decreases by l, and the adjoint variables β_l = const. In the case of "false resonances", the averaged Hamiltonian does not depend on some components of the vectors $\alpha^{(1)}, \dots, \alpha^{(l)}$, and the respective momenta are also constant in the first approximation.

The problem of studying multi-frequency systems remains a challenge. More subtle cases than those considered above cause significant difficulties for the application of the averaging method. The system can "pass resonances" in some frequencies but "stick at resonances" with respect to others [15, 38, 69, 115, 123, 124, 150], these conditions being able to change in time. The situation is significantly simplified when investigating controlled oscillatory motions of multi-frequency systems on a relatively short time interval $t \in [t_0, \Theta\varepsilon^{-1/2}]$ ([150] and Section 4.4).

2.5.3. EXAMPLE OF TWO-FREQUENCY SYSTEM

To illustrate the developed technique, we give the solution of the control problem for a nonlinear oscillatory system acted on by a harmonic external force, both in resonance and nonresonance cases.

(1) *Model example*. Consider the controlled system on a fixed time interval in the absence of constraints on the control:

$$\begin{aligned} &\ddot{x} + \nu^2 x = \varepsilon u + \varepsilon\mu \sin t, \qquad x(0) = x^0, \quad \dot{x}(0) = \dot{x}^0, \\ &|u| < \infty, \quad t \in \left[0, \Theta\varepsilon^{-1}\right], \quad \mu, \nu, \Theta = \text{const}, \\ &x(T) = x^*, \quad \dot{x}(T) = \dot{x}^*, \quad J[u] = \frac{1}{2}\int_0^\Theta u^2\, d\tau, \quad \tau = \varepsilon t. \end{aligned} \tag{2.5.14}$$

Problem (14) admits the exact solution. To make a comparison of the exact and the approximate solutions more convenient, we introduce Van der Pohl's variables a and b, see (1.1.13)

$$\begin{aligned} &x = a\cos\nu t + b\sin\nu t, \qquad \dot{x} = \nu(-a\sin\nu t + b\cos\nu t), \\ &\dot{a} = -\varepsilon(u + \mu\sin t)\sin\nu t, \qquad \dot{b} = \varepsilon(u + \mu\sin t)\cos\nu t, \\ &a^0 = x^0, \qquad a^* = x^*\cos\nu T - \dot{x}^*\nu^{-1}\sin\nu T, \\ &b^0 = \dot{x}^0\nu^{-1}, \qquad b^* = x^*\sin\nu T + \dot{x}^*\nu^{-1}\cos\nu T. \end{aligned} \tag{2.5.15}$$

According to the maximum principle, the optimal control is (p_a and p_b are adjoint variables)

$$u^* = -p_a\sin\nu t + p_b\cos\nu t, \qquad p_a, p_b = \text{const}. \tag{2.5.16}$$

The constants p_a and p_b are determined from the boundary conditions (15):

$$\begin{aligned} &p_0 = \varepsilon\,\frac{f_{cc}(T)\left[\Delta a + \varepsilon\mu f_s(T)\right] + f_{sc}(T)\left[\Delta b - \varepsilon\mu f_c(T)\right]}{\varepsilon^2\left[f_{ss}(T)f_{cc}(T) - f_{sc}^2(T)\right]}, \\ &p_b = \varepsilon\,\frac{f_{sc}(T)\left[\Delta a + \varepsilon\mu f_s(T)\right] + f_{ss}(T)\left[\Delta b - \varepsilon\mu f_c(T)\right]}{\varepsilon^2\left[f_{ss}(T)f_{cc}(T) - f_{sc}^2(T)\right]}, \\ &\Delta a = a^* - a^0, \qquad \Delta b = b^* - b^0, \end{aligned}$$

$$f_{ss}(t) = \frac{1}{2}\Big(t - \frac{1}{2\nu}\sin 2\nu t\Big), \qquad f_{cc}(t) = \frac{1}{2}\Big(t + \frac{1}{2\nu}\sin 2\nu t\Big), \tag{2.5.17}$$
$$f_{ss}(t)f_{cc}(t) - f_{sc}^2(t) = \frac{1}{4}\Big(t^2 - \frac{1}{\nu^2}\sin^2 \nu t\Big) > 0,$$
$$f_{sc}(t) = \frac{1}{2\nu}\sin^2 \nu t, \qquad f_s(t) = \frac{1}{2}\Big[\frac{\sin(1-\nu)t}{1-\nu} - \frac{\sin(1+\nu)t}{1+\nu}\Big],$$
$$f_c(t) = \frac{1}{2}\Big[\frac{1-\cos(1-\nu)t}{1-\nu} + \frac{1-\cos(1+\nu)t}{1+\nu}\Big].$$

It is worth noting that the order of both numerators and denominators in expressions (17) for p_a and p_b, in terms of the parameter ε, is zero, i.e. $p_a = O(1)$, $p_b = O(1)$, since $\varepsilon T = \Theta \sim 1$. The expressions for $a(t,\varepsilon)$ and $b(t,\varepsilon)$ are written easily using the expressions introduced in (15)

$$a(t,\varepsilon) = a^0 + \varepsilon\big[p_a f_{ss}(t) - p_b f_{sc}(t)\big] - \varepsilon\mu f_s(t),$$
$$b(t,\varepsilon) = b^0 - \varepsilon\big[p_a f_{sc}(t) - p_b f_{cc}(t)\big] + \varepsilon\mu f_c(t). \tag{2.5.18}$$

Since $T = \Theta\varepsilon^{-1}$ is asymptotically large, one can assume, for the sake of simplicity, that $T = N(2\pi/\nu)$, where $N \gg 1$ is a natural number. Then, according to (15), $f_{ss}(T) = f_{cc}(T) = T/2$, $f_{sc}(T) = 0$, and the expressions for p_a and p_b are considerably simplified:

$$p_a = \frac{2}{\Theta}\big[\Delta a + \varepsilon\mu f_s(T)\big], \qquad p_b = \frac{2}{\Theta}\big[\Delta b - \varepsilon\mu f_s(T)\big]. \tag{2.5.19}$$

The control $u_p^*(t)$ to be found is obtained after substitution of p_a and p_b from (17) or (19) into (16).

In the nonresonant case, for $(1-\nu) \sim 1$, one immediately establishes from (17) and (18), using the averaging method, that to the first approximation with respect to ε, the external force has no influence:

$$u_p^*(t) = -\frac{2}{\Theta}(\Delta a \sin \nu t - \Delta b \cos \nu t),$$
$$a = a^0 + \Delta a\frac{\tau}{\Theta}, \qquad b = b^0 + \Delta b\frac{\tau}{\Theta}, \qquad \tau \in [0,\Theta], \tag{2.5.20}$$
$$J_0 = \big(\Delta a^2 + \Delta b^2\big)\Theta^{-1}, \qquad p_a = 2\Delta a\Theta^{-1}, \qquad p_b = 2\Delta b\Theta^{-1}.$$

In the resonant case, for $1 - \nu = \varepsilon\gamma$, $\gamma \sim 1$, one can obtain from the exact solution, using the averaging method, the expressions for p_a, p_b and u_p^*, J_0 to the first approximation with respect to ε:

$$p_a = \frac{1}{\Theta}\left(2\Delta a + \frac{\mu}{\gamma}\sin\gamma\Theta\right), \qquad p_b = \frac{1}{\Theta}\left[2\Delta b - \frac{\mu}{\gamma}(1-\cos\theta)\right],$$
$$u_p^*(t) = -p_a\sin\nu t + p_b\cos\nu t, \qquad J_0 = \frac{\Theta}{4}(p_a^2 + p_b^2). \tag{2.5.21}$$

As has been mentioned above, the optimal control in the feedback form ($\Theta \to \Theta - \tau$, $a^0 \to a$, $b^0 \to b$) has a singularity as $\tau \to \Theta$, $\tau \in [\tau_0, \Theta]$.

In the general case, expressions (21) significantly differ from the respective expressions (20), since in the case of resonance, small external forces give rise to a significant change of slow variables, thus "producing significant work". Under the exact resonance ($\gamma = 0$), the expressions for p_a and p_b are simplified: $p_a = 2\Delta a\Theta^{-1} + \mu$, $p_b = 2\Delta b\Theta^{-1}$.

Below, we consider an example of the problem with constrained control which illustrates the influence of periodic perturbations on the qualitative behaviour of the solution.

(2) *Time-optimal control problem*. Consider the time-optimal criterion for the two-point problem (15) (T is not fixed beforehand):

$$J[u] = \varepsilon T = \Theta \longrightarrow \min_u, \qquad |u| \leqslant u^0. \tag{2.5.22}$$

The exact solution for problem (15) and (22) is not known. To a first approximation with respect to ε, the optimal control and the Hamiltonian (averaged) are

$$u^*(t) = u^0\,\mathrm{sign}(-p_a\sin\nu t + p_b\cos\nu t), \qquad p_a, p_b = \mathrm{const},$$
$$h_0 = \varepsilon^{-1}\langle H^*\rangle = -1 + (2/\pi)u^0(p_a^2 + p_b^2)^{1/2} - \mu p_a\varphi_s(\tau),$$
$$\varphi_s(\tau) \equiv 0, \qquad (1-\nu) \sim 1;$$
$$\varphi_s(\tau) = (1/2)\cos\gamma\tau, \qquad 1 - \nu = \varepsilon\gamma, \quad \gamma \sim 1. \tag{2.5.23}$$

In the nonresonant case, the approximate solution is determined uniquely and has the form (the action of the external force is not developed in the first approximation)

$$a(\tau) = a^0 + \Delta a\frac{\tau}{\Theta}, \qquad b(\tau) = b^0 + \Delta b\frac{\tau}{\Theta},$$
$$\Theta = (\pi/2)(\Delta/u^0), \qquad u^*(t) = u^0\,\mathrm{sign}\cos(\nu t + \theta), \tag{2.5.24}$$
$$\cos\theta = \Delta b/\Delta, \qquad \sin\theta = \Delta a/\Delta, \qquad \Delta = (\Delta a^2 + \Delta b^2)^{1/2}.$$

In the case of resonance, according to (23), the averaged variables $a = a(\tau)$, $b = b(\tau)$ and the optimal time Θ are given by

$$\begin{aligned} &a(\tau) = a^0 + (2/\pi)u^0\tau\cos\theta - (\mu/2\gamma)\sin\gamma\tau, \qquad a(\Theta) = a^*, \\ &b(\tau) = b^0 + (2/\pi)u^0\tau\sin\theta, \qquad b(\Theta) = b^*, \quad \Theta > 0, \\ &\left[\Delta a + (\mu/2\gamma)\sin\gamma\Theta\right]^2 + \Delta b^2 = (2/\pi)^2\left(u^0\Theta\right)^2. \end{aligned} \tag{2.5.25}$$

Here, θ and Θ are the unknowns. Having defined the optimal time $\Theta > 0$ from the last transcendental equation (25) we can easily express $\cos\theta$ and $\sin\theta$ from the boundary conditions for a and b. It is easy to prove that for $\gamma \neq 0$, this equation always has a solution $\Theta > 0$, this solution being unique and close to (24) as $\mu \to 0$. For sufficiently large values of $|\mu|$, the solution is not unique and, moreover, multiple roots are possible. If $|\mu| \to \infty$, then the number of the roots also unlimitedly increases as $|\mu|$, with the maximal values of these roots increasing by the same rate. The solution of the time-optimal control problem is defined by the minimal root $\Theta^* = \min \Theta_i \geqslant 0$. The construction of a complete picture reflecting the behaviour of the roots as functions of the system parameters causes no difficulties.

Let the exact resonance take place: $\gamma = 0$. Then the transcendental equation for Θ (25) becomes quadratic:

$$\begin{aligned} &\left(\Delta a + \frac{1}{2}\mu\Theta\right)^2 - (2/\pi)^2\left(u^0\Theta\right)^2 + \Delta b^2 = 0, \\ &\Theta_{1,2} = \frac{\mu\Delta a \pm \left\{\mu^2\Delta a^2 + 4\Delta^2\left[(2/\pi)^2u^{02} - \mu^2/4\right]\right\}^{1/2}}{2\left[(2/\pi)^2u^{02} - \mu^2/4\right]}. \end{aligned} \tag{2.5.26}$$

The analysis of the roots (26) shows that, provided a sufficiently small external force $|\mu| < (4/\pi)u^0$, the solution of the problem exists and is unique: $\Theta = \Theta_1 > 0$. If the inequality $|\mu\Delta b| > (4/\pi)u^0\Delta$ holds, then the control problem is not solvable.

It may happen that the equation has two positive roots $\Theta_{1,2} > 0$:

$$\mu\Delta a < 0, \qquad (4/\pi)u^0\Delta|\Delta b|^{-1} > |\mu| > (4/\pi)u^0.$$

In this case, $\Theta^* = \Theta_2$; in the limit for $|\mu| = (4/\pi)u^0\Delta|\Delta b|^{-1}$ both roots coincide, i.e. Θ is a double root. If $\Delta b = 0$, then the control problem is solvable under the condition

$$\Theta_{1,2} = \frac{\Delta a}{2}\,\frac{\mu \pm 4u^0/\pi}{(4u^0/\pi)^2 - \mu^2} = -\frac{\Delta a}{2}\,\frac{1}{\mu \mp 4u^0/\pi} > 0.$$

However, if $\Delta a = 0$, then the solution exists and is unique under $|\mu| < 4u^0/\pi$. In the critical case $|\mu| = 4u^0/\pi$, the solution also exists and is unique if $\mu\Delta a < 0$, see above. Therefore, the presence of resonant terms qualitatively changes the control scheme for problems with constraints on the control function compared with unconstrained problems. The optimal control in the feedback form is found by using similar constructions if one takes $t = t_0$ as the initial time and then makes the change $t_0 \to t$, $a^0 \to a$, $b^0 \to b$ in the obtained expressions for Θ, p_a, p_b, and u^*.

2.5.4. AVERAGING IN THE PROBLEM OF CONTROL OF MOTION FOR MORE GENERAL SYSTEMS

The methods of the approximate solution of optimal control problems developed above are applicable to weakly controlled systems of a more general form than (1.1.6), namely,

$$\begin{aligned} \dot{a} &= A(\tau, a, \psi, z, u, \varepsilon), & a(t_0) &= a^0, \\ \dot{\psi} &= \nu(\tau) + \Psi(\tau, a, \psi, z, u, \varepsilon), & \psi(t_0) &= \psi^0, \\ \dot{z} &= \Lambda(\tau)z + Z(\tau, a, \psi, z, u, \varepsilon), & z(t_0) &= z^0. \end{aligned} \tag{2.5.27}$$

Here, a, ψ, and z are vectors of arbitrary dimensions; $\Lambda(\tau)$ is an asympotically stable matrix, i.e. its eigenvalues $\lambda_i(\tau)$, $\tau \in [\tau_0, \Theta]$, have negative real parts. Perturbing functions A, Ψ and Z are 2π-periodic in $\psi = (\psi_1, \dots, \psi_{n_\psi})$ and sufficiently smooth in $(a, z) \in D_a \times D_z$, where D_z is a neighbourhood of the point $z = 0$. Let us suppose, in addition, that the following estimates are valid

$$A, \Psi, Z = O(\varepsilon + |z|^2), \quad \varepsilon \in [0, \varepsilon_0], \quad (a, z) \in D_a \times D_z, \quad \psi \in R^{n_\psi}. \tag{2.5.28}$$

Under appropriate assumptions concerning the smallness of ε_0, $|z|$ and the properties (boundedness and smoothness) of the control function $u = u(t, \varepsilon)$ or $u = u(t, a, \psi, z, \varepsilon)$, the system (25), with the estimates (28), will move within an ε-neighbourhood of the asymptotically stable local integral manifold $a \in D_a$, $\psi \in R^{n_\psi}$, $z = 0$ [38, 74, 120] when

$$T \geqslant t \geqslant t^0 = t_0 + \Delta t_0, \qquad T = \Theta\varepsilon^{-1}, \qquad \Delta t_0 \sim \ln \varepsilon^{-1}.$$

Equations (1.1.1) are reduced to the set (27) and (28), if, provided $\varepsilon = 0$, $\tau, x = \text{const}$, the generating family of solutions (1.1.3) is asymptotically stable for the considered values of $\tau \in [\tau_0, \Theta]$ and $a = (c, x) \in D_a$. The dimension n_c of the vector c is considered to satisfy the condition $n_c \leqslant n_y - n_\psi$. The case $n_c = n_y - n_\psi$, when the vector z is absent, has been considered in Chapters 1 and 2; if $n_c < n_y - n_\psi$, then, based on the Lyapunov–Floquet theory [75, 115], the original system (1.1.1) is reduced to the form (27), (28) by a nonsingular transformation $(x, y) \to (a, \psi, z)$ [38, 120, 150].

For the system (27) and (28), the following optimal control problem with respect to the variables a and ψ is stated:

$$M(\tau, a, \psi)\big|_T = 0, \qquad J[u] = g(\tau, a, \psi)\big|_T \longrightarrow \min_u, \qquad u \in U. \tag{2.5.29}$$

In what follows, we assume that $t \in [t_0, T]$, i.e. $z(t) = O(\varepsilon)$; then the influence of the vector z on the variables a and ψ will be of the order of ε for $t - t^0 \sim 1/\varepsilon$. Hence, in the first approximation given by the averaging method, one can put $z = 0$ and consider the optimal control problem to be independent of z. This assertion is proved by means of estimates for averages of A, Ψ, Z and the variable r, adjoint with the vector z. These estimates have the form

$$\begin{aligned}
&\lim_{T\to\infty} \frac{1}{T} \int_{t^0}^{T} \Phi^*\left(\tau, \alpha, \psi_0(t), z_0(t), p, q, r, \varepsilon\right) dt \\
&\quad = \lim_{T\to\infty} \frac{1}{T} \int_{t^0}^{T} \Phi^*\left(\tau, \alpha, \psi_0(t), 0, p, q, 0, 0, \varepsilon\right) dt + O(\varepsilon^2), \\
&\psi_0(t) = \int_{t^0}^{t} \nu(\tau')\, dt' + \psi^0, \qquad z_0(t) = z^0 \exp\left[\int_{t^0}^{t} \Lambda(\tau')\, dt'\right], \\
&r(t) = O(\varepsilon), \qquad \Phi = (A, \Psi, Z), \qquad \Phi^* \equiv \Phi\big|_{u^*}.
\end{aligned} \tag{2.5.30}$$

Here, A^*, Ψ^*, and Z^* are 2π-periodic functions of ψ obtained by the substitution of $u = u^*(\tau, a, \psi, z, p, q, r, \varepsilon)$ into the perturbations A, Ψ, and Z. For this, one should remember that

$$\begin{aligned}
&\Phi = \varepsilon\Phi^{(1)}(\tau, a, \psi, z, u, \varepsilon) + \Phi^{(2)}(\tau, a, \psi, z, \varepsilon), \\
&\Phi^{(2)} = O\left(|z|^2\right).
\end{aligned} \tag{2.5.31}$$

The estimate (30) for $r(t)$ under $t - t^0 \sim 1/\varepsilon$ is obtained from the equation and terminal condition

$$\dot r = -r\Lambda(\tau) - \frac{\partial}{\partial z}\left[(p \cdot A) + (q \cdot \Psi) + (r \cdot Z)\right]^*, \qquad r(T) = 0,$$

taking into account the estimates (31). Thus, to a first approximation with respect to ε, one can confine oneself to the solution of the optimal control problem (20) for the system

$$\begin{aligned} &\dot{a} = \varepsilon A^{(1)}(\tau, a, \psi, 0, u, 0), \qquad a(t^0) = a^0, \\ &\dot{\psi} = \nu(\tau) + \varepsilon \Psi^{(1)}(\tau, a, \psi, 0, u, 0), \qquad \psi(t^0) = \psi^0, \\ &\varepsilon A^{(1)} \equiv A(\tau, a, \psi, 0, u, \varepsilon), \qquad \varepsilon \Psi^{(1)} \equiv \Psi(\tau, a, \psi, 0, u, \varepsilon). \end{aligned} \tag{2.5.32}$$

If it is required to construct the solution of the optimal control problem (27) and (28) with a higher accuracy with respect to ε, then the averaging procedure (the successive change of variables) for the respective boundary-value problem should be generalized by using the scheme of the averaging method developed by Volosov [38, 149, 150]. The solution of problem (29) and (32) found on the basis of the approach given in Chapters 1 and 2 is assumed to be the solution of the first approximation.

2.6. Analysis of Some Special Aspects of Using the Averaging Method for Controlled Systems

2.6.1. AVERAGING OF QUASILINEAR SYSTEMS IN THE CASE OF SMALL FREQUENCIES

Consider the single-frequency system (1.2.13). If the frequency $\nu = \nu(\tau, \varepsilon)$ is asymptotically small with respect to ε or can become arbitrarily small for $t \in [t_0, \Theta\varepsilon^{-1}]$, $\Theta = \text{const}$, then the application of different averaging schemes given in Chapters 1 and 2 is significantly complicated. The difficulties arise as a result of both singularities in the right-hand sides of equations in the standard form (see (1.1.10), (1.1.14), etc.) and the violation of the basic condition for the existence of the uniform average with respect to t [38, 149, 150]. The schemes of averaging given above, as well as the error estimates, become inconsistent because corresponding expressions in the formulae for changes of the variables contain $\nu(\tau, \varepsilon)$ in the denominators (see (1.2.16), (2.1.7), (2.1.8), etc.). Simple examples show [12] that, in such cases, the proximity of the solution of the system formally averaged with respect to the phase and the solution of the original system, as a rule, does not take place for $t - t_0 \sim 1/\varepsilon$. The averaging method either becomes unsuitable for the approximate analysis of the motion on the interval $t \in [t_0, \Theta\varepsilon^{-1}]$ and one should use other asymptotic approaches, for example those given in Chapters 7 and 8, or more accurate error estimates are required which can turn out to be acceptable for the solution of applied problems. Below, we consider such cases and give some error estimates for the averaging method. For more detailed considerations, see [12]. For the sake of convenience, one can write down the system (1.2.13) using variables of

Van der Pohl's type $z = (a, \theta, p, q)$, having made a change of phase $\psi = \varphi + \theta$, $\dot{\psi} = \nu$ analogously to (1.1.11):

$$\begin{aligned} &\dot{z} = \varepsilon g(\tau, z, \varphi), \qquad \tau = \varepsilon t, \qquad z(t_0) = z^0, \\ &\dot{\psi} = \nu(\tau, \varepsilon), \qquad \varphi(t_0) = \varphi^0, \qquad t \in \left[t_0, \Theta\varepsilon^{-1}\right]. \end{aligned} \tag{2.6.1}$$

Here, z is a vector of an arbitrary dimension, φ is a scalar, and g is a function, 2π-periodic or quasiperiodic in φ and smooth in z.

(1) *Asymptotically small frequency.* Let the quantity ν satisfy the estimate $\nu = O(\varepsilon^\alpha)$, $\alpha > 0$, on the time interval $t \in [t_0, \Theta\varepsilon^{-1}]$, i.e. there exist constants $c_{1,2}$, $c_2 > c_1 > 0$, such that $c_1\varepsilon^\alpha \leqslant \nu(\tau, \varepsilon) \leqslant c_2\varepsilon^\alpha$. If $\alpha \geqslant 1$, then the variable φ is slow and formal averaging with respect to φ in general gives the error $O(1)$ for the variable z. For $\alpha < 1$, the changes

$$\begin{aligned} &\omega(\tau, \varepsilon) = \nu(\tau, \varepsilon)\varepsilon^{-\alpha}, \qquad c_2 \geqslant \omega \geqslant c_1 > 0, \qquad \theta = \varepsilon^\alpha t, \\ &\theta \in \left[\theta_0, \Theta\mu^{-1}\right], \qquad \mu = \varepsilon^{1-\alpha}, \qquad \mu \ll 1, \qquad \theta_0 = \varepsilon^\alpha t_0, \end{aligned} \tag{2.6.2}$$

reduce system (1) to the form (the prime denotes the derivative with respect to θ)

$$\begin{aligned} &z' = \mu g(\tau, z, \varphi), \qquad \tau = \varepsilon t = \mu\theta \in [\tau_0, \Theta], \\ &\varphi' = \omega(\tau, \varepsilon), \qquad \theta \in \left[\theta_0, \Theta\mu^{-1}\right]. \end{aligned} \tag{2.6.3}$$

One can apply to system (3) the above described schemes of averaging with respect to the phase φ, leading to errors $O(\mu^{j_*+1})$ on the interval $\theta \in [\theta_0, \Theta\mu^{-1}]$, i.e. $t \in [t_0, \Theta\varepsilon^{-1}]$; j_* being the order of the procedure of separation of motions, see Section 2.1. Variations of the slow variable z and the phase φ are of the order $O(1)$ and $O(\mu^{-1})$, respectively. This approach is justified if the expression for g does not contain the quantity ν in the denominator, e.g., has the form (1.1.27), and the initial and terminal conditions for z lead to values of the order $O(1)$.

Another situation is also possible (see (1.1.14) and others) when the function g has the form

$$g = g_1\nu^{-1} + g_2\nu'\nu^{-1} + g_3\left(Y\nu^{-2}\right)', \qquad g_{2,3} = O(1).$$

System (1) is reduced to the form (3) if the estimates

$$g_1 \sim \nu = O(\varepsilon^\alpha), \qquad \nu'\nu^{-1} = O(1), \qquad (Y\nu^{-2})' = O(1).$$

are valid.

(2) *Variable frequency; passing the zero value.* Suppose the frequency $\nu(\tau,\varepsilon)$ becomes asymptotically small, vanishes, or even changes its sign in the course of the control of the motion. Here, we assume that the function $\nu(\tau,\varepsilon)$ is approximated by one of the expressions

$$\begin{aligned}
&\nu(\tau,\varepsilon) = \nu_0(\tau)|\tau_* - \tau|^\alpha + \varepsilon^\beta \sigma(\tau), \qquad \alpha,\beta = \text{const} > 0;\\
&\nu(\tau,\varepsilon) = \nu_0(\tau)|\tau_* - \tau|^\alpha \operatorname{sign}(\tau_* - \tau) + \varepsilon^\beta \sigma(\tau);\\
&\nu_0(\tau) \geqslant \nu^0 > 0, \qquad |\sigma(\tau)| \leqslant \sigma^0,\\
&\nu^0, \sigma^0 = \text{const} > 0, \qquad \tau, \tau_* \in [\tau_0, \Theta].
\end{aligned} \tag{2.6.4}$$

Then the averaging in the system (1) and (4) leads to power-type singularities in the function of the change and in the estimates. It is worth noting that the second expression (4) becomes negative for $\tau > \tau_*$; such a situation can be encountered in control problems for rotations of a rigid body (for instance, (1.1.2), Sections 1.4.2 and 6.3). For definiteness, we consider in what follows the first expression (4) for $\sigma(\tau) \equiv 0$ or, what is the same, $\beta \geqslant 1$ (for $\beta < 1$, $\sigma \geqslant \sigma^0$, the estimates of Section 2.6.1, (1) are valid). A more general case, when $\sigma(\tau) \not\equiv 0$ and $\beta < 1$ in the second expression (4), is analyzed analogously to that given below.

Let $\zeta(\tau, z^0)$ be the known general solution of the formally averaged (with respect to φ) system (1) and (4) when $\nu = \nu_0(\tau)|\tau_* - \tau|^\alpha$. Then the estimate of the error $(z - \zeta)$ is obtained, as usual (see [35, 38]), by using Gronwall's lemma. Using the substitution

$$\begin{aligned}
&z = \zeta + v + \delta,\\
&v(t,\tau,\zeta,\varepsilon) = \varepsilon \int_{t_0}^{t} [g(\tau,\zeta,\varphi(s,\varepsilon)) - g_0(\tau,\zeta)]\, ds,
\end{aligned} \tag{2.6.5}$$

where g_0 is the average of g over φ and, when integrating, the variables τ and ζ are considered as parameters, we obtain the estimate for the unknown δ

$$\begin{aligned}
|\delta| &\leqslant e^{L\Theta} \max_{t\in[t_0,\Theta\varepsilon^{-1}]} \int_{t_0}^{t} \left(L|v| + M\left\|\frac{\delta v}{\partial \zeta}\right\| + \left|\frac{\partial v}{\partial \tau}\right| \right) ds\\
&\leqslant K\Theta e^{L\Theta} \max_{t\in[t_0,\Theta\varepsilon^{-1}]} \left(|v| + \left\|\frac{\delta v}{\partial \zeta}\right\| + \left|\frac{\partial v}{\partial \tau}\right| \right).
\end{aligned} \tag{2.6.6}$$

Here, $K > 0$ is a constant determined in terms of L (the Lipschitz constant of the function g in z) and M (the maximum of $|g_0(\tau, \zeta(\tau, z^0))|$ over τ, $\tau \in [\tau_0, \Theta]$). Thus, according to (6), the error $(z-\zeta)$ of the averaging method for $t \in [t_0, \Theta\varepsilon^{-1}]$ is determined by estimates of the function v and its derivatives, i.e. $|v|$, $\|\partial v/\partial\zeta\|$ and $|\partial v/\partial\tau|$. Below, we give a method for obtaining such estimates, suitable for the estimation of all three quantities.

Suppose we wish to estimate the quantity

$$w = \varepsilon \max_{t\in[0,\Theta\varepsilon^{-1}]} \left| \int_0^t f\big(\varphi(s,\varepsilon)\big)\mathrm{d}s \right|, \qquad \dot\varphi = \nu_0(\tau)|\tau_* - \tau|^\alpha, \tag{2.6.7}$$

in terms of the parameter ε. Here, f is a quasiperiodic bounded function of φ with zero average, see (5). Without loss of generality, we can assume $t_0 = 0$, $\varphi(0,\varepsilon) = 0$, $\tau_* \geqslant 0$ and $\nu_0 = \text{const} > 0$; the connection between τ and φ reduces to an expression of the form

$$\dot\varphi = \nu_0|\tau_* - \tau|^\alpha = \nu_0\left|\tau_*^{1+\alpha} - \frac{1+\alpha}{\nu_0}\gamma\right|^{\alpha/(1+\alpha)}, \qquad \gamma = \varepsilon\varphi. \tag{2.6.8}$$

Then, taking into account relationship (8), we write expression (7) for the quantity w to be estimated in the form

$$\begin{gathered} w = \frac{\varepsilon}{\nu_0} \max_{\varphi\in[0,\Gamma\varepsilon^{-1}]} \left| \int_0^\varphi f(\psi)\left|\tau_*^{1+\alpha} - \frac{1+\alpha}{\nu_0}\varepsilon\psi\right|^{-\alpha/(1+\alpha)} \mathrm{d}\psi \right|, \\ \Gamma = \frac{\nu_0}{1+\alpha}\left|\tau_*^{1+\alpha} - |\tau_* - \Theta|^{1+\alpha}\right|, \qquad \varphi \in \left[0, \Gamma\varepsilon^{-1}\right]. \end{gathered} \tag{2.6.9}$$

The quantity $\gamma = \gamma_* = \nu_0(1+\alpha)^{-1}\tau_*^{1+\alpha}$, for which the frequency $\tilde\nu(\gamma_*) = \nu(\tau_*) = 0$ according to (8) when $0 \leqslant \tau_* \leqslant \Theta$, satisfies the inequalities $0 \leqslant \gamma_* \leqslant \Gamma$. Using simple transformations and substitutions one can obtain the estimate for the expression of w (9)

$$\begin{gathered} |w| \leqslant D\varepsilon^\rho, \qquad \rho = (1+\alpha)^{-1}, \quad 0 < \rho \leqslant 1, \quad t \in \left[0, \Theta\varepsilon^{-1}\right], \\ D = \nu_0^{-\rho}\rho^{\alpha\rho}\left[\left| \int_0^{\gamma_*\varepsilon^{-1}} \frac{f(\gamma_*\varepsilon^{-1} - y)}{y^{\alpha\rho}}\mathrm{d}y \right| + \right. \\ \left. + \max_{\gamma\in[\gamma_*,\Gamma]} \left| \int_0^{(\gamma-\gamma_*)\varepsilon^{-1}} \frac{f(\gamma_*\varepsilon^{-1} - y)}{y^{\alpha\rho}}\mathrm{d}y \right|\right]. \end{gathered} \tag{2.6.10}$$

Since $F(y,\varepsilon) = f(\gamma_*\varepsilon^{-1} - y)$ is a quasiperiodic function of y, which has a zero average uniformly with respect to ε, and where the quantity $\alpha\rho$ satisfies the inequalities $0 \leqslant \alpha\rho < 1$, one can obtain uniform estimates for the integrals under the sign of absolute value in (10). Thus, it is required to estimate the improper integral I of the form

$$I = \int_0^{\varkappa\varepsilon^{-1}} \frac{F(y,\varepsilon)}{|y|^{\alpha\rho}}\, dy, \qquad |\varkappa| \leqslant \Gamma. \tag{2.6.11}$$

The integral I (11) has singularities when $y = 0$ and when $\varepsilon \to 0$, since the upper limit becomes infinite. The estimate of I is carried out in the following way: $I = I_1 + I_2$, then for $h > 0$ or $h < 0$ ($\varkappa > 0$ or $\varkappa < 0$) we have

$$\begin{aligned}
&|I| \leqslant |I_1| + |I_2|, \qquad I_1 = \int_0^h \frac{F}{|y|^{\alpha\rho}}\, dy, \quad I_2 = \int_0^{\varkappa\varepsilon^{-1}} \frac{F}{|y|^{\alpha\rho}}\, dy, \\
&|I_1| \leqslant a|h|^{\rho}, \qquad a = (1+\alpha) \max_{|y|<\infty} |F|, \\
&|I_2| \leqslant b|h|^{-\alpha\rho}, \qquad b = 2 \max_{|h|\leqslant|H|} \left| \int_h^H F\, dy \right|, \quad |H| < \infty.
\end{aligned} \tag{2.6.12}$$

The minimum over h for the estimate of I (12) is achieved under $|h| = \alpha b a^{-1}$; as a result, we obtain the following uniform estimate

$$|I| \leqslant a^{\alpha\rho} b^{\rho} \rho^{-1} \alpha^{-\alpha\rho}. \tag{2.6.13}$$

Therefore, the estimate to be found for the error of the averaging method for system (1) and (8) is reduced to the form

$$\begin{aligned}
&|z - \zeta| \leqslant C\Theta e^{L\Theta} \varepsilon^{\rho}, \qquad t \in \left[t_0, \Theta\varepsilon^{-1}\right], \\
&L, C = \text{const} > 0, \quad \rho = (1+\alpha)^{-1}, \quad \rho \in (0,1], \quad \alpha \in [0,\infty),
\end{aligned} \tag{2.6.14}$$

on the base of (5), (6), (10)–(13). Here, $C > 0$ is a constant determined in terms of the parameters K, ν_0, and α, introduced above, and corresponding constants similar to a and b, which characterize the functions v, $\partial v/\partial\zeta$, and $\partial v/\partial\tau$.

The analysis of system (1), with the frequency ν changing its sign according to the second expression (4), leads to estimates of the error of the averaging method, similar to (14). The question of the justification of the averaging method in the case of small and vanishing frequencies for more general controlled systems with slowly varying parameters requires further study.

(3) *Example.* Consider the problem of optimal stabilization of an axial rotation of a dynamically symmetric apparatus controlled by small torques. We study the problem of quenching the equatorial component of the vector ω of angular velocity of a rigid body, when the variable rate of axial rotation ω_3 is given. The equations of controlled rotations in the coordinate system based on the principal axes of inertia at the mass centre, and the corresponding boundary-value problem, take the form (see (1.1.24))

$$\begin{aligned} &I\dot{\omega}_1 + (I_3 - I)\omega_3\omega_2 = M_1, \qquad \omega_1(0) = \omega_1^0, \quad \omega_1(T) = 0, \\ &I\dot{\omega}_2 - (I_3 - I)\omega_3\omega_1 = M_2, \qquad \omega_2(0) = \omega_2^0, \quad \omega_2(T) = 0, \\ &I_3\dot{\omega}_3 = M_3, \qquad \omega_3(0) = \omega_3^0, \quad \omega_3(T) = \omega_3^T. \end{aligned} \tag{2.6.15}$$

Furthermore, we introduce, for convenience, the following dimensionless variables: the time t', phase coordinates $l_{1,2}$ and ν, the controls $u_{1,2}$, v, and a small parameter ε, and we write down the corresponding initial and terminal values:

$$\begin{aligned} &t' = \omega^0 t, \quad l_{1,2} = I\omega_{1,2}/L^0, \quad \nu = \left(I_3\omega_3/L^0\right)N, \\ &\varepsilon\alpha_{1,2}u_{1,2} = M_{1,2}/\left(L^0\omega^0\right), \quad \varepsilon v = \left(M_3/\left(L^0\omega^0\right)\right)N, \quad T' = \omega^0 T, \\ &l_{1,2}(0) = I\omega_{1,2}^0/L^0 \equiv l_{1,2}^0, \quad l_{1,2}(T') = 0, \\ &\nu\big|_{t'=0,T'} = \left(I_3\omega_3^{0,T}/L^0\right)N \equiv \nu^{0,T}, \quad N = (I_3 - I)L^0/\left(II_3\omega^0\right). \end{aligned} \tag{2.6.16}$$

Here, $\varepsilon\alpha_{1,2}$ ($\varepsilon \ll 1$) are small quantities characterizing the effectiveness of the controls, ω^0 and L^0 are initial values of the angular velocity vector ω and angular momentum $L = J\omega$, respectively, and the primes accompanying t' and T' will be omitted from now on. Using relation (16), we reduce the boundary-value problem (15) to the form

$$\begin{aligned} &\dot{l}_{1,2} \pm \nu l_{2,1} = \varepsilon\alpha_{1,2}u_{1,2}, \qquad l_{1,2}(0) = l_{1,2}^0, \qquad l_{1,2}(T) = 0, \\ &\dot{\nu} = \varepsilon v, \qquad \nu(0) = \nu^0, \qquad \nu(T) = \nu^T. \end{aligned} \tag{2.6.17}$$

We further assume that the time $t = T$ of termination of the control process is determined by varying the magnitude of $\nu(t)$, characterizing the angular velocity ω_3 of the axial rotation of the apparatus. This may depend on the constraints imposed on the control v, $|v| \leqslant v_0$, (i.e. on the axial torque M_3) and on the quantity $|\nu^T - \nu^0|$

(i.e. on the difference $(\omega_3^T - \omega_3^0)$). Thus, the control $v = v^*(t)$, the variable $\nu = \nu(\tau)$ and time T are as follows:

$$\begin{aligned} &v = v^*(t) = v_0 \operatorname{sign}\left(\nu^T - \nu^0\right), \qquad \nu = \nu(\tau) = \nu^0 + v^*\tau, \\ &\tau = \varepsilon t \in [0, \Theta], \qquad \Theta = \varepsilon T = \left|\nu^T - \nu^0\right| v_0^{-1} \qquad \left(v^* \equiv 0,\ t > T\right). \end{aligned} \tag{2.6.18}$$

We consider here a more general case, when the axial component of the angular velocity may change its sign, i.e. when there exists $\tau = \tau_* \in [0, \Theta]$ such that $\nu(\tau_*) = 0$. We pose the problem of control, optimal with respect to the energy consumed, over the variable $l_{1,2}$, with the instant T of the termination of the process fixed according to (18).

We introduce the osculating variables $a_{1,2}$ by means of the substitution

$$l_1 = a_1 \cos\varphi - a_2 \sin\varphi, \qquad l_2 = a_1 \sin\varphi + a_2 \cos\varphi, \qquad \dot\varphi = \nu,$$

where the phase φ is a known function of time. As a result of the transformation, the problem of optimal control is reduced to the standard form

$$\begin{aligned} &\dot a_1 = \varepsilon(\alpha_1 u_1 \cos\varphi + \alpha_2 u_2 \sin\varphi), \qquad a_1(0) = l_1^0, \quad a_1(T) = 0, \\ &\dot a_2 = \varepsilon(-\alpha_1 u_1 \sin\varphi + \alpha_2 u_2 \cos\varphi), \qquad a_2(0) = l_2^0, \quad a_2(T) = 0, \\ &\dot\varphi = \nu(\tau) = \nu^0 + v^*\tau, \qquad \varphi(0) = 0, \\ &\Phi[u_1, u_2] = \frac{1}{2}\int_0^\Theta (u_1^2 + u_2^2)\,d\tau \longrightarrow \min_{u_i, u_2}. \end{aligned} \tag{2.6.19}$$

Here, we assume that the quantity Θ is sufficiently large, therefore the possible constraints which can be imposed on the controls u_1 and u_2 are not attained and do not appear in the formulation of problem (19). According to Pontryagin's maximum principle, the momenta p_1 and p_2, adjoint with a_1 and a_2, respectively, are retained, i.e. $p_{1,2} = \text{const}$, and the optimal control is equal to

$$u_1 = \alpha_1(p_1 \cos\varphi - p_2 \sin\varphi), \qquad u_2 = \alpha_2(p_1 \sin\varphi + p_2 \cos\varphi). \tag{2.6.20}$$

Substituting the expressions for u_1 and u_2 (20) into (19) and taking the initial conditions into account only, we arrive at the special case of the initial-value problem, since its right-hand sides are known 2π-periodic functions of the phase φ. The solution of the boundary-value problem can also be obtained by elementary methods and leads to very bulky expressions. Averaging over φ enables us to obtain simple approximate expressions for the phase variables $\xi_{1,2}$ and momenta $\eta_{1,2}$

$$\begin{aligned} \xi_{1,2} &= l_{1,2}^0 + (1/2)\eta_{1,2}(\alpha_1^2 + \alpha_2^2) = l_{1,2}^0(1 - \tau\Theta^{-1}), \\ \eta_{1,2} &= -2l_{1,2}^0\Theta^{-1}(\alpha_1^2 + \alpha_2^2)^{-1}. \end{aligned} \tag{2.6.21}$$

Substituting $\eta_{1,2}$ (21) into the controls $u_{1,2}$ (20) in place of $p_{1,2}$ gives approximately optimal open-loop and feed-back control for $a_{1,2}$:

$$\begin{aligned} u_1^* &= -2\alpha_1(\alpha_1^2 + \alpha_2^2)^{-1}\Theta^{-1}(l_1^0 \cos\varphi - l_2^0 \sin\varphi), \\ u_2^* &= -2\alpha_2(\alpha_1^2 + \alpha_2^2)^{-1}\Theta^{-1}(l_1^0 \sin\varphi + l_2^0 \cos\varphi), \\ &\quad (l_{1,2}^0 \to l_{1,2},\ \Theta \to \Theta - \tau). \end{aligned} \tag{2.6.22}$$

The open-loop control $u_{1,2}^*(\varphi, l_1^0, l_2^0)$ (22) leads to the functions $a_{1,2}^*(t, l_1^0, l_2^0, \varepsilon)$, whose values at $t = T$ are in the $\sqrt{\varepsilon}$-proximity of zero. The value of the functional Φ (19) for $u_{1,2}^*$ (22) differs by an amount of the order of $O(\sqrt{\varepsilon})$ from the exact minimum value. It should be noted that the approximate solution (18), (21), and (22) of the problem in question holds also in the more general case: $|I_1 - I_2| = O(\varepsilon)$, $|I_{1,2} - I_3| = O(1)$. The proposed approach can also be used with a wider class of perturbations acting on system (15), depending on the vector ω and the slow time τ.

2.6.2. AVERAGING IN MULTI-FREQUENCY SYSTEMS WHEN PASSING THROUGH RESONANCES

The models considered in Section 2.5 correspond to special cases when, during the entire motion, either commensurability of frequencies (resonance is possible) or incommensurability (no resonance) takes place. In real conditions, the controlled system can "pass through resonances" [123]. Such situations are of considerable practical interest. The development of asymptotic methods for the investigation of such systems, and primarily the method of averaging, is of great theoretical and applied value. The general situation which can arise in a multi-frequency system is rather complicated and is not amenable to detailed analysis. We consider, in what follows, the case of the quasilinear system reducible to the two-frequency one, where passage through a resonance takes place when $t = t_* \in [t_0, T]$:

$$\begin{aligned} &\dot z = \varepsilon Z(\tau, z, \psi, \varphi), \qquad \dot\psi = \omega(\tau) + \varepsilon\Psi(\tau, z, \psi, \varphi), \\ &\dot\varphi = \nu(\tau), \qquad \nu(\tau) \geqslant \nu_0 > 0, \quad \omega n - \nu m = \Omega, \quad \Omega(\tau_*) = 0. \end{aligned} \tag{2.6.23}$$

Here, $m \geqslant 0$ and $n \geqslant 1$ are relatively simple not so large integers, and $\Omega(\tau)$ is a smooth function. Independent averaging of the function Z will lead to an error in determining the slow variable z. This error is defined by the properties of the function $\Omega(\tau)$ in the neighbourhood of the point $\tau = \tau_*$, i.e. by the rate of passing through the resonance. Let the power-type approximation be valid, see (4)

$$\Omega(\tau) = k(\tau)\left|\frac{\tau - \tau_*}{\Theta}\right|^{\gamma}, \qquad |k| \geqslant k_0 > 0, \quad 0 < \gamma < \infty.$$

Then the limit case $\gamma \to 0$ corresponds to the nonresonant case while $\gamma \to \infty$ satisfies the resonance relationship ("sticking at the resonance"). In the nonresonant case, independent averaging gives the error $O(\varepsilon)$ for $t - t_0 \sim 1/\varepsilon$, see above. For commensurable frequencies ω and ν, it is easy to show by simple examples that independent averaging can lead to the error $O(1)$. In the intermediate case $0 < \gamma < \infty$, an estimate of the type (14) takes place. In particular, based on the results [123], one can establish that when $\gamma = 1$ the error is $O(\varepsilon^{1/2})$, since the basic condition guaranteeing passage through resonances is valid: $|\omega'\nu - \nu'\omega| \geqslant \nu_0^2 k_0/n > 0$. Therefore, the inequality

$$\left|\omega'\nu - \nu'\omega\right| \geqslant \varkappa > 0, \qquad \tau \in [\tau_0, \Theta],$$

and the corresponding smoothness conditions imply the estimate $O(\varepsilon^{1/2})$ for the error of the averaged solution of system (23) with respect to the slow vector variable z and the phase ψ.

2.6.3. METHOD OF AVERAGING IN PROBLEMS WITH BANG-BANG CONTROLS

As has been mentioned in Section 1.2, the control problems reducing to equations with discontinuous right-hand sides cause substantial difficulties for the application of asymptotic methods. These difficulties are aggravated by the fact that one has to solve two-point boundary-value problems. In the case of bang-bang controls, right-hand sides of the equations have discontinuities of the first kind, the number of which is of the order of $1/\varepsilon$ on the time interval considered. Moreover, in real problems, optimal and quasioptimal trajectories, as a rule, slide along discontinuity surfaces which can be time-varying. This makes available results inapplicable [1, 60, 128], since the motion on the discontinuity surfaces may have an arbitrary (nonoscillating) behaviour. Sliding modes may appear due to singular controls which also correspond to discontinuity surfaces for the control functions.

It is convenient to illustrate the features of studying boundary-value problems of the maximum principle and application of the averaging technique for bang-bang systems by an example of the linear (relative to the control) system reduced to a standard form:

$$\begin{aligned}
&\dot z = \varepsilon g(t,z) + \varepsilon G(t,z)u, \qquad z(t_0) = z^0, \qquad z \in D_z,\\
&N(z)\big|_T = 0, \qquad J[u] = \varphi(z)\big|_T \longrightarrow \min_u, \qquad u \in U,\\
&U = \{u\colon\ |u_k| \leqslant u_k^0,\ k = 1,\dots,m\}, \qquad t \in [t_0, T], \quad T = \Theta\varepsilon^{-1},\\
&z = (z_1,\dots,z_n), \quad N = (N_1,\dots,N_l), \quad G - (n\times m).
\end{aligned} \tag{2.6.24}$$

Here, g, G, N, and φ are sufficiently smooth functions of z; g and G having uniform averages with respect to t. According to the maximum principle, the optimal control u^* and the Hamiltonian function H^* are (p is the adjoint vector corresponding to z)

$$\begin{aligned}
&u_k^* = u_k^0 \operatorname{sign}\left(\sum_{i=1}^{n} p_i G_{ik}\right), \qquad k = 1,\dots,m,\\
&H^*(t,z,p,\varepsilon) = \varepsilon(p\cdot g) + \varepsilon\big(p\cdot Gu^*\big).
\end{aligned} \tag{2.6.25}$$

The function H^* has a uniform average over t. It is required to find the solution for the boundary-value problem of the maximum principle:

$$\begin{aligned}
&\dot z = \varepsilon g(t,z) + \varepsilon G(t,z)u^*(t,z,p),\\
&\dot p = -\varepsilon p\,\frac{\partial g}{\partial z} - \varepsilon p\,\frac{\partial G}{\partial z}\,u^*(t,z,p),\\
&z(t_0) = z^0, \qquad N(z)\big|_T = 0, \qquad p(T) = \left(-\frac{\partial\varphi}{\partial z} + \lambda\frac{\partial N}{\partial z}\right)\bigg|_T.
\end{aligned} \tag{2.6.26}$$

In the extended space of variables (t,z,p), the right-hand sides of system (26) may have discontinuities on the surfaces $\Pi_k = 0$, which are defined by discontinuities of the functions $u_k^*(t,z,p)$:

$$\Pi_k(t,z,p) = \sum_{i=1}^{n} p_i G_{ik}(t,z) = 0, \qquad k = 1,\dots,m. \tag{2.6.27}$$

Sections $(t,z) = \text{const}$ of surfaces (27) are n-dimensional hyperplanes in the $(2n+1)$-dimensional space of points $S = (t,z,p)$. If, during the motion, the point

$$S(t) = (t,z,p) \notin \bigcup_{k=1}^{m} \Pi_k,$$

then the averaging technique developed above is applicable. But if the point $S(t) = (t, z(t), p(t))$ can intersect some discontinuity surfaces, then the averaging method for its applicability requires, as follows from [31, 60, 62, 64, 65, 101, 128, 155], that the trajectories "transfix" the surface (i.e. are transverse to it), i.e. scalar products of the tangential vector s and normal vectors ρ_k to discontinuity surfaces at the intersection points at the time instants $t = t_k^* - 0$ and $t = t_k^* + 0$ are not equal to zero:

$$\begin{aligned}
&|(s \cdot \rho_k)| \geqslant \chi > 0, \qquad S_k^* = S(t_k^*) = (t_k^*, z(t_k^*), p(t_k^*)),\\
&s = \left(1, \varepsilon g + \varepsilon G u^*, -\varepsilon p g_z' - \varepsilon p G_z' u^*\right),\\
&\rho_k = (\partial\Pi_k/\partial t, \partial\Pi_k/\partial z, \partial\Pi_k/\partial p).
\end{aligned} \tag{2.6.28}$$

Here, k is the ordinal number of the discontinuity surface Π_k (27) intersected by the trajectory $(t, z(t), p(t))$ at time $t = t_k^*$. At $t_k^* - 0$ and $t_k^* + 0$, the scalar product $(s \cdot \rho_k)$, see (28), must keep its sign unchanged. Otherwise, sliding motion along the discontinuity surface $\Pi_k = 0$ (27) can occur. When $\varepsilon > 0$ is sufficiently small, the scalar product $(s \cdot \rho_k)$, according to (28), is equal to

$$(s \cdot \rho_k) = \partial\Pi_k/\partial t + O(\varepsilon), \qquad \frac{\partial\Pi_k}{\partial t} = \sum_{i=1}^{n} p_i \frac{\partial G_{ik}(t, z)}{\partial t}. \tag{2.6.29}$$

It arises from (26)–(29) that the situation when $p \equiv 0$ on some interval (the case of singular controls [31, 62, 64, 83, 105, 131, 155]) is critical, since, certainly, $(s \cdot \rho_k) = 0$. It is worth noting that such situations are natural for multi-dimensional systems of the type (24), and asymptotic analysis requires taking specific features of the controlled system into account. Investigation of concrete examples shows that singular controls lead, as a rule, to the nonuniqueness of the solution of the problem in terms of both the trajectory and control. In typical situations, the absolute minimum of the functional is achieved, as in the example of Section 1.2 (the case of a terminal functional), or some part of the state coordinates of the system reaches the terminal manifold earlier than the other (problems analogous to the minimum-time ones, see Chapters 5 and 6), and so on. Solution of concrete control problems for mechanical oscillatory systems shows that the averaging method, combined with other asymptotic and numerical methods, can also be a sufficiently effective tool for the simplification of the analysis in the aforenamed complicated cases of degeneration of necessary optimality conditions.

CHAPTER 3

Averaging Method in Optimal Control Problems for Single-Frequency Essentially Nonlinear Systems

On the basis of the maximum principle, we develop an asymptotic approach to achieving optimal control for nonlinear rotatory-oscillatory systems reducible to a standard form with a scalar rotating phase. Some concrete problems of the optimal control of mechanical systems are solved to a first approximation.

3.1. Nonlinear Systems Reducible to the Standard Form of Controlled Systems with a Rotating Phase

3.1.1. REDUCTION OF EQUATIONS OF CONTROLLED NONLINEAR OSCILLATIONS TO A STANDARD FORM

Consider a weakly controlled system similar to (1.1.1):

$$\begin{aligned} \dot{x} &= \varepsilon X(x, y, u), \qquad x(t_0) = x^0, \qquad u(t) \in U, \\ \dot{y} &= Y_0(x, y) + \varepsilon Y(x, y, u), \qquad y(t_0) = y^0. \end{aligned} \tag{3.1.1}$$

The variables x and y are vectors of any dimension, and the functions X, Y_0, and Y possess the properties defined in Section 1.1. The difference lies in the assumption that the unperturbed phase ψ is scalar and has the form (see (1.1.3)):

$$\psi = \omega(a)(t - t_0) + \psi^0, \qquad a = (x, c), \qquad \omega(a) \geqslant \omega^0 > 0. \tag{3.1.2}$$

Here, the frequency ω depends on the controlled slow variables x and c. Using the transformation, similar to (1.1.3), one can arrive at the system of the form (1.1.4), (1.1.5), which can then be represented in a form similar to (1.1.6):

$$\begin{aligned} \dot{a} &= \varepsilon f(a, \psi, u), \qquad a(t_0) = a^0, \qquad a, a^0 \in D_a, \\ \dot{\psi} &= \omega(a) + \varepsilon F(a, \psi, u), \qquad \psi(t_0) = \psi^0, \qquad |\psi| < \infty. \end{aligned} \tag{3.1.3}$$

The functions f and F are supposed to be 2π-periodic in the scalar phase ψ.

3.1.2. EXAMPLES OF NONLINEAR SYSTEMS. REDUCTION TO THE STANDARD FORM

As an illustration, we give below some examples of controlled systems of type (1) and describe the ways of reducing them to the standard form (3). Such systems were studied for some particular types of control functions $u = u(t, a, \psi)$, for instance, those periodic in t and ψ. In recent years, methods for the analysis of a wider class of nonlinear oscillatory systems of the form (3), where the phase ψ is a vector of any dimension, have been developed intensively. Many publications have been devoted to asymptotic methods of analysis of essentially nonlinear oscillations [1, 4, 5, 8, 20, 35, 36, 38, 45, 46, 53, 60, 61, 69, 71, 73, 74, 77–82, 91, 94, 98, 100, 104, 115, 119, 120, 123, 124, 130, 133, 137, 147, 149, 150].

(1) *Rotatory-oscillatory system.* In applied mechanics, one often encounters a system having a single degree of freedom y and slow parameters x in the form [1, 150]

$$\begin{aligned} &\ddot{y} + Q(x, y, \dot{y}) = \varepsilon q(x, y, \dot{y}, u), \qquad y(t_0) = y^0, \qquad \dot{y}(t_0) = \dot{y}^0, \\ &\dot{x} = \varepsilon X(x, y, \dot{y}, u), \qquad x(t_0) = x^0, \qquad u \in U. \end{aligned} \tag{3.1.4}$$

We suppose that, provided $\varepsilon = 0$, $x = \text{const}$, and Equation (4) for y admits a solution describing rotations

$$\begin{aligned} &y = y_0(x, c, \psi) = \psi + \varphi(x, c, \psi), \\ &\dot{y} = \omega \frac{\partial y_0}{\partial \psi} = \omega\left(1 + \frac{\partial \varphi}{\partial \psi}\right), \end{aligned} \tag{3.1.5}$$

or oscillations

$$\begin{aligned} &y = y_0(x, c, \psi) = \varphi(x, c, \psi), \\ &\dot{y} = \omega \frac{\partial y_0}{\partial \psi} = \omega \frac{\partial \varphi}{\partial \psi}. \end{aligned} \tag{3.1.6}$$

Here, $\psi = \omega(x, c)(t - t_0) + \psi^0$ is a rotating phase, φ is a 2π-periodic function of ψ, c is the amplitude or the energy. In the case of rotations, and the functions Q, q, and X are 2π-periodic in y.

Regarding (5) or (6) to describe the change of y, $\dot{y}$, and x for c, ψ, and x and differentiating with respect to t, one can obtain, according to (4), the following system:

$$\frac{\partial y_0}{\partial c}\dot{c} + \frac{\partial y_0}{\partial \psi}\dot{\psi} + \varepsilon\left(\frac{\partial y_0}{\partial x} \cdot X\right) = \omega \frac{\partial y_0}{\partial \psi}, \qquad \dot{x} = \varepsilon X(x, y_0, \omega \partial y_0/\partial \psi, u),$$

$$\frac{\partial}{\partial c}\left(\omega\frac{\partial y_0}{\partial \psi}\right)\dot{c}+\omega\frac{\partial y_0^2}{\partial \psi^2}\dot{\psi}+\varepsilon\left(\frac{\partial}{\partial x}\left(\omega\frac{\partial y_0}{\partial \psi}\right)\cdot X\right)+Q\left(x,y_0,\omega\frac{\partial y_0}{\partial \psi}\right) \qquad (3.1.7)$$
$$=\varepsilon q\left(x,y_0,\omega\frac{\partial y_0}{\partial \psi},u\right).$$

Using the identity $\omega^2\partial^2 y_0/\partial\psi^2+Q\equiv 0$ (see Section 1.1) and then solving (7) with respect to $\dot{c}$ and $(\dot{\psi}-\omega)$ one obtains the system

$$\dot{c}=-\frac{\varepsilon}{W}\left[\frac{q}{\omega}\frac{\partial y_0}{\partial \psi}+\left(\left(\frac{\partial^2 y_0}{\partial \psi^2}\frac{\partial y_0}{\partial x}-\frac{\partial y_0}{\partial \psi}\frac{\partial^2 y_0}{\partial \psi\partial x}\right)\cdot X\right)\right],$$
$$\dot{\psi}=\omega+\frac{\varepsilon}{W}\left[\frac{q}{\omega}\frac{\partial y_0}{\partial c}-\left(\left(\frac{\partial y_0}{\partial c}\frac{\partial^2 y_0}{\partial \psi\partial x}-\frac{\partial^2 y_0}{\partial \psi\partial c}\frac{\partial y_0}{\partial x}\right)\cdot X\right)\right], \qquad (3.1.8)$$
$$c(t_0)=c^0,\qquad \psi(t_0)=\psi^0(y_0(x^0,c^0,\psi^0)=y^0\quad (\omega\partial y_0/\partial\psi)^0=\dot{y}^0),$$
$$\dot{x}=\varepsilon X\left(x,y_0(x,c,\psi),\omega\partial y_0/\partial\psi,u\right),\qquad x(t_0)=x^0.$$

The derivative of y_0 with respect to c in (8) is taken in a partial way. Here, the Wronskian W is a 2π-periodic function of ψ:

$$W=W(x,c,\psi)=\frac{1}{\omega}\det\frac{\partial(y_0,\dot{y}_0)}{\partial(c,\psi)}=\frac{\partial y_0}{\partial c}\frac{\partial^2 y_0}{\partial \psi^2}-\frac{\partial y_0}{\partial \psi}\frac{\partial^2 y_0}{\partial \psi\partial c}, \qquad (3.1.9)$$
$$W(x,c,\psi)\neq 0,\qquad x\in D_x,\quad c\in D_c,\quad |\psi|<\infty.$$

The system of the form (4) arises when studying mechanical systems having an unperturbed Lagrangian function L of the form $L=(1/2)m(x,y)\dot{y}^2-\Pi(x,y)$, where $m\geqslant m_0>0$ is the inertia characteristic, y and $\dot{y}$ are, respectively, the generalized coordinate and velocity, x is a vectorial parameter, and Π is the potential energy. In this case, the function Q in (4) has the form

$$Q(x,y,\dot{y})=\frac{1}{m}\left(\frac{1}{2}\frac{\partial m}{\partial y}\dot{y}^2+\frac{\partial \Pi}{\partial y}\right) \qquad (3.1.10)$$

and, for rotations, it has to be periodic in y with a zero mean value. If the function Q has the form (10), then the derivation of system (8) and its averaging can be performed by using known integrals of the unperturbed system, see below.

(2) *A system close to the conservative one.* Consider now a simpler variant of system (4):

$$\ddot{y} + Q(x, y) = \varepsilon q(x, y, \dot{y}, u), \qquad y(t_0) = y^0, \qquad \dot{y}(t_0) = y^0,$$
$$\dot{x} = \varepsilon X(x, y, \dot{y}, u), \qquad x(t_0) = x^0. \tag{3.1.11}$$

The standard system (8) is constructed on the basis of the integrals

$$\frac{1}{2}\dot{y}^2 + \Pi(x, y) = c, \qquad \dot{y}(x, c, y) = \pm\sqrt{2}(c - \Pi)^{1/2},$$
$$\psi = \omega(x, c)\int \dot{y}^{-1}(x, c, y)\,\mathrm{d}y \equiv \Psi(x, c, y), \tag{3.1.12}$$
$$\Pi(x, y) = \int Q(x, y)\,\mathrm{d}y, \quad \omega = \frac{2\pi}{T^0}, \quad T_0(x, c) = \oint \dot{y}^{-1}(x, c, y)\,\mathrm{d}y.$$

Differentiation of c and ψ in (12) with respect to t according to (11) gives

$$\dot{c} = \varepsilon\left[q\dot{y} + (\partial\Pi/\partial x \cdot X)\right], \qquad \dot{x} = \varepsilon X,$$
$$\dot{\psi} = \omega + \varepsilon\left\{q\dot{y}\,\frac{\partial\Psi}{\partial x} + \left(\left[\frac{\partial\Psi}{\partial c} + \frac{\partial\Psi}{\partial c}\frac{\partial\Pi}{\partial x}\right]\cdot X\right)\right\}. \tag{3.1.13}$$

The right-hand sides of Equations (13) are expressed through x, c, ψ, and u, and the variable y is connected with the phase ψ. For the rotational mode, the functions $\partial\Psi/\partial x$ and $\partial\Psi/\partial c$ turn out to be periodic in y and, by this, in ψ, as in the case of oscillations [1, 50].

A pendulum with a controlled equilibrium position experiencing small acceleration of the suspension point (see (1.1.17) and (1.1.18)) can serve as an example of system (11). The pendulum length l (or σ) plays the role of the parameter x. It should be noted that if the length l is controlled, the system will be essentially nonlinear even in the case of small oscillations.

(3) *The case of fast rotations.* Consider a rotatory system of the form (11) without a small parameter ($\varepsilon = 1$) under the assumption that the initial angular velocity of rotation $\dot{y}^0$ is large, i.e. the kinetic energy $(1/2)\dot{y}^2$ ($m = 1$) significantly exceeds the potential energy, which is periodic and bounded in y. In this case, the motion is close to uniform rotation, $y \approx \theta = \dot{y}^0 t$ [1, 150]. Having introduced the fast independent variable θ and divided the equation for y by $\dot{y}^{02}$ and the equation for x by $\dot{y}^0$, we

obtain, under appropriate conditions imposed on the magnitudes of the perturbations q and X and on the order of their growth with respect to $\dot{y}$, the following system:

$$\begin{aligned} &\ddot{y} = \varepsilon q(x, y, \dot{y}, u), && y(t_0) = y^0, && \dot{y}(t_0) = \dot{y}^0, \\ &\dot{x} = \varepsilon X(x, y, \dot{y}, u), && x(t_0) = x^0. \end{aligned} \tag{3.1.14}$$

System (14) is easily reduced to the standard form (13): $y = \psi$, $\dot{y} = c$. Systems of the form (13) arise, for instance, when studying some problems concerning the control of motion in weak force fields.

(4) *A nonlinear oscillator with a controlled equilibrium position*. Consider a system governed by equations of the form

$$m\ddot{y} + F(y - s) = 0, \qquad \dot{s} = \varepsilon u. \tag{3.1.15}$$

Here, y is the absolute coordinate of the mass m, s is a velocity-controlled equilibrium position, and $-F(z)$ is a restoring force; $F(0) = 0$, $zF(z) > 0$ for $z \neq 0$. Equations (15) are easily reduced to the standard form (13), see Section 3.5 below. Some control problems for pendulum-type systems with moving suspension points (Chapter 5) and elastic systems with controlled boundary conditions [1, 8, 13, 16, 23, 24, 34, 41, 42, 46, 67, 90, 103, 117, 146, 156] give rise to the analysis of equations of the type (15).

It should be noted that in the more general case, the oscillatory system with controlled equilibrium position is described by the Hamiltonian function

$$H = (1/2)(p - v)^2 + \Phi(\varphi), \qquad v = v(\varphi, \mathbf{u}),$$

where φ is a coordinate, p is a momentum, u is a control, $v(\varphi, 0) \equiv 0$. The equations of motion imply that, under appropriate assumptions and the condition $\dot{\mathbf{x}} = \mathbf{u} \equiv 0$, the system performs oscillating ($\Phi'(0) = 0$, $\varphi\Phi'(\varphi) > 0$, $\varphi \neq 0$) and rotating ($\Phi(\varphi + 2\pi) \equiv \Phi(\varphi)$, $v(\varphi + 2\pi, \mathbf{u}) \equiv v(\varphi, \mathbf{u})$) motions. Moreover, it can move as a whole, without oscillations if the conditions $\mathbf{u} = \mathbf{u}_0 = \text{const}$, $\varphi \equiv 0$, $p = v(0, \mathbf{u}_0)$ hold. In particular, for plane oscillations and rotations of a pendulum ($m = l = 1$), the following expressions are valid:

$$\dot{\varphi} = \frac{\partial H}{\partial p} = p - v, \qquad \dot{p} = -\frac{\partial H}{\partial \varphi} = (p - v)\frac{\partial v}{\partial \varphi} - g \sin \varphi,$$

$$H = (1/2)(p - v)^2 - g \cos \varphi, \qquad v = \dot{x}_0 \cos \varphi + \dot{y}_0 \sin \varphi,$$

$$\mathbf{x} = (x_0, y_0), \qquad \mathbf{u} = (u_x, u_y), \qquad \dot{x}_0 = u_x, \qquad \dot{y}_0 = u_y.$$

(5) *Control of elements of a plane elliptic orbit.* If the eccentricity e is not small ($0 < e_1 \leqslant e \leqslant e_2 < 1$), simultaneous Equations (1.1.23) in the absence of perturbations admit the general periodic solution which is found in implicit form as a complete set of integrals [1, 46, 71, 72, 108, 116, 130]. The application of the technique using integrals of motion allows one to construct and to analyse the standard system in implicit form on an asymptotically large time interval, see Section 3.5 and Chapter 6.

(6) *Control of rotations of a rigid body.* The class of optimal control problems for rotational motions of a rigid body with a fixed point (the centre of mass), taking perturbations of a different physical nature into account, seems to be of practical importance and theoretical interest. Within this class, there are problems of stabilization, orientation, twist, etc. The mathematical model for the analysis of such systems is a set of equations of motion in the coordinate frame fixed within the body (of the Euler type) [1, 6, 7, 9, 11, 21, 46, 84, 91, 97, 111, 112, 127, 138], or in the coordinate frame connected with the angular momentum vector [1, 46, 91, 97, 127, 138]. Other types of equations of motion are possible as well . These equations are reducible to the standard form (3). For a special case of an axially symmetrical body, the problem of controlling the equatorial component of the angular velocity vector has been considered in Section 1.4.2 in a quasilinear interpretation. The asymptotic technique developed below allows control problems to be investigated for such objects in a nonlinear consideration. In Chapter 6, we will give solutions for some problems concerning the optimal control of rigid body rotations.

3.2. Averaging in Problems with a Fixed Time of Termination of the Control Process

3.2.1. THE STATEMENT OF THE PROBLEM AND THE MAXIMUM PRINCIPLE

The terminal set and the functional are specified by general relationships of the type (1.3.6)

$$\begin{gathered} M(a)\big|_T = 0, \qquad M = (M_1, \dots, M_l), \qquad 0 \leqslant l \leqslant n-1, \\ J[u] = g(a)\big|_T \longrightarrow \min_u, \qquad u(t) \in U, \qquad T = \Theta\varepsilon^{-1}. \end{gathered} \tag{3.2.1}$$

The functions M, g and the domain U are defined analogously to Section 1.2 and are such that there exists a solution of the optimal control problem (1.3) and (1) for $\varepsilon \in (0, \varepsilon_0]$. It is supposed that singular controls are absent and the system is sufficiently smooth. For each concrete case of singular controls, a special consideration is required which can be simplified using the approach described.

As has been mentioned in Section 1.3, the functional containing the integral part can be reduced to the form (1).

$$J[u] = g_0(a)\big|_T + \varepsilon \int_{t_0}^{T} G(a, \psi, u)\, dt,$$
$$J[u] = g_0(a)\big|_T + a_{n+1}\big|_T, \quad \dot{a}_{n+1} = \varepsilon G(a, \psi, u), \quad a_{n+1}(t_0) = 0. \tag{3.2.2}$$

Analogously to Section 1.3, the functions M and g in (1) must not depend essentially on the phase ψ. The necessity of such a restriction of the control problem statement follows from the analysis of the boundary-value problem of the maximum principle. We suppose that the optimal control u^* is determined uniquely from the maximality conditions for the Hamiltonian function H and, moreover, that the control is smooth in all arguments and 2π-periodic in ψ, i.e.

$$H \longrightarrow \max_u, \qquad u \in U, \quad H = \varepsilon(p \cdot f) + q(\omega + \varepsilon F),$$
$$u^* = u(a, \psi, p, q) = \arg\max_u H, \qquad u \in U, \tag{3.2.3}$$
$$H^* = H\big|_{u^*} = \varepsilon(p \cdot f^*) + q(\omega + \varepsilon F^*).$$

Here, p and q are the variables adjoint with a and ψ. The application of the technique given in Section 1.2 results in the following boundary-value problem:

$$\dot{a} = \varepsilon f^*(a, \psi, p, q), \qquad \dot{\psi} = \omega(a) + \varepsilon F^*(a, \psi, p, q),$$
$$\dot{p} = -q\frac{\partial \omega}{\partial a} - \varepsilon \frac{\partial}{\partial a}\left[(p \cdot f^*) + qF^*\right], \qquad \dot{q} = -\varepsilon \frac{\partial}{\partial \psi}\left[(p \cdot f^*) + qF^*\right],$$
$$a(t_0) = a^0, \qquad \psi(t_0) = \psi^0, \qquad M(a)\big|_T = 0, \tag{3.2.4}$$
$$p(T) = \frac{\partial}{\partial a}\left[(\lambda \cdot M) - g\right]\Big|_T, \qquad q(T) = 0.$$

It is important to note that, unlike the quasilinear case (1.1.6) leading to the standard set of equations for the boundary-value problem of the maximum principle (1.2.9) or (1.3.3), the set of differential equations (4) obtained for (1.3) does not have the standard form. As follows from (4), the adjoint variable p is not formally slow: $\dot{p} = -q\omega' \not\equiv 0$ when $\varepsilon = 0$. The equation for the vector p is similar to that for the vectorial amount off-resonance arising when investigating multi-frequency nonlinear oscillatory systems in the neighbourhood of the resonance point [1, 4, 5, 20, 23, 46, 60, 77, 94, 115, 119, 120, 123, 124, 150]. Such cases cause significant difficulties for asymptotic analysis. It turns out, in addition, that the solution of the boundary-value problem of the maximum principle, as a rule, is essentially nonunique, analogously to Section 1.3. The two circumstances just mentioned are connected with each other and substantially complicate the analysis of the boundary-value problem and the construction of the asymptotic solution of the optimal control problem.

3.2.2. DERIVATION OF THE STANDARD SYSTEM WITH ROTATING PHASE

The modulus of the adjoint vector (p, q) is assumed to be bounded by the quantity of the order unity. It can be made by using a proper normalization. In what follows, we use the boundary condition $q(T) = 0$ from (4) and the canonical property of the time-invariant system consisting in nonvariability of H^* along the trajectories

$$H^* = \varepsilon(p \cdot f^*) + q(\omega + \varepsilon F^*) = C = \text{const.} \tag{3.2.5}$$

Under the assumptions of the smoothness and boundedness of the functions f^* and F^*, as well as their derivatives, on the time interval $t \in [t_0, T]$, $T = \Theta\varepsilon^{-1}$, the vector a is also bounded: $a \in D_a$ when $a^0 \in D_{a^0} \subseteq D_a$. Then it follows from (5) at $t = T$ that $C = \varepsilon(p \cdot f^*)|_T \sim \varepsilon$.

Assuming $C = \varepsilon h$, where the parameter h is equal to ($h \sim 1$ when $\varepsilon \to 0$)

$$h = (p \cdot f^*)\big|_T, \qquad f^* = f^*(a, \psi, p, q), \qquad q(T) = 0, \tag{3.2.6}$$

one can rewrite relationship (5), taking (6) into account, in the form

$$q = \varepsilon\omega^{-1}\big[h - (p \cdot f^*) - qF^*\big]. \tag{3.2.7}$$

Relationship (7) should be considered as an equation relative to the variable q. Within the conditions of boundedness of partial derivatives of the functions f^* and F^* with respect to q in a neighbourhood of $q = 0$, the solution exists and has the form

$$\begin{aligned} &q = \varepsilon\omega^{-1}\big[h - (p \cdot f_0^*)\big] + \varepsilon^2 q_2(h, a, \psi, p, \varepsilon), \\ &f_0^* = f^*(a, \psi, p, 0); \qquad q = O(\varepsilon). \end{aligned} \tag{3.2.8}$$

The function $q = q(h, a, \psi, p, \varepsilon)$ can be constructed by using series expansions in terms of powers of ε, the successive approximations technique or Newton's method (the method of tangents [1, 11, 49, 147]), q_2 being smooth. Substituting (3) into (4) we obtain a family of boundary-value problems of the following form (h is the family parameter)

$$\begin{aligned}
&\dot{a} = \varepsilon f_0^*(a, \psi, p) + \varepsilon^2 f_2(h, a, \psi, p, \varepsilon),\\
&\dot{p} = -\varepsilon \frac{\omega'(a)}{\omega(a)}\left[h - (p \cdot f_0^*)\right] - \varepsilon \frac{\partial}{\partial a}(p \cdot f_0^*) + \varepsilon^2 P_2(h, a, \psi, p, \varepsilon),\\
&\dot{\psi} = \omega(a) + \varepsilon F_1(h, a, \psi, p, \varepsilon),\\
&a(t_0) = a^0, \quad \psi(t_0) = \psi^0, \quad M(a)\big|_T = 0, \quad p(T) = \frac{\partial}{\partial a}\left[(\lambda \cdot M) - g\right]\Big|_T.
\end{aligned} \tag{3.2.9}$$

Thus, we have constructed the standard system (9), where f_2, P_2, and F_1 are known functions, periodic in ψ. The solution of the boundary-value problem (9) for fixed $h \in I_h$ and $\varepsilon > 0$ exists and is equal to ($t \in [t_0, T]$)

$$a = a(t, h, \varepsilon), \qquad \psi = \psi(t, h, \varepsilon), \qquad p = p(t, h, \varepsilon). \tag{3.2.10}$$

Corresponding sufficient conditions have been established in Sections 2.3 and 2.4. However, relationship (6) determining the parameter h admits, as a rule, many roots. Provided some general conditions hold, similar to those stated in Section 1.3, the number of these roots is of the order of ε^{-1}, and the distance between adjacent roots has the order of ε. To prove this, we rewrite Equation (6), taking (8) into account, in the form

$$h = \left(p \cdot f_0^*(a, \psi, p)\right)\big|_T, \tag{3.2.11}$$

where a, ψ, and p are substituted according to (10). Since the solution of the optimal control problem exists, Equation (11) admits at least one real root $h_* \in I_h$. Hence, for sufficiently small $\varepsilon > 0$, other roots must exist in an ε-vicinity of $h = h_*$. It is conditioned by fast oscillations of the function $f_0^*|_T$ in h due to the periodicity in ψ. The frequency of these oscillations is of the order of ε^{-1} and the amplitude has the order of unity. Indeed, the first property stems from the estimate of the quantity $\partial\psi/\partial h|_T$, since in the general case

$$\frac{\partial \psi}{\partial h}\Big|_T = \frac{\partial}{\partial h}\int_{t_0}^{T} \dot{\psi}\, dt \sim \frac{\partial}{\partial h}\int_{t_0}^{T} \omega(a)\, dt = \int_{t_0}^{T}\left(\omega'(a) \cdot \frac{\partial a}{\partial h}\right) dt \sim \varepsilon^{-1}. \tag{3.2.12}$$

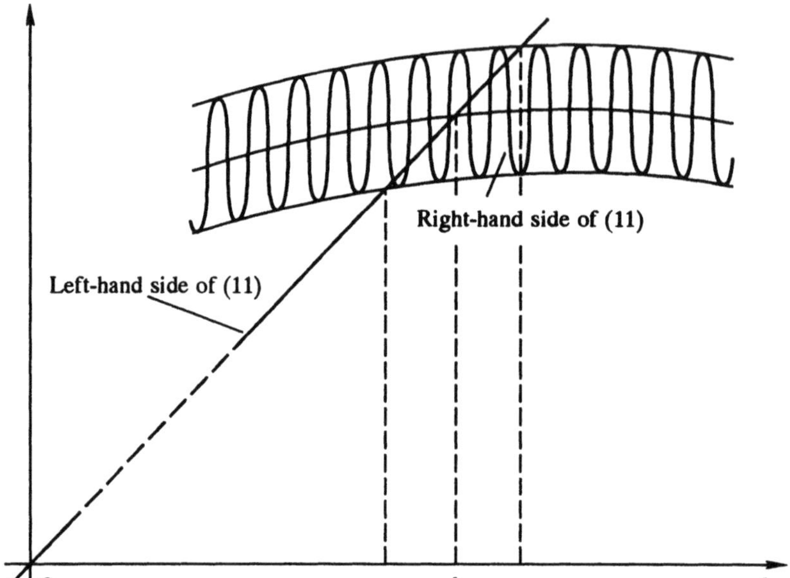

Fig. 3.1.

The amplitude of the oscillations is characterized by the magnitude of the swing of the function f_0^* in ψ, i.e. it has the order of unity. The typical behaviour of the right-hand side of Equation (11) as a function of h is shown in Fig. 3.1. This behaviour qualitatively coincides with that presented in Figs. 1.6 and 1.7. Sufficient conditions for such behaviour of the roots are established by using the solution of the problem of the first approximation, as was done in Section 1.3. These conditions will be discussed below in more detail.

Thus, we can regard the fact of the essential nonuniqueness of the solution of the maximum principle boundary-value problem (4), or (6), (8)–(11), to be established in the general case. The set of solutions is defined by the set of roots $\{h_\nu\}$ of Equation (11). The optimal value of h^* is chosen from the minimality condition for the functional (1)

$$J^* = \min_{u \in U} J[u] = \min_h \tilde{J}(h, \varepsilon), \qquad h \in \{h_\nu\},$$
$$\tilde{J}(h, \varepsilon) = g\big(a(T, h, \varepsilon)\big). \tag{3.2.13}$$

The typical qualitative behaviour of the function $\tilde{J}(h, \varepsilon)$ for small ε is given in Fig. 3.2. Having determined h^* from (13) we find uniquely the variables a^*, ψ^*,

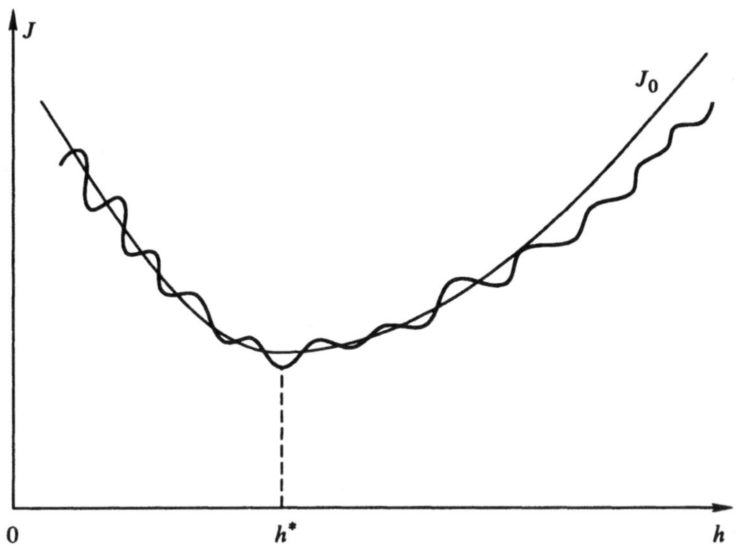

Fig. 3.2.

p^*, and q^*, according to (10) and (8), and then the optimal control u^* (3), both open-loop and feedback, is determined. The aforenamed difficulties, associated with the solution of boundary-value problem (9) and transcendental Equation (11) with subsequent optimization over h according to (13), can be overcome by using the asymptotic method developed below. This approach, similar to that given in Section 1.3, essentially uses the solution of the problem of the first approximation.

3.2.3. BOUNDARY-VALUE PROBLEM TO A FIRST APPROXIMATION

Having omitted the terms of order ε^2 in the standard system (9) and averaging over the phase ψ we can obtain the following boundary-value problem in the slow time $\tau = \varepsilon t$

$$\begin{aligned} &\dot{\xi} = f_0(\xi, \eta), \qquad \dot{\eta} = -\frac{\omega'}{\omega}[h - (\eta \cdot f_0)] - \frac{\partial}{\partial \xi}(\eta \cdot f_0), \\ &\xi(\tau_0) = a^0, \qquad M(\xi)\big|_\Theta = 0, \qquad h(\Theta) = \frac{\partial}{\partial \xi}[(\lambda \cdot M) - g]\Big|_\Theta; \\ &\dot{\varphi} = \varepsilon^{-1}\omega(\xi), \qquad \varphi(\tau_0) = \psi^0 \qquad \Big(f_0(\xi, \eta) = \langle f_0^*(\xi, \psi, \eta)\rangle\Big). \end{aligned} \tag{3.2.14}$$

There is no reason for keeping the terms of order unity in the equation for φ (14), since the error of the same order in determining the phase ψ is conditioned by the error $O(\varepsilon)$ in determining the quantity a: $a = \xi + O(\varepsilon)$ [1, 38, 46, 149, 150]. It is known [1, 35, 38, 149, 150] that the solution of the initial-value problem for system (9) with specified values of the variables a^0, p^0, and ψ^0 and parameters h, t_0, and Θ is close to the solution of the averaged system (14), provided the initial values ξ^0, η^0, and φ^0 and the values of the parameter $\bar{h}$ are sufficiently close, i.e.

$$\begin{aligned}
&|a-\xi| \leqslant c\varepsilon, \qquad |p-\eta| \leqslant c\xi, \qquad |\psi-\varphi| \leqslant c,\\
&|a^0-\xi^0| \leqslant d\varepsilon, \qquad |p^0-\eta^0| \leqslant d\varepsilon,\\
&|\psi^0-\varphi^0| \leqslant d, \qquad |h-\bar{h}| \leqslant d\varepsilon \qquad (c, d = \text{const},\ t \in [t_0, \Theta\varepsilon^{-1}]).
\end{aligned} \tag{3.2.15}$$

To justify proximity estimates similar to (15) for solutions of boundary-value problems (9) and (14), we suppose that the solution of the averaged problem exists, is unique and possesses the isolatedness property, formulated in Section 1.3, when $h \in I_h$. In this case, the addition of small terms $O(\varepsilon)$ to right-hand sides of the equations and the boundary conditions (14) leads to a similar small change in the solution of the perturbed problem.

The investigation of the averaged boundary-value problem enables the following to be established. The set of equations for ξ, η admits the integral

$$\omega^{-1}(\xi)[h - (\eta \cdot f_0(\xi, \eta))] = \beta = \text{const}. \tag{3.2.16}$$

This can be verified immediately by differentiation of (16) according to Equation (14), and using the equality $\partial(\eta \cdot f_0)/\partial\eta = f_0$ which stems from (1.2.7). Comparison of (10) and (8) shows that $\beta = \varepsilon^{-1}\langle q\rangle$. Using the integral (16), we can represent the equations for ξ and η (14) in a more convenient form

$$\dot{\xi} = f_0(\xi, \eta), \qquad \dot{\eta} = -\omega'(\xi)\beta - \frac{\partial}{\partial\xi}(\eta \cdot f_0), \qquad (\dot{\ }) \equiv \frac{\mathrm{d}}{\mathrm{d}\tau}. \tag{3.2.17}$$

If β is considered as a fixed parameter, then system (17), as is easily established by differentiation, is Hamiltonian, with the following Hamiltonian function

$$\lim_{\varepsilon\to 0} \varepsilon^{-1}\langle H^*(\xi, \psi, \eta, \varepsilon\beta)\rangle = \omega(\xi)\beta + (\eta \cdot f_0(\xi, \eta)) = h = \text{const}. \tag{3.2.18}$$

The constancy of h is in agreement with the relationship (16). Parallel with h-family of Equations (14) we consider in what follows β-family (17). The parameters

β and h are connected by relationship (18) or (16). We suppose that in some range of values $h \in I_h$ and $\beta \in I_\beta$, this connection is one-to-one and, besides, $\beta = 0 \in I_\beta$. Below, we establish, analogously to Section 1.3, that the last condition holds under some additional assumptions. It is connected with the sufficient conditions of the existence of the root $h_* \in I_h$ for the averaged (over ψ) relationship (11) and the behaviour of the roots $\{h_\nu\}$ of the exact equation, as is depicted in Fig. 3.1. As a result, the β-family of boundary-value problems for system (17) has a unique solution which can be represented in the form:

$$\begin{gathered} \xi = \xi(\tau - \tau_0, \Theta - \tau_0, a^0, \beta), \qquad \eta = \eta(\tau - \tau_0, \Theta - \tau_0, a^0, \beta), \\ \tau \in [\tau_0, \Theta], \quad a^0 \in D_{a^0} \subset D_a, \quad \xi \in D_a, \quad \beta \in I_\beta. \end{gathered} \tag{3.2.19}$$

3.2.4. THE CONNECTION OF THE AVERAGED AND ORIGINAL BOUNDARY-VALUE PROBLEMS

The parameter β must satisfy Equation (11), where one has substituted the expressions for a, ψ, and p:

$$a = \xi + O(\varepsilon), \quad p = \eta + O(\varepsilon), \quad \psi = \psi^0 + \frac{1}{\varepsilon}\int_{\tau_0}^{\tau} \omega(\xi)\,d\theta + O(1), \tag{3.2.20}$$

and the expression for h, according to (16) and (18). The following relationship is obtained

$$\begin{gathered} \Phi\left(\xi, \eta, \psi^0 + \frac{1}{\varepsilon}\int_{\tau_0}^{\tau} \omega(\xi)\,d\theta + O(1)\right)\bigg|_{\Theta} = \beta + O(\varepsilon), \\ \Phi(\xi, \eta, \psi) \equiv \omega^{-1}(\xi)\big(\eta \cdot [f_0^*(\xi, \psi, \eta) - f_0(\xi, \eta)]\big). \end{gathered} \tag{3.2.21}$$

The function Φ, introduced in (21), is periodic in ψ and has a zero mean value. Let the condition

$$\frac{d}{d\beta}\int_{\tau_0}^{\Theta} \omega\big(\xi(\tau - \tau_0,\ \Theta - \tau_0, a^0, \beta)\big)\,d\tau\bigg|_{\beta=0} = \gamma_0 \neq 0 \tag{3.2.22}$$

hold. Then for sufficiently small $\varepsilon > 0$, the left-hand side of relationship (21) in some neighbourhood of the value $\beta = 0 \in I_\beta$ is a rapidly oscillating function of β with a frequency of order $\gamma_0\varepsilon^{-1}$, an amplitude of order unity and the average $O(\varepsilon)$. Hence, Equation (21), analogously to Equation (11), has $O(\varepsilon^{-1})$ roots within some interval

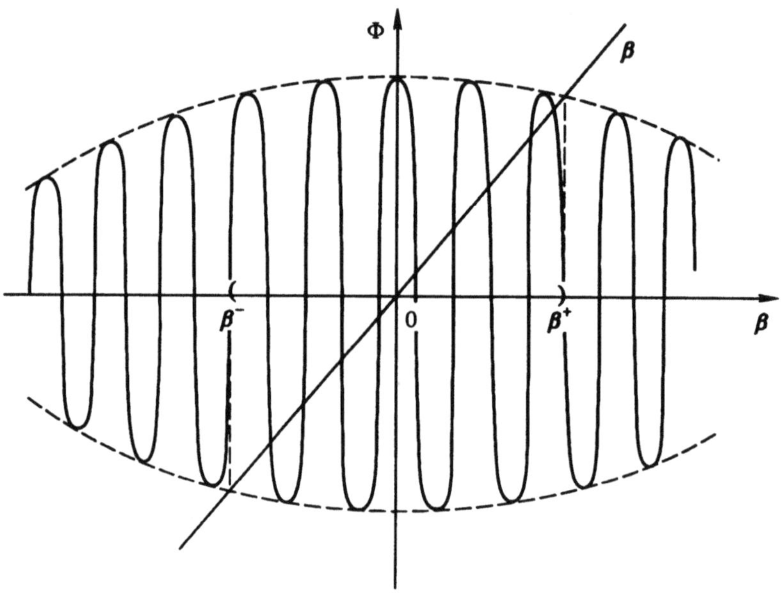

Fig. 3.3.

of values β, including $\beta = 0$, adjacent roots differing from each other by $O(\varepsilon)$. Any point of this interval possesses the property that, in an ε-vicinity of this point there exists a root β_ν of the exact equation (21) corresponding to the root h_ν of Equation (11). The connection between these values is one-to-one, and is specified according to (16) and (18). The qualitative behaviour of roots β_ν of Equation (21) is shown in Fig. 3.3. More profound reasons for the introduction of the parameter β are developed in Section 4.1, when constructing higher approximations.

3.2.5. THE CHOICE OF THE OPTIMAL SOLUTION

For determining the optimal value $\beta \in I_\beta$, we use condition (13), where the functional J is represented in the form

$$\begin{aligned} J &= \tilde{J}(h,\varepsilon) = J_0[\beta] + O(\varepsilon), \qquad h \in I_h, \quad \varepsilon \in (0, \varepsilon_0], \\ J_0[\beta] &= g\big(\xi\big(\Theta - \tau_0, \Theta - \tau_0, a^0, \beta\big)\big), \qquad \beta \in I_\beta. \end{aligned} \tag{3.2.23}$$

Here, the parameters h and β are connected according to (16) and (18), and the function J_0 is supposed to be smooth in $\beta \in I_\beta$. We consider the following minimization problem for a smooth function $J_0[\beta]$:

$$J_0[\beta] \longrightarrow \min_{\beta}, \qquad \beta \in I_\beta; \qquad J_0^* = J_0\left[\beta_0^*\right]. \tag{3.2.24}$$

It is important to note that here the parameter β is to be chosen from the continuous set of admissible values.

The proximity of $\widetilde{J}(h,\varepsilon)$ and $J_0[\beta]$ and the smoothness of J_0 in β imply that the minima (13) and (24) differ by an amount $O(\varepsilon)$, i.e. $J^* - J_0^* = O(\varepsilon)$. Hence, the solution β_0^* and J_0^* of problem (24) gives the approximate solution of the original optimal control problem (1) and (13) in the following sense. In an ε-neighbourhood of the determined β_0^*, there exists β_ν corresponding to some root h_0 of Equation (11), for which, due to the established properties, the following proximity estimates are valid

$$\widetilde{J}(h_\nu,\varepsilon) = J_0[\beta_\nu] + O_1(\varepsilon) = J_0\left[\beta_0^*\right] + O_2(\varepsilon) = J^* + O_3(\varepsilon). \tag{3.2.25}$$

Here, $O_{1,2,3}(\varepsilon)$ are small quantities of the order of ε. Therefore, the difference between the exact and approximate solutions in terms of the functional is $O(\varepsilon)$ and there is an exact root β_ν of Equation (21) in an ε-neighbourhood of β_0^*. The ε-proximity to the optimal value β^* corresponding to h^* is not guaranteed. Due to this, the ε-proximity of solutions of the boundary-value problems in slow variables for all $t \in [t_0, \Theta\varepsilon^{-1}]$ is also not guaranteed. But the ε-proximity takes place in terms of the fulfillment of the boundary conditions, since the boundary and transversality conditions are satisfied identically in $\beta \in I_\beta$ or $h \in I_h$.

Thus, to a first approximation, we may confine ourselves to considering the minimization of the smooth function $J_0[\beta] \to \min_\beta$, $\beta \in I_\beta$, according to (24). It is shown below that there exists a point within the set I_β suspected to be the extremum. The following statement is valid.

LEMMA 3.1. *The point $\beta = 0$ is a point of a local extremum when $\gamma_0 \neq 0$; the value $J_0[0]$ at the point $\beta = 0$ is a local minimum of $J_0[\beta]$ for $\gamma_0 > 0$ and is a maximum for $\gamma_0 < 0$.*

Proof. Differentiation of the function $J_0[\beta]$ from (23) with respect to β gives

$$J_0'[\beta] = (\partial g/\partial\xi \cdot \partial\xi/\partial\beta)\big|_\Theta. \tag{3.2.26}$$

Taking into account the boundary conditions (14) for ξ and η, as well as the equalities $(\partial\xi/\partial\beta)|_{\tau_0} = \partial a^0/\partial\beta = 0$, enables (26) to be represented in the form

$$\begin{aligned} J_0' &= \left(\left[\frac{\partial}{\partial\xi}(\lambda \cdot M) - \eta\right] \cdot \frac{\partial\xi}{\partial\beta}\right)\bigg|_\Theta = -\left(\eta \cdot \frac{\partial\xi}{\partial\beta}\right)\bigg|_\Theta \\ &= -\int_{\tau_0}^{\Theta} \frac{\mathrm{d}}{\mathrm{d}\tau}\left(\eta \cdot \frac{\partial\xi}{\partial\beta}\right)\mathrm{d}\tau. \end{aligned} \tag{3.2.27}$$

In the second equality (27) we have used the fact that $\partial(\lambda \cdot M)/\partial\beta|_\Theta = 0$, since the condition $M|_\Theta = 0$ holds identically in β, see above. Differentiating the expression under the sign of integral (27) and using the equation for η (17) we obtain

$$J_0' = \int_{\tau_0}^{\Theta} \left\{ \left(\left[\beta \frac{\partial \omega}{\partial \xi} + \frac{\partial}{\partial \xi}(\eta \cdot f_0) \right] \cdot \frac{\partial \xi}{\partial \beta} \right) - \left(\eta \cdot \frac{\mathrm{d}}{\mathrm{d}\tau} \frac{\partial \xi}{\partial \beta} \right) \right\} \mathrm{d}\tau. \tag{3.2.28}$$

The expression for the derivative $(\partial\xi/\partial\beta)$ included in (28) is obtained form the differentiability of the function $f_0(\xi, \eta)$ (14):

$$\frac{\mathrm{d}}{\mathrm{d}\tau} \frac{\partial \xi}{\partial \beta} = \frac{\partial}{\partial \beta} \dot{\xi} = \frac{\partial f_0}{\partial \xi} \frac{\partial \xi}{\partial \beta} + \frac{\partial f_0}{\partial \eta} \frac{\partial \eta}{\partial \beta}.$$

The contribution of the second term after substitution of the obtained expression into (28) is equal to zero, on the basis of the equality $\partial(\eta \cdot f_0)/\partial\eta = f_0$. As a result, we have

$$J_0'[\beta] = \beta \int_{\tau_0}^{\Theta} \left(\frac{\partial \omega}{\partial \xi} \cdot \frac{\partial \xi}{\partial \beta} \right) \mathrm{d}\tau = \beta \frac{\mathrm{d}}{\mathrm{d}\beta} \int_{\tau_0}^{\Theta} \omega(\xi)\, \mathrm{d}\tau. \tag{3.2.29}$$

The equality to zero of the derivative $J_0'[\beta]$ (29) implies that $\beta = 0$ is the point of the possible extremum for $J_0[\beta]$ if $\beta = 0 \in I_\beta$ (which has been suggested above). Under the condition of boundedness of the second derivative of $J_0[\beta]$ with respect to β, the value $\beta = 0$ corresponds to a local minimum, provided the strict inequality

$$J_0''[0] = \frac{\mathrm{d}}{\mathrm{d}\beta} \int_{\tau_0}^{\Theta} \omega(\xi)\, \mathrm{d}\tau \Big|_{\beta=0} = \gamma_0 > 0 \tag{3.2.30}$$

holds. The quantity γ_0 is defined in (22) and, by assumption, is not equal to zero. Therefore, inequality (30), implying conditions of rapid oscillations (22) and the behaviour of the roots β_ν of Equation (21) qualitatively shown in Fig. 3.3, is sufficient for local optimality of the value $\beta_0^* = 0$ as well as of the corresponding variables ξ and η and the functional J_0.

To derive the sufficient conditions for a global minimum on some interval $\beta \in I_\beta$ we calculate the function $J_0[\beta]$ by integrating its derivative (29) by parts:

$$J_0[\beta] - J_0[0] = \int_{\tau_0}^{\Theta} \left[\beta\omega(\xi(\tau, \beta)) - \int_0^{\beta} \omega(\xi, (\tau, \beta'))\, \mathrm{d}\beta' \right] \mathrm{d}\tau. \tag{3.2.31}$$

The condition $J_0[\beta] - J_0[0] \geqslant 0$, according to (31), implies the inequality

$$\beta \int_{\tau_0}^{\Theta} \omega\big(\xi(\tau,\beta)\big)\,\mathrm{d}\tau \geqslant \int_0^{\beta} \mathrm{d}\beta' \int_{\tau_0}^{\Theta} \omega\big(\xi(\tau,\beta')\big)\,\mathrm{d}\tau, \qquad \beta \in I_\beta. \tag{3.2.32}$$

Another equivalent condition follows from the definition of $J_0[\beta]$ in (23):

$$g\big(\xi(\Theta,\beta)\big) \geqslant g\big(\xi(\Theta,0)\big), \qquad \beta \in I_\beta. \tag{3.2.33}$$

The choice of the form for the optimality condition is determined by reasons of convenience. It is important to note that the sufficient conditions of local optimality (29) and (30) can be verified on the basis of the known general solution of Equations (17) under $\beta = 0$. Indeed, the functions ξ and η, which are required for the verification, can be constructed by series expansions in β when $|\beta| \leqslant \beta^0$. To do this, one has to solve some linear boundary-value problems with known fundamental matrices. These constructions are similar to those made in Section 2.2, see (2.2.1)–(2.2.6). In this case, the homogeneous part of the set of equations is the system for variations related to (17) under $\beta = 0$; its fundamental matrix is found by differentiating the general solution mentioned above with respect to parameters (constants of integration), see the example in Section 3.2.7 and the problem in Section 6.2.

3.2.6. CONCLUDING REMARKS

The procedure of constructing optimal control, as in Section 1.3, includes the construction of a β-family of solutions for the boundary-value problem (16) and (14) and subsequent minimization of the functional of the first approximation $J_0[\beta]$ (24) with respect to the family parameter β over the set I_β of its values for which the solution of the boundary-value problem exists in the considered domain D_a. As has been established above, the value $\beta_0^* = 0$, as a rule, is optimal for smooth systems, provided the solution of the optimal control problem is unique. The state and adjoint variables, according to (8), (19), and (20), are expressed as

$$a = \xi^*, \qquad p = \eta^*, \qquad q = 0, \qquad \psi = \psi^0 + \varepsilon^{-1}\varphi_{-1}, \tag{3.2.34}$$

where $\beta = \beta_0^*$ is substituted. Using (34) one can obtain the approximate solution of the optimal control problem possessing the property of ε-proximity with respect to initial and boundary conditions, as well as the ε-proximity of the control with respect to the functional. Provided that the envelopes of the function J_1

$$J_1(\beta, y, \varepsilon) = J_\varepsilon[\beta] - J_0[\beta], \qquad y = \varphi_{-1}(\Theta,\beta) = \int_{\tau_0}^{\Theta} \omega\,\mathrm{d}\tau, \tag{3.2.35}$$

are sufficiently smooth with respect to y, considerations according to the scheme of Section 1.3.4 allow us to establish, that the value β_0^* gives rise to ε-proximity of the exact and approximate optimal motions in slow variables (15), see Fig. 3.2. A more detailed investigation of this issue can be found in Section 2.4.

One cannot change the fast phase $\psi = \varphi + O(1)$ for φ in the control u^* (3) without loss of accuracy. Therefore, it is possible to represent the approximate control u^* in the feedback form u_s or in the form of the "partial" programme u_p:

$$u_s = u^*\left(a, \psi, \eta^*(0, \Theta - \tau, a, \beta_0^*), 0\right),$$
$$u_p = u^*\left(\xi^*, \psi, \eta^*, 0\right). \tag{3.2.36}$$

One should bear in mind that the feedback control may have a singularity at $\tau = \Theta$, since for the specified terminal manifold $M(a)$ (1), the adjoint variable $\eta^* \to \infty$ as $\tau_0 \to \Theta$.

It should be noted that, in applications, the solution of the averaged boundary-value problem (14) or (17) is usually substantially simpler than that of the original problem (4) or (9). This fact is due to the reduced dimension of the averaged system and the absence of rapid oscillations. This enables the slow time to be introduced when integrating the system. Optimal control problems for systems with one degree of freedom are reduced, as a rule, to quadratures and finite equations. Solutions for some problems of optimal control of oscillations in mechanical systems can be found in Chapters 5 and 6.

3.2.7. MODEL EXAMPLE

To illustrate the approach described above, consider an example of velocity control in a dynamic system having one degree of freedom, y:

$$\ddot{y} = \varepsilon\left[u + F(y) - \chi\dot{y}\right], \quad \chi = \text{const}, \quad y(t_0) = y^0, \quad \dot{y}(t_0) = \dot{y}^0 > 0. \tag{3.2.37}$$

Here, F is a function, 2π-periodic in y with the mean value F_0 and u is a control. Equation (37) describes controlled motions in weak periodic fields, fast rotations (see (1.14)), etc. The change $y = \psi$, $\dot{y} = a$ reduces Equation (37) to the standard form (1.3):

$$\dot{a} = \varepsilon\left[u + F(\psi) - \chi a\right], \quad \dot{\psi} = a; \quad a(t_0) = \dot{y}^0 > 0, \quad \psi(t_0) = y^0. \tag{3.2.38}$$

For system (38), we pose the optimal control problem:

$$a(T) = \dot{y}^* \geqslant 0, \quad J[u] = \varepsilon \int_{t_0}^{T} u^2 \, dt, \qquad |u| < \infty, \quad T = \Theta/\varepsilon. \tag{3.2.39}$$

By the mode (2), the integral functional $J[u]$ (39) can be represented in the terminal form. According to (14) and (17), the boundary-value problem of the first approximation is reduced to the form

$$\begin{gathered}\dot{\xi} = \frac{1}{2}\eta + F_0 - \chi\xi, \quad \xi(\tau_0) = \dot{y}^0, \quad \xi(\Theta) = \dot{y}^*, \\ \dot{\eta} = \chi\eta - \beta, \quad \eta(\Theta) = \lambda, \quad \tau = \varepsilon t \in [\tau_0, \Theta], \quad \tau_0 = \varepsilon t_0.\end{gathered} \tag{3.2.40}$$

When $\beta = 0$, the solution of the linear boundary-value problem (40) is

$$\begin{gathered}\xi(\tau) = \left[\dot{y}^0 - \frac{F_0}{\chi} - \frac{\lambda}{4\chi}e^{-\chi(\Theta-\tau_0)}\right]e^{-\chi(\Theta-\tau_0)} + \frac{F_0}{\chi} + \frac{\lambda}{4\chi}e^{-\chi(\Theta-\tau)}, \\ \eta(\tau) = \lambda e^{-\chi(\Theta-\tau)}, \\ \lambda = 4\chi\left\{\dot{y}^* - \dot{y}^0 e^{-\chi(\Theta-\tau_0)} - \frac{F_0}{\chi}\left[1 - e^{-\chi(\Theta-\tau_0)}\right]\right\}\left[1 - e^{-2\chi(\Theta-\tau_0)}\right]^{-1}.\end{gathered} \tag{3.2.41}$$

The approximate optimal control and the functional, taking (41) into account, are found in the form

$$\begin{gathered}u_p(\tau) = \frac{1}{2}\eta(\tau), \qquad u_s(\Theta - \tau, y) = \frac{1}{2}\eta\Big|_{\tau_0\to\tau,\ \dot{y}^0\to a=\dot{y}}, \\ J_0^* = J_0[0] = (\lambda^2/2\chi)\left[1 - e^{-2\chi(\Theta-\tau_0)}\right].\end{gathered} \tag{3.2.42}$$

It is easy to show that the value $\beta_0^* = 0$ gives the minimum of $J_0[\beta]$, according to (24). Indeed, the linearity of system (40) implies that the solution of the boundary-value problem is linear in β, $|\beta| < \infty$:

$$\xi = \xi_0 + \beta\xi_1, \qquad \eta = \eta_0 + \beta\eta_1, \qquad \lambda = \lambda_0 + \beta\lambda_1. \tag{3.2.43}$$

Having substituted (43) into (42) we obtain the expression

$$J_0[\beta] = \frac{1}{4}\int_{\tau_0}^{\Theta} (\eta_0^2 + 2\beta\eta_0\eta_1 + \beta^2\eta_1^2)\, d\tau. \tag{3.2.44}$$

Since $\eta_1 \not\equiv 0$, it follows from (44) that $\min_\beta J_0[\beta]$ exists and is unique. On the other hand, formula (29) implies that $\beta = 0$ is the minimum point for $J_0[\beta]$. Therefore, the integral of $\eta_0\eta_1$ in (44) must be equal to zero, which is easy to verify by immediate integration. Then, from the form of the functional (44), using inequality (33), we ascertain the strict global optimality of the point $\beta_0^* = 0$ and the minimum $J_0[0]$.

3.3. Asymptotic Analysis of the Optimal Control Problem with Free Time of the Process Termination

3.3.1. STATEMENT OF THE CONTROL PROBLEM AND THE ANALYSIS OF THE BOUNDARY-VALUE PROBLEM OF THE MAXIMUM PRINCIPLE

Consider a nonlinear controlled (in slow variables) system (1.3), (2.1), where $T = \Theta\varepsilon^{-1}$ is a nonfixed instant of termination of the control process which is to be found. In this case [1, 5, 28, 40, 110, 131], the transversality conditions at the right endpoint ($t = T$) are supplemented with the equality (see (1.3.5) and (2.4), (2.5))

$$H^*\big|_T = \varepsilon\big(p \cdot f_0^*(a, \psi, p)\big)\big|_T = 0. \tag{3.3.1}$$

Here, the functions H^* and f_0^* are defined by formulae (2.3) and (2.8). The construction of an asymptotic solution under nonfixed Θ ($\Theta \sim 1$ when $\varepsilon \to 0$) is conducted following a scheme similar to that for a specified Θ, according to Section 3.2 (see also Section 1.3). For this reason, we focus below mainly on differences from the constructions of Section 3.2.

It follows from (1) that in corresponding relationships of the boundary-value problem of the maximum principle (2.6)–(2.11), one has to put the parameter h equal to zero. However, the solution of the boundary-value problem (2.9) will include another unknown parameter Θ and this solution can be represented in the form, similar to (2.10):

$$a = a(t, \Theta, \varepsilon), \qquad \psi = \psi(t, \Theta, \varepsilon), \qquad p = p(t, \Theta, \varepsilon). \tag{3.3.2}$$

We suppose that the solution of the form (2) exists in the domain considered for the values Θ from some set I_Θ. Substituting (2) into condition (1) one can obtain the equation for determining $\Theta = \varepsilon T$:

$$\left(p\left(\frac{\Theta}{\varepsilon}, \Theta, \varepsilon\right) \cdot f_0^*\left(a\left(\frac{\Theta}{\varepsilon}, \Theta, \varepsilon\right), \psi\left(\frac{\Theta}{\varepsilon}, \Theta, \varepsilon\right), p\left(\frac{\Theta}{\varepsilon}, \Theta, \varepsilon\right)\right)\right) = 0. \tag{3.3.3}$$

As with the analysis of Equation (2.1), we establish that Equation (3), due to the rapid oscillation of $f_0^*|_T$ in Θ, admits, as a rule, a number ($\sim \varepsilon^{-1}$) of roots, the distance between adjacent roots being $O(\varepsilon)$. To determine the optimal value $\Theta \in \{\Theta_\nu\}$ it is required, analogously to (2.13), to solve the minimization problem

$$J^* = \min_{\Theta} J(\Theta, \varepsilon), \quad \Theta \in \{\Theta_\nu\}; \qquad J(\Theta, \varepsilon) = g\left(a\left(\frac{\Theta}{\varepsilon}, \Theta, \varepsilon\right)\right). \tag{3.3.4}$$

The qualitative behaviour pattern of the function $J(\Theta, \varepsilon)$ from (4) in the neighbourhood of the minimum for small ε is similar to that shown in Fig. 3.2 (with the change $h \to \Theta$).

3.3.2. AVERAGED BOUNDARY-VALUE PROBLEM OF THE FIRST APPROXIMATION

In the first approximation with respect to ε, the boundary-value problem has the form (2.14), where $h = 0$ and Θ is an unknown parameter to be determined. The solution ξ and η is supposed to satisfy the isolatedness condition, formulated in Section 3.2.3, with the change of h for Θ. The boundary-value problem is transformed to the form (2.17) by using the integral (2.16) where $h = 0$. Relationships (2.16) and (2.18) define the dependence $\Theta(\beta)$. In what follows, we consider $\Theta = \Theta(\beta)$, $\beta \in I_\beta$, and the problem of determining the optimal value for the parameter β is stated.

Substitution of the dependence $\Theta(\beta)$ into solution (2.19) and (2.20) and then into relationship (3) gives a transcendental equation relative to β. This equation coincides with (2.21), but here $\Theta = \Theta(\beta)$. Supposing that the following condition of rapid oscillation, similar to (2.22),

$$\frac{\mathrm{d}}{\mathrm{d}\beta}\int_{\tau_0}^{\Theta(\beta)} \omega\big(\xi\big(\tau - \tau_0, \Theta - \tau_0, a^0, \beta\big)\big)\mathrm{d}\tau\Big|_{\beta=0} = \mu_0 \neq 0 \tag{3.3.5}$$

holds, we arrive at the conclusion that the parameter β is defined by the minimality condition for the function $J_0[\beta]$, $\beta \in I_\beta$, $\beta = 0 \in I_\beta$. As a result, we obtain the minimization problem for the functional of the first approximation, similar to (2.24):

$$J_0[\beta] = g\big(\xi\big(\Theta(\beta) - \tau_0, \Theta(\beta) - \tau_0, a^0, \beta\big)\big) \longrightarrow \min_\beta, \qquad \beta \in I_\beta. \tag{3.3.6}$$

The investigation and the solution of problem (6) are carried out analytically, analogously to Section 3.2.5; see also Section 1.3.3, item (2).

3.3.3. DETERMINING THE OPTIMAL SOLUTION

Under the adopted assumptions relative to the smoothness of the functions f, ω, F, and g, the functional $J_0[\beta]$ (6) is differentiable with respect to $\beta \in I_\beta$. The derivative is calculated following the technique of Section 3.2.5:

$$J_0'[\beta] = \left(\frac{\partial g}{\partial \xi}\Big|_{\Theta(\beta)} \cdot \frac{\mathrm{d}\xi_*}{\mathrm{d}\beta}\right), \qquad \xi_*(\beta) = \xi\big|_{\Theta(\beta)}. \tag{3.3.7}$$

Here, $\Theta(\beta)$ is a root of Equation (2.16) where $h = 0$ and

$$\xi = \xi\big(\tau - \tau_0, \Theta - \tau_0, a^0, \beta\big), \qquad \eta = \eta\big(\tau - \tau_0, \Theta - \tau_0, a^0, \beta\big).$$

This equation is reduced to the form

$$\left[\omega(\xi)\beta + (\eta \cdot f_0(\xi, \eta))\right]\Big|_{\Theta(\beta)} = 0 \quad \left(= \omega(\xi)\beta + (\eta \cdot f_0(\xi, \eta))\right). \tag{3.3.8}$$

We take the expression for $\partial g/\partial \xi|_{\Theta(\beta)}$ from (2.14) and substitute it into (7). Then we find, analogously to (2.27), that

$$J_0'[\beta] = \left(\left[\frac{\partial(\lambda \cdot M)}{\partial \xi}\Big|_{\Theta} - \eta_*\right] \cdot \frac{d\xi_*}{d\beta}\right) = -\left(\eta_* \cdot \frac{d\xi_*}{d\beta}\right). \tag{3.3.9}$$

Here, the first term in the first equality equals zero, since

$$M\Big|_{\Theta} \equiv 0, \qquad \beta \in I_\beta, \qquad \Theta = \Theta(\beta).$$

Then we transform expression (9), differentiating $\xi_*(\beta)$ with respect to β as a composite function, according to (7):

$$J_0'[\beta] = -(\eta \cdot \dot{\xi})\big|_{\Theta}\Theta'(\beta) - (\eta \cdot \partial\xi/\partial\beta)\big|_{\Theta}. \tag{3.3.10}$$

In the expression (10), $\dot{\xi} = d\xi/d\tau$ is replaced with $f_0(\xi, \eta)$, according to (2.17), and the second term is represented as an integral of the derivative, as in (2.27). Taking into account the equalities $\partial\xi/\partial\beta|_{\tau_0} = \partial a^0/\partial\beta = 0$ we can obtain the expression

$$J_0'[\beta] = -(\eta_* \cdot f(\xi_*, \eta_*))\Theta'(\beta) - \int_{\tau_0}^{\Theta\beta} \frac{d}{d\tau}\left(\eta \cdot \frac{\partial\xi}{\partial\beta}\right) d\tau. \tag{3.3.11}$$

The first term in (11) is replaced according to (8), and the integral is calculated using the mode of Section 3.2.5, see (2.28) and (2.29). As a result, we obtain the expression for the derivative $J_0'[\beta]$, $\beta \in I_\beta$, in appearance identical to (2.29):

$$J_0'[\beta] = \beta \frac{d}{d\beta} \int_{\tau_0}^{\Theta} \omega\big(\xi(\tau - \tau_0, \Theta - \tau_0, a^0, \beta)\big) d\tau, \qquad \Theta = \Theta(\beta). \tag{3.3.12}$$

The substantial difference is that $\Theta = \Theta(\beta)$, $\beta \in I_\beta$. Since, according to (5), the point $\beta = 0 \in I_\beta$, the subsequent conclusions of Section 3.2.5 about the optimality of the value $\beta_0^* = 0$ remain valid. The following assertion takes place.

LEMMA 3.2. *The sufficient condition of the local optimality, including requirement* (5) *of the aforenamed behaviour of roots* $\{\beta_\nu\}$, *has the form similar to* (2.30): $\mu_0 > 0$.

The global optimality conditions coincide with (2.31)–(2.33) when $\Theta = \Theta(\beta)$. The procedure of constructing the state and adjoint variables, as well as the optimal control, and substantiation of the proximity of the asymptotic solution obtained rely on analogous reasonings. It is worth noting that the feedback control u_s (2.36) turns out to depend only on a and ψ. The dependence on τ is absent due to time-invariance (the quantity $\Theta - \tau_0$ is determined from (8) as a function of a^0 and β).

3.3.4. MODEL EXAMPLE

To illustrate the technique expounded in Section 3.3, we consider the control problem for fast rotations of the pendulum-type system (see (2.37)):

$$\begin{gathered} \ddot{y} = \varepsilon u + \varepsilon F(y), \quad y(0) = y^0, \quad \dot{y}(0) = \dot{y}^0 > 0, \\ \dot{y}(T) = \dot{y}^* \geqslant 0, \quad |u(t)| < \infty, \quad t \in [0, T], \quad T = \Theta\varepsilon^{-1}, \\ J[u] = k\Theta + \int_0^\Theta u^2 \, d\tau \longrightarrow \min_u, \qquad k > 0, \quad \tau = \varepsilon t. \end{gathered} \tag{3.3.13}$$

Here, F is a 2π-periodic function, $F_0 = 0$. The phase in the system of the type (1.3) is $\psi = y$, and the frequency $\omega(a)$ is the velocity $a = \dot{y}$. According to (2.2), the functional from (13) is written as

$$J[u] = b(T), \qquad \dot{b} = \varepsilon\left(k + u^2\right), \qquad b(0) = 0. \tag{3.3.14}$$

On the basis of (2.17) and (2.14), the averaged boundary-value problem for ξ and η is written in the form

$$\dot{\xi} = \frac{1}{2}\eta, \quad \dot{\eta} = -\beta, \quad \xi(0) = \dot{y}^0, \quad \xi(\Theta) = \dot{y}^*. \tag{3.3.15}$$

The solution of (15) is

$$\begin{gathered} \xi(\tau) = -\frac{1}{4}\beta(\Theta - \tau)^2 + \Theta^{-1}(\Theta - \tau)\left(\Delta + \frac{1}{4}\beta\Theta^2\right) + \dot{y}^*. \\ \eta(\tau) = \beta(\Theta - \tau) - 2\Theta^{-1}\left(\Delta + \frac{1}{4}\beta\Theta^2\right), \qquad \Delta = \dot{y}^0 - \dot{y}^*, \end{gathered} \tag{3.3.16}$$

and the connection between Θ and β is given by Equation (8) which takes the form

$$\left(\beta\xi + \frac{1}{2}\eta^2 - k\right)\Big|_\Theta = 0 \quad \left(= \beta\xi + \frac{1}{2}\eta^2 - k\right). \tag{3.3.17}$$

Substitution of ξ and η from (16) into (17) leads to the relationship

$$\beta\dot{y}^* + \Theta^{-2}\left(\Delta + \frac{1}{4}\beta\Theta^2\right)^2 - k = 0. \tag{3.3.18}$$

Functional (14) in the first approximation is

$$J_0[\beta] = 2k\Theta - (\beta^2/24)\Theta^3 - \frac{1}{2}\Delta\beta\Theta - \dot{y}^*\beta\Theta. \tag{3.3.19}$$

The derivatives $J_0'[0]$, $\Theta'(0)$, and the quantity $\Theta(0)$ are given by

$$\begin{aligned} J_0'[0] &= 2k\Theta'(0) - \left(\dot{y}^* + \frac{1}{2}\Delta\right)\Theta(0), \\ \Theta'(0) &= \frac{1}{2}k^{-1}\left(\dot{y}^* + \frac{1}{2}\Delta\right)\Theta(0), \qquad \Theta(0) = k^{-1/2}|\Delta|. \end{aligned} \tag{3.3.20}$$

Expressions (20) in accordance with the general expression (12) imply that $J_0'[0] = 0$. Therefore, the necessary optimality condition is fulfilled. To calculate the second derivative $J_0''[0]$ one can use formula (12) for $J_0'[\beta]$. As a result, we obtain

$$J_0''[0] = (1/24)\Theta^3(0) + \frac{1}{2}k^{-1}\left(\dot{y}^* + \frac{1}{2}\Delta\right)^2\Theta(0) > 0. \tag{3.3.21}$$

Inequality (21) implies local optimality of the value $\beta_0^* = 0$. The values $\beta = 0$ and $\Theta(0)$, having been substituted into (16) and (19), define the optimal solution of the control problem (13) and (14) in the first approximation with respect to ε:

$$\begin{aligned} &\xi(\tau) = \dot{y}^0 - \tau k^{1/2}\operatorname{sign}\Delta, \qquad \Theta(0) = k^{-1/2}|\Delta|, \\ &u^* = \frac{1}{2}\eta, \qquad u_p^* = -k^{1/2}\operatorname{sign}\Delta, \qquad u_s^* = -k^{1/2}\operatorname{sign}(\dot{y} - \dot{y}^*), \\ &J_0^*[0] = 2k^{1/2}|\Delta| = 2k\Theta(0). \end{aligned}$$

It is worth noting that the contribution of each term in the functional (13) in the first approximation is the same and equals $k^{1/2}|\Delta|$. The global minimality of $J_0[0]$ at $\beta_0^* = 0$ is established on the basis of expressions (18) and (19).

3.4. Averaging of Essentially Nonlinear Systems on the Basis of Unperturbed Integrals

3.4.1. THE BACKGROUND OF THE PROBLEM

Direct application of methods of separation of motions (the averaging method in particular) for nonlinear systems often gives rise to substantial analytical and computational difficulties. These difficulties are due to the fact that the solution of the generating system, as a rule, is known only in implicit form as a complete set of first integrals. As examples of such systems, important for applications in mechanics, we can consider a nonlinear oscillator, a pendulum-type system, rotational motions of a rigid body about a fixed point (the centre of mass, Euler's or Lagrange's cases), the motion of a particle along an elliptic orbit (Kepler's problem), and many others. Motions of such systems are specified either in implicit form or through special functions, which is essentially the same. Therefore, it becomes necessary to construct averaging schemes on the basis of integrals of the unperturbed system. For rotatory-oscillatory pendulum-type systems, which admit a simple set of first integrals, implicit averaging schemes of the first and second approximations are developed in [1, 20, 150]. For problems of rigid body dynamics, the averaging scheme based on the Euler–Poinsot motion has been used in [1, 46]; the averaging based on the Lagrange motion can be carried out analogously, see below. The averaging over the angular variable (the true anomaly) in Kepler's problem is performed in [1, 5, 46, 72, 86, 88]. In recent years, direct averaging schemes based on an approximate representation of the generating motion are being developed. Below, we give various averaging schemes based on known integrals of unperturbed systems.

3.4.2. IMPLICIT AVERAGING SCHEMES FOR OSCILLATORY SYSTEMS

Consider the controlled system (1.1) having the special form (x and y are scalars, and z is a vector)

$$\begin{aligned}\dot{x} &= y\nu(x, y, z) + \varepsilon f(x, y, z, u),\\ \dot{y} &= -x\nu(x, y, z) + \varepsilon g(x, y, z, u),\\ \dot{z} &= \varepsilon h(x, y, z, u).\end{aligned} \tag{3.4.1}$$

The "frequency" ν is supposed to be positive, i.e. $0 < \nu_1 \leqslant \nu \leqslant \nu_2 < \infty$. For $\varepsilon = 0$, system (1) is completely integrable:

$$\begin{aligned}&x^2 + y^2 = c = \text{const} \geqslant 0, \qquad z = \text{const},\\ &t = \int \frac{d\varphi}{\nu(\sqrt{c}\cos\varphi, \sqrt{c}\sin\varphi, z)} \qquad (x = \sqrt{c}\cos\varphi,\ y = \sqrt{c}\sin\varphi).\end{aligned} \tag{3.4.2}$$

According to (2), the period of oscillations T_0 for the unperturbed system (1) is calculated by integrating the following expression:

$$
\begin{aligned}
T_0 = T_0(c, z) &= \oint_{x^2+y^2=c} \frac{ds}{\sqrt{c}\,\nu(x, y, z)} \\
&= \int_0^{2\pi} \frac{d\varphi}{\nu(\sqrt{c}\cos\varphi, \sqrt{c}\sin\varphi, z)} \\
&\left(ds = \left(dx^2 + dy^2\right)^{1/2},\ x = \sqrt{c}\cos\varphi,\ y = \sqrt{c}\sin\varphi\right).
\end{aligned}
\tag{3.4.3}
$$

The equation for the slow variable c is obtained by differentiating expression (2) according to the perturbed system (1) and easily expressed through the original variables:

$$
\dot{c} = 2\varepsilon(xf + yg), \qquad x^2 + y^2 = c \geqslant 0. \tag{3.4.4}
$$

Using (2), the expression for the phase of oscillations ψ can be represented in the following form:

$$
\begin{aligned}
\dot{\psi} &= \frac{2\pi}{T_0(c, z)} + \varepsilon\Psi(c, z, \psi, u) \\
&= \frac{d}{dt}\left(\frac{2\pi}{T_0(c, z)} \int \frac{d\varphi}{\nu(\sqrt{c}\cos\varphi, \sqrt{c}\sin\varphi, z)}\right).
\end{aligned}
\tag{3.4.5}
$$

The differential constraint for the phase ψ and the variables x, y, and φ allows averaging any function $F(x, y, z, u)$, where u is a control, depending on x, y, z and other slow variables (for instance, the adjoint ones). The averaging scheme can be taken in one of the following forms:

$$
\begin{aligned}
\langle F \rangle &= \frac{2\pi}{T_0(c, z)} \int_0^{T_0} F\,dt = \frac{1}{T_0} \oint_{x^2+y^2=c} \frac{F\,ds}{\sqrt{c}\,\nu(x, y, z)} \\
&= \frac{1}{T_0} \int_0^{2\pi} \frac{F\,d\varphi}{\nu(\sqrt{c}\cos\varphi, \sqrt{c}\sin\varphi, z)}.
\end{aligned}
\tag{3.4.6}
$$

The specific choice is defined by reasons of convenience. Other transformations of Equations (1) for x, y, and z, convenient for averaging, are possible as well. The change of (x, y) for variables of amplitude-phase type (a, ψ) leads to the equivalent equations (when $a \neq 0$):

$$\begin{aligned}
&\dot{a} = \varepsilon(f \cos\psi - g \sin\psi) \qquad (\dot{z} = \varepsilon h),\\
&\dot{\psi} = \nu - \varepsilon a^{-1}(f \sin\psi + g \cos\psi),\\
&x = a\cos\psi, \quad y = -a\sin\psi, \quad a \neq 0; \quad \nu = \nu(x, y, z).
\end{aligned} \tag{3.4.7}$$

The averaging is conducted in the following way (F is the right-hand side of system (7)):

$$\begin{aligned}
&\langle F\rangle = \frac{1}{T_0}\int_0^{2\pi} \frac{F\,\mathrm{d}\psi}{\nu(a\cos\psi, -a\sin\psi, z)},\\
&T_0(a, z) = \int_0^{2\pi} \frac{\mathrm{d}\psi}{\nu(a\cos\psi, -a\sin\psi, z)}.
\end{aligned} \tag{3.4.8}$$

The transition from the original variables (x, y) to the variables (a, b, φ) of Van der Pohl's type gives the following averaging scheme:

$$\begin{aligned}
&\dot{a} = \varepsilon(f\cos\varphi - g\sin\varphi), \qquad x = a\cos\varphi + b\sin\varphi,\\
&\dot{b} = \varepsilon(f\sin\varphi + g\cos\varphi), \qquad y = -a\sin\varphi + b\cos\varphi,\\
&\dot{\varphi} = \nu(x, y, z), \qquad T_0(a, b, z) = \int_0^{2\pi} \frac{\mathrm{d}\varphi}{\nu},\\
&\langle F\rangle = \frac{1}{T_0(a, b, z)}\int_0^{2\pi} \frac{F\,\mathrm{d}\varphi}{\nu(a\cos\varphi + b\sin\varphi, -a\sin\varphi + b\cos\varphi, z)}.
\end{aligned} \tag{3.4.9}$$

Here, F is the right-hand side of system (9).

Let us consider a system more general than (1), in which

$$\begin{aligned}
&\dot{x} = y\nu(x, y, z) + \varepsilon f(x, y, z, u),\\
&\dot{y} = -Q(x, z)\nu(x, y, z) + \varepsilon g(x, y, z, u),\\
&\dot{z} = \varepsilon h(x, y, z, u).
\end{aligned} \tag{3.4.10}$$

The function Q is supposed to admit a stable equilibrium position $x = x^*(z)$ for all considered values of z; without any loss of generality, we can assume $x^*(z) = 0$ and $Q'(0, z) > 0$. Then, in some vicinity of the origin $x = y = 0$, $z = \text{const}$, the unperturbed system (10) is integrable and has the oscillating solution $x = x_0(c, z, \psi)$ and $y = y_0(c, z, \psi)$, which is defined by

$$
\begin{aligned}
&\frac{1}{2}y^2 + \Pi(x, z) = c = \text{const} \geqslant 0, \qquad z = \text{const},\\
&\Pi(x, z) = \int_0^x Q(\varkappa, z)\,\mathrm{d}\varkappa, \qquad \Pi > 0, \quad x \neq 0,\\
&y = y_{1,2}(x, c, z) = \pm\sqrt{2}\,[c - \Pi(x, z)]^{1/2}, \qquad y^2 \leqslant 2c,\\
&x_1 \leqslant x \leqslant x_2, \qquad x_{1,2}(c, z) = \arg\left(c - \Pi(x, z)\right),\\
&t + \tau = \int_0^x y_{1,2}^{-1}(\varkappa, c, z)\nu^{-1}(\varkappa, y_{1,2}, z)\,\mathrm{d}\varkappa \equiv \theta(x, c, z),\\
&\psi = \frac{2\pi}{T_0(c, z)}(t + \tau) = \frac{2\pi}{T_0(c, z)}\theta(x, c, z),\\
&T_0(c, z) = \oint_{y^2/2+\Pi=c} \nu^{-1}(y^2 + Q^2)^{-1/2}\,\mathrm{d}s\\
&\qquad = \int_{x_1}^{x_2} |y|^{-1}\left(\nu^{-1}(x, y_1, z) + \nu^{-1}(x, y_2, z)\right)\mathrm{d}x.
\end{aligned}
\tag{3.4.11}
$$

Based on the generating solution specified by the set of integrals (11), the averaging of the perturbation F in the equations for c and z is carried out as follows:

$$
\begin{aligned}
\langle F\rangle &= \frac{1}{T_0}\int_0^{T_0} F\,\mathrm{d}t\\
&= \frac{1}{T_0(c, z)}\int_{x_1(c,z)}^{x_2(c,z)} \frac{F}{|y(x, c, z)|}\left(\frac{1}{\nu(x, y_1, z)} + \frac{1}{\nu(x, y_2, z)}\right)\mathrm{d}x.
\end{aligned}
\tag{3.4.12}
$$

Analogously, one can generalize the averaging scheme, based on the generating oscillating motion $x = x_0(c, z, \psi)$, $y = y_0(c, z, \psi)$ defined by the relationships ($\varepsilon = 0$):

$$
\begin{aligned}
&\dot{x} = G(y, z)\nu(x, y, z), \qquad \dot{y} = -Q(x, z)\nu(x, y, z),\\
&\Gamma(y, z) + \Pi(x, z) = c \geqslant 0, \qquad \Gamma(y, z) = \int_0^y G(\chi, z)\,\mathrm{d}\chi,\\
&T_0(c, z) = \oint_{\Gamma+\Pi=c} \nu^{-1}(G^2 + Q^2)^{-1/2}\,\mathrm{d}s,
\end{aligned}
\tag{3.4.13}
$$

$$\langle F\rangle = \frac{1}{T_0}\int_0^{T_0} F\,\mathrm{d}t = \frac{1}{T_0}\oint_{\Gamma+\Pi=c} F\nu^{-1}\left(G^2+Q^2\right)^{-1/2}\mathrm{d}s.$$

To perform integration in the expressions for T_0 and $\langle F\rangle$ it is required to make a parametrization for a family of closed curves $x(s,c,z)$, $y(s,c,z)$: $\Gamma(y,z)+\Pi(x,z)=c$, where c and z are parameters of the family.

The averaging is performed analogously to (13), also in the case when the generating solution is governed by general equations admitting a stable equilibrium position $x=y=0$:

$$\begin{gathered}\dot{x}=G(x,y,z), \qquad \dot{y}=-Q(x,y,z), \qquad z=\mathrm{const}; \\ G(x,0,z)=Q(0,y,z)\equiv 0; \qquad \partial G/\partial y>0, \qquad \partial Q/\partial x>0.\end{gathered} \tag{3.4.14}$$

It is supposed that in some vicinity of the point $x=y=0$, the first integral $E(x,y,z)=c\geqslant 0$ is known, where the function E has a local minimum. Then, according to (13), we have

$$\begin{gathered}T_0(c,z)=\oint_{E=c}\left(G^2+Q^2\right)^{-1/2}\mathrm{d}s, \\ \langle F\rangle = \frac{1}{T_0}\int_0^{T_0}F\,\mathrm{d}t=\frac{1}{T_0(c,z)}\oint_{E=c}F\left(G^2+Q^2\right)^{-1/2}\mathrm{d}s.\end{gathered} \tag{3.4.15}$$

In expressions (15), as above, it is required to make a parametrization of the family of closed curves $E(x,y,z)=c$ along which the integration is carried out: $x=x(s,c,z)$, $y=y(s,x,z)$.

Let us consider the system with variable parameters in the following form:

$$\begin{gathered}\dot{x}=X(x,z)+\varepsilon f(x,z,u), \qquad x\in D_x, \\ \dot{z}=\varepsilon h(x,z,u), \qquad z\in D_z.\end{gathered} \tag{3.4.16}$$

In system (16), x is a vector of any dimension $n\geqslant 1$ for which the generating motion is expressed as

$$\begin{gathered}x=x_0(\varphi,c,z), \qquad c=(c_1,\ldots,c_{n-1}), \quad c\in D_c, \\ \dot{\varphi}=\Phi(\varphi,c,z), \qquad |\varphi|<\infty,\end{gathered} \tag{3.4.17}$$

where x_0 and Φ are 2π-periodic functions of the scalar angular variable φ ($0 < \Phi_1 \leqslant \Phi \leqslant \Phi_2 < \infty$). The averaging of the perturbing function F is performed analogously to (3) and (9):

$$T_0(c, z) = \int_0^{2\pi} \Phi^{-1}(\varphi, c, z)\,\mathrm{d}\varphi, \qquad \langle F\rangle = \frac{1}{T_0}\int_0^{2\pi} F\Phi^{-1}\,\mathrm{d}\varphi. \tag{3.4.18}$$

The averaging schemes developed above can be transferred to systems (10), (13), (14), and (16) admitting oscillating and rotating motions.

If $c = (c_1, \dots, c_l)$, $l < n - 1$, and the motion of the generating system takes place in the vicinity of an asymptotically stable local integral manifold, then the averaging is carried out according to scheme (18), in which the variables corresponding to the manifold are assumed to equal zero (Sections 2.5 and 4.5). The case when φ is a vector is rather complicated for analysis and requires reduction to the standard form [1, 38, 74, 115, 119, 120, 123, 128, 149, 150].

3.4.3. AVERAGING OF ROTATORY-OSCILLATORY SYSTEMS

Consider a system of the form (10), where Q is a 2π-periodic function of x with the mean value equal to zero. Then, as follows from the expression (11) for the first integral, under $c > \max_x \Pi(x, z)$, the variable y retains the sign determined by the initial condition. Let $0 < y_1 \leqslant y \leqslant y_2 < \infty$. The expressions to be found and the averaging scheme then have the form:

$$\begin{aligned}
&\frac{1}{2}y^2 + \Pi(x, z) = c > 0, \qquad y = \sqrt{2}(c - \Pi)^{1/2} > 0,\\
&x = \psi + v(\psi, c, z), \quad y = \frac{2\pi}{T_0}\left(1 + \frac{\partial v}{\partial \psi}\right),\\
&t + \tau = \int_0^x \frac{\mathrm{d}\varkappa}{y(\varkappa, c, z)} \equiv \theta(x, c, z), \qquad \psi = \frac{2\pi}{T_0}(t + \tau) = \frac{2\pi}{T_0}\theta,\\
&T_0(c, z) = \int_0^{2\pi} y^{-1}(x, c, z)\nu^{-1}(x, y(x, c, z), z)\,\mathrm{d}x,\\
&\langle F\rangle = \frac{1}{T_0}\int_0^{T_0} F\,\mathrm{d}t = \frac{1}{T_0}\int_0^{2\pi} F y^{-1}\nu^{-1}\,\mathrm{d}x.
\end{aligned} \tag{3.4.19}$$

To apply the scheme of averaging over x to the rotatory-oscillatory system it is necessary that the functions F and ν are 2π-periodic in x. The rotating variable $x_0(\psi, c, z)$ has the sense of an angular variable φ in the averaging scheme (17) and (18). If the first integral of the unperturbed rotatory-oscillatory system (14) has the general form

$$E(x, y, z) = c > 0, \qquad E(x + 2\pi, y, z) \equiv E(x, y, z), \tag{3.4.20}$$

and $y(x, c, z) = \arg(c - E)$ is a strictly sign-definite (for example, positive-definite) function for x: $|x| < \infty$, then the averaging leads to integration with respect to the variable x:

$$\begin{aligned} &T_0(c, z) = \int_0^{2\pi} G^{-1}(x, y(x, c, z), z)\, dx, \\ &c \in [c_1, c_2], \quad c_{1,2} > 0; \quad z \in D_z, \\ &\langle F \rangle = \frac{1}{T_0} \int_0^{2\pi} FG^{-1}(x, y(x, c, z), z)\, dx. \end{aligned} \tag{3.4.21}$$

In particular, let the first integral (20) of the system of the form (14) be equal to

$$\begin{aligned} &\frac{1}{2} m(x, z) y^2 + \Pi(x, z) = c > 0. \qquad |x| < \infty, \\ &0 < m_1 \leqslant m(x, z) \leqslant m_2 < \infty, \qquad y \in [y_1, y_2], \quad z \in D_z \end{aligned} \tag{3.4.22}$$

(m is the inertia characteristic). Then, provided m and Π are 2π-periodic in x, the averaging scheme (21) is simplified, taking into account the known expression for y

$$y = y(x, c, z) = \sqrt{2} m^{-1/2}(x, z) \left[c - \Pi(x, z)\right]^{1/2}.$$

The most general averaging scheme (17) and (18) turns out to be applicable also to rotatory-oscillatory single-frequency systems.

It is worth noting that the construction of averaging schemes of higher approximations in terms of powers of ε gives rise to rather bulky uninteresting expressions [150].

3.5. The Control of a Nonlinear Oscillator with Regulated Equilibrium Position

3.5.1. NONLINEAR SYSTEM WITH AN ELASTIC ELEMENT

Consider, as an example, a nonlinear model of the mechanical oscillatory system shown in Fig. 3.4. Having neglected transient processes, we assume that the particle P, controlled by the velocity v, and the particle G of the mass m can slide along parallel guides, separated by a distance d. The control action is transmitted through a linear massless spring with a stiffness coefficient c and the length l_0, when

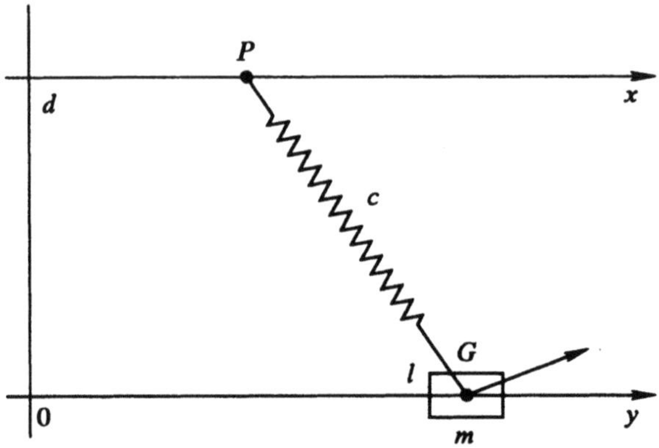

Fig. 3.4.

unextended. The potential energy of the spring Π appears to depend only on the difference $(y-x)$, where y and x are absolute coordinates of the particles G and P:

$$\Pi = \Pi(y-x) = \frac{1}{2}c(l-l_0)^2, \qquad l^2 = d^2 + (y-x)^2. \tag{3.5.1}$$

The equilibrium position $y = x$ is stable when $d \geqslant l_0$ and unstable when $d < l_0$. However, in the last case the system has two additional stable equilibrium positions $y_{1,2} = x \pm (l_0^2 - d^2)^{1/2}$. If $d = 0$, then the system is "linear" and is described by equations of the type (5.3.2). But if $d = l_0$, then the term $O((y-x)^4)$ dominates in the function Π (1) and the system is essentially nonlinear. Its generating solution is described in terms of elliptic functions [46, 87]. The equations of motion in Lagrange's form for this system are

$$\begin{aligned} &\ddot{y} + \nu^2\left(1 - \frac{l_0}{l}\right)(y-x) = 0, \\ &\dot{x} = v, \qquad v \in [v_1, v_2], \qquad \nu^2 = \frac{c}{m}. \end{aligned} \tag{3.5.2}$$

Introducing dimensionless variables and parameters, and assuming v to be small, we obtain a system of the form (1.15) which maintains this form in the uniform force field. System (2) is a generalization of the model investigated in [150] (for the uncontrolled model $x = \text{const}$).

3.5.2. STATEMENT OF OPTIMAL CONTROL PROBLEMS

For the nonlinear oscillatory system

$$\begin{aligned} &\ddot{y} + F(y - x) = 0, \qquad y(0) = y^0, \qquad \dot{y}(0) = \dot{y}^0, \\ &\dot{x} = \varepsilon u, \qquad x(0) = x^0, \qquad u \in [u_1, u_2], \end{aligned} \tag{3.5.3}$$

we assume that $y = x$ is an isolated point of minimum of the potential energy, i.e. $F(0) = 0$ and $zF(z) > 0$ for $z \neq 0$ within some neighbourhood of the point $z = 0$. Having assumed that $u = 0 \in [u_1, u_2]$, we pose the following optimal control problems for system (3).

A. *Minimum-time quenching relative oscillations.* Terminal conditions and the functional have the form

$$y(T) = x(T), \quad \dot{y}(T) = 0; \qquad T \longrightarrow \min_u, \quad u \in [u_1, u_2]. \tag{3.5.4}$$

For $t > T$ and $u \equiv 0$, system (3) remains at a state of rest (4). Terminal values of the variables y and x are not specified here.

B. *Acceleration of the system as a whole.* Terminal conditions and the functional have the form

$$y(T) = x(T), \quad \dot{y}(T) = \varepsilon u^0; \qquad T \longrightarrow \min_u, \quad u \in [u_1, u_2]. \tag{3.5.5}$$

When $t > T$ and $u = u^0$, the system moves with the velocity εu^0 as a whole, without relative oscillations. Problem (4) can be considered as a special case of statement (5) with $u^0 = 0$.

C. *Transferring to a given point.* It is required to satisfy the following conditions at the end of the control process:

$$y(T) = x(T) = x^*, \quad \dot{y}(T) = 0; \qquad T \longrightarrow \min_u, \quad u \in [u_1, u_2]. \tag{3.5.6}$$

Here, x^* is a given quantity; for $t > T$, the system remains at the state (6), provided $u \equiv 0$.

Statements A, B, and C of optimal control problems have a definite applied aspect for the class of mechanical oscillatory systems in question whose equilibrium positions are controlled by the velocity of their displacement (3), see also Chapter 5.

3.5.3. THE REDUCTION TO THE STANDARD FORM WITH ROTATING PHASE

When $\varepsilon = 0$, system (3) admits complete integration (see (1.12)). Let $z(h,\psi)$ be the general solution of the equation $\ddot{z} + F(z) = 0$. Then, using the change $(y,\dot{y},x) \to (h,\psi,x)$:

$$y = x + z(h,\psi), \qquad \dot{y} = \omega(h)\frac{\partial z}{\partial \psi}, \qquad x = x, \tag{3.5.7}$$

we obtain the standard system of the form (1.13). Here, h is the integral of the "energy of relative oscillations", ψ is a phase, ω is a frequency, see (1.12) and z is a function 2π-periodic in ψ:

$$h = \frac{1}{2}\omega^2(h)\left(\frac{\partial z}{\partial \psi}\right)^2 + \Pi(z), \quad \omega = \frac{2\pi}{T_0}, \quad \Pi(z) = \int_0^z F(s)\,ds. \tag{3.5.8}$$

Indeed, differentiating the change (7) according to (3) we obtain the system of the standard form with the rotating phase ψ and the terminal conditions A, B, and C corresponding to (4)–(6)

$$\begin{aligned} &\dot{h} = \varepsilon\frac{\omega}{W}\frac{\partial^2 z}{\partial \psi^2}u, \qquad \dot{x} = \varepsilon u; \quad h(0) = h^0, \quad x(0) = x^0, \\ &\dot{\psi} = \omega - \frac{\varepsilon}{W}\left(\omega'\frac{\partial z}{\partial \psi} + \omega\frac{\partial^2 z}{\partial h \partial \psi}\right)u, \quad \psi(0) = \psi^0; \\ &W = \omega'\left(\frac{\partial z}{\partial \psi}\right)^2 + \omega\frac{\partial z}{\partial \psi}\frac{\partial^2 z}{\partial h \partial \psi} - \omega\frac{\partial z}{\partial h}\frac{\partial^2 z}{\partial \psi^2} = \frac{1}{\omega(h)}; \\ &\text{A, B: } h(T) = 0; \qquad \text{C: } h(T) = 0, \ x(T) = x^*. \end{aligned} \tag{3.5.9}$$

Here, W is the Wronskian determinant for the corresponding system in variations; the property $W = W(h)$ follows from the fact that the original system is conservative, and the equality $W = 1/\omega$ is established by differentiating the first relationship with respect to h.

3.5.4. PROCEDURE OF ASYMPTOTIC SOLUTION

Let us denote by p, q, and r the variables adjoint with h (or $a(h)$), ψ, and x, respectively. Since x is a cyclic variable, $r = \text{const}$. Moreover, in problems A and B (9), the quantity $r = 0$. The Hamiltonian function and the adjoint system are linear in $|r|$. This enables this constant to be normalized in the following way:

$$\text{A, B: } r = 0; \qquad \text{C: } r \in \{-1, 0, 1\}. \tag{3.5.10}$$

In the first approximation, the optimal control is determined from the condition of the type (2.3), where $q = 0$:

$$u^* = \frac{1}{2}(u_1 + u_2) + \frac{1}{2}(u_2 - u_1)\operatorname{sign}(r - pF). \tag{3.5.11}$$

We keep the original denotations of variables for their averaged values and assume, as usual, $\langle q \rangle = \varepsilon\beta$. As a result, we obtain the following expression for the averaged Hamiltonian K in the first approximation:

$$\varepsilon^{-1}K = \frac{1}{2}(u_1 + u_2)r + \frac{1}{2}(u_2 - u_1)\langle |r - pF| \rangle + \beta\omega = \text{const}. \tag{3.5.12}$$

Boundary-value problems of the first approximation for statements A, B, and C are described by corresponding equations and conditions:

$$\begin{aligned}
&\dot{h} = -\frac{1}{2}(u_2 - u_1)\langle F \operatorname{sign}(r - pF)\rangle, \qquad h(0) = h^0, \quad h(\Theta) = 0,\\
&\dot{p} = -\frac{1}{2}(u_2 - u_1)(\partial/\partial h)\langle |r - pF| \rangle - \beta\omega' \qquad (\Theta = \varepsilon T),\\
&\dot{x} = \frac{1}{2}(u_1 + u_2) + \frac{1}{2}(u_2 - u_1)\langle \operatorname{sign}(r - pF)\rangle, \qquad x(0) = x^0\\
&\quad (\text{C: } x(\Theta) = x^*).
\end{aligned} \tag{3.5.13}$$

The third equation (13) is reduced to the quadrature after the first two equations have been integrated.

Let us consider the solution of problem (13) for statements A and B. Then, assuming $r = 0$ according to (10), we can find that the decrease rate of h is maximal when $\operatorname{sign} p = -1$. If $\beta = 0$, then the second equation (13) for p implies that its non-trivial solution has a constant sign, provided $|\partial F/\partial h|$ is bounded. Thus, assuming $r = \beta = 0$ and $\operatorname{sign} p = -1$, we can obtain the optimal solution. The terminal condition (13) for h will be satisfied, i.e. the value Θ will be bounded if for small h the value $\langle |F| \rangle = O(h^\alpha)$, where $\alpha < 1$.

As a result, we obtain the desired solution of the first approximation:

$$\begin{aligned}
&\tau = \frac{2}{u_2 - u_1}\int_h^{h^0} \frac{dl}{F(l)},\\
&F_0 = \langle |F| \rangle = \frac{1}{2\pi}\int_0^{2\pi} |F|\, d\psi = \frac{\omega^2}{2\pi}\int_0^{2\pi} \left|\frac{\partial^2 z}{\partial \psi^2}\right| d\psi,\\
&x(\tau) = x^0 + \frac{1}{2}(u_1 + u_2)\tau, \qquad \Theta = \tau\big|_{h=0}.
\end{aligned} \tag{3.5.14}$$

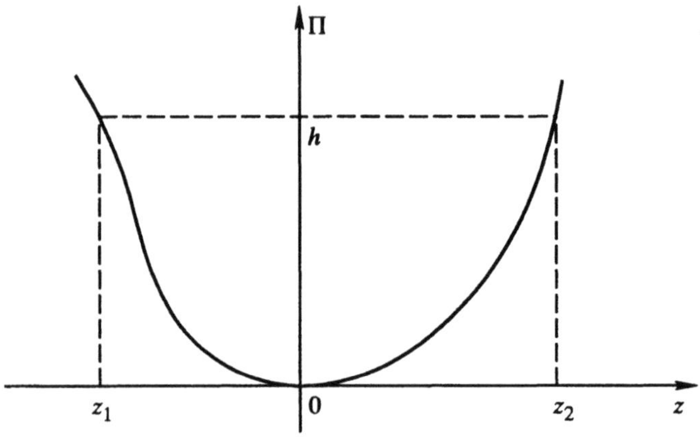

Fig. 3.5.

If the constraint on u is symmetric ($u_2 = -u_1$), then, according to (14), the "drift" of the equilibrium position is not observed in the first approximation: $x(\tau) = x^0$. Due to the property of the function F in (3): $\operatorname{sign} F(z) = \operatorname{sign} z$, we have for u^*

$$u^* = \frac{1}{2}(u_1 + u_2) + \frac{1}{2}(u_2 - u_1)\operatorname{sign}(y - x). \tag{3.5.15}$$

The physical sense of the control u^* is clear: this control reduces the difference $|y - x|$ by moving the suspension point.

It is worth noting that one can perform the averaging over ψ in (14) on the basis of integrals of the unperturbed motion. If the function $\Pi(z)$ has an ordinary shape, given in Fig. 3.5, then (see Section 3.4)

$$\begin{aligned} F_0(h) &= 2\frac{\omega(h)}{2\pi}\int_{z_1(h)}^{z_2(h)} \frac{|F|\,dz}{\sqrt{2}[h - \Pi(z)]^{1/2}} = (2\sqrt{2}/\pi)h^{1/2}\omega(h) \\ &= 4\sqrt{2}h^{1/2}/T_0(h). \end{aligned} \tag{3.5.16}$$

Here, $z_{1,2}(h)$ are turning points defined by the equation $\Pi(z) = h$, the relationship $F = d\Pi/dz$ having been taken into account. As a result, the expressions (14) for τ and Θ are reduced to integration of the function $h^{-1/2}T_0(h)$ within the indicated limits. The condition $F_0(h) = O(h^\alpha)$, $\alpha < 1$, is reduced to the estimate $T_0(h) = O(h^\sigma)$, $\sigma > -1/2$.

3.5.5. EXAMPLES

Based on (14), we construct the solution of the linear problem $F(z) = z$ which has a period $T_0 = 2\pi$. The dependence $x(\tau)$ has the form (14), and the quantities Θ and $a(\tau)$ are given by

$$\Theta = \frac{\pi a^0}{u_2 - u_1}, \quad a^0 = \left(2h^0\right)^{1/2}, \quad a = (2h)^{1/2} = a^0\left(1 - \frac{\tau}{\Theta}\right). \tag{3.5.17}$$

Consider the case of an arbitrary power-type nonlinearity

$$\Pi(z) = \mu|z|^\gamma, \qquad \mu, \gamma = \text{const} > 0. \tag{3.5.18}$$

Then for the turning points $z = z_{1,2}(h)$ and the oscillation period $T_0(h)$ we obtain the following expressions (Γ is Euler's gamma function [87]):

$$\begin{aligned} z_2(h) &= -z_1(h) = (h/\mu)^{1/\gamma}, \qquad h \geqslant 0, \\ T_0(h) &= 2\sqrt{2}\mu^{-1/\gamma}h^\varkappa \int_0^1 \left(1 - x^\gamma\right)^{-1/2} \mathrm{d}x, \\ &= \left(2\sqrt{2}\pi/\gamma\right)\Gamma(1/\gamma)\Gamma^{-1}(1+\varkappa)\mu^{-1/\gamma}h^\varkappa, \\ T_0 &\sim h^\varkappa, \qquad \varkappa = 1/\gamma - 1/2 > -1/2. \end{aligned} \tag{3.5.19}$$

Substituting expressions (19) into (16) and (14), we can obtain the dependences to be determined:

$$h(\tau) = h^0\left(1 - \frac{\tau}{\Theta}\right)^\gamma, \qquad \Theta = \frac{\sqrt{\pi}}{u_2 - u_1}\frac{\Gamma(1/\gamma)}{\Gamma(1+\varkappa)}\left(\frac{h^0}{\mu}\right)^{1/\gamma}. \tag{3.5.20}$$

In the case of a linear system when $\gamma = 2$, $\mu = 1/2$, and $h = a^2/2$, expressions (19) and (20) turn into (17).

The period T_0 of oscillations in the vicinity of the equilibrium position $y = x$ for the nonlinear system with an elastic element (1) and (2) in dimensionless variables ($\nu t \to t$, $z/d \to z$, $l_0/d \to l_0 < 1$) is expressed by

$$\begin{aligned} T_0(h, l_0) &= 2\int_{-z^*}^{z^*} \left[2h - \left(\left(1 + z^2\right)^{1/2} - l_0\right)^2\right]^{-1/2} \mathrm{d}z, \\ z^* &= z^*(h, l_0) = \left[\left(l_0 + \sqrt{2h}\right)^2 - 1\right]^{1/2} > 0. \end{aligned}$$

For small amplitudes of relative oscillations $z^* \ll 1$, the quantity T_0 is calculated approximately in terms of powers of z^*.

3.5.6. DISPLACEMENT OF A NONLINEAR SYSTEM

Solution of the boundary-value problem C (13) is reduced to integration of equations for $h(\tau)$ and $p(\tau)$ with a known integral $K = \text{const}$, calculation of a quadrature to determine $x(\tau)$ and calculation of unknown parameters Θ, r, and β. This can be done using the boundary condition (13) for x, relationships (9), and the minimality condition for $\Theta(\beta)$ (see Section 3.3). Averaging and solution of the boundary-value problem in the general case is performed numerically. One succeeds in the analytical integration of the boundary-value problem (9) for a linear system and the graphic construction of the dependence of the adjoint variables on the parameters and state variables, i.e. in finding an approximate time-optimal control in the feedback form (see Section 5.3 and 5.4).

CHAPTER 4

The Foundation of Asymptotic Methods of the Separation of Motions in Essentially Nonlinear Controlled Systems

Here we consider methods of the canonical separation of motions and the construction of higher approximations to the solution of the maximum principle boundary-value problem. We consider some generalizations of asymptotic procedures: the two-parameter averaging scheme, averaging in multi-frequency systems and more general systems, and averaging of motions in a neighbourhood of a separatrix.

4.1. Reduction to the Standard Form and Canonical Averaging

4.1.1. THE DIMENSION REDUCTION OF THE MAXIMUM PRINCIPLE SET OF EQUATIONS AND WHITTAKER'S EQUATIONS

We pose the problem of constructing the solution of Equations (3.2.4) with an arbitrary specified accuracy in terms of ε. Unlike the case of the quasilinear system (1.2.8) or (1.3.3), the canonical system (3.2.4) is not of the standard form, and the set of $(2n+1)$ equations (3.29) reduced to the standard form is not Hamiltonian. It turns out that this system can be reduced to the canonical form, which enables the effective technique of canonical averaging (see Section 2.1) to be applied.

As in Section 3.2, we use the constancy of H^*:

$$H^* = \omega(a)q + \varepsilon\big(p \cdot f^*(a, \psi, p, q)\big) + \varepsilon q F^*(a, \psi, p, q) = \varepsilon h. \tag{4.1.1}$$

Here, f^* and F^* are sufficiently smooth functions of the slow variables a, p, and q. The constant h is determined by the transversality conditions (3.2.11) when T is fixed, and by the condition $h = 0$ when T is free. Relationship (1) is regarded as an equation with respect to q which can be solved either exactly or with sufficient accuracy in terms of ε:

$$\begin{aligned} q &= \varepsilon Q(h, a, \psi, p, \varepsilon) \\ &= \frac{\varepsilon}{\omega}[h - (p \cdot f_0^*)]\left\{1 - \frac{\varepsilon}{\omega}\left[\frac{\partial}{\partial q}(p \cdot f_0^*) + F_0^*\right]\right\} + \varepsilon^2 + \cdots . \end{aligned} \tag{4.1.2}$$

After substitution of expression (2) into (3.2.4), we obtain the following equations for a, ψ, and p:

$$\begin{aligned} \dot{a} &= \varepsilon f^*(a,\psi,p,\varepsilon Q), \qquad \dot{\psi} = \omega(a) + \varepsilon F^*(a,\psi,p,\varepsilon Q), \\ \dot{p} &= -\varepsilon Q\frac{\partial\omega}{\partial a} - \varepsilon\frac{\partial}{\partial a}[(p\cdot f^*) + qF^*]\Big|_{q=\varepsilon Q}. \end{aligned} \tag{4.1.3}$$

The initial and terminal conditions for a and ψ in (3.2.4) have the form

$$a(t_0) = a^0, \qquad \psi(t_0) = \psi^0, \qquad M(a)\big|_T = 0. \tag{4.1.4}$$

In the case of fixed time T of the control process, the transversality conditions, after having obtained the solution of system (3) as functions of t and h, are given by

$$p(T) = \frac{\partial}{\partial a}[(\lambda\cdot M) - g]\Big|_T, \qquad Q(h,a,\psi,p,\varepsilon)\big|_T = 0. \tag{4.1.5}$$

Due to (1) and (3.2.4), the second equation of (5), determining the unknown parameter h, can be represented in another form:

$$\left(p\cdot f_0^*(a,\psi,p)\right)\big|_T = h. \tag{4.1.6}$$

When the time T is not fixed, the boundary-value problem of the maximum principle retains the form (2)–(6), but in all expressions one should put $h = 0$. The unknown parameter $\Theta = \varepsilon T$ is determined as a root of the second equation (5) or Equation (6) where $h = 0$.

Thus, it is required to find the set of solutions of the boundary-value problem which (as is established under some additional general conditions) is determined by the set of roots $\{h_\nu\}$ or $\{\Theta_\nu\}$ and then to determine the optimal value for the root by solving the minimization problem (see (3.2.13)):

$$J_\nu \longrightarrow \min_\nu; \qquad J_\nu = J(h_\nu,\varepsilon), \quad J_\nu = J(\Theta_\nu,\varepsilon). \tag{4.1.7}$$

To simplify the construction of the approximate solution by using the averaging method we make the phase ψ, $\psi \in [\psi^0,\psi_T]$ ($\psi_T \sim \varepsilon^{-1}$) to be an independent variable in (3) (dividing by $\dot{\psi}$). It is admissible for sufficiently small $\varepsilon > 0$, since

$\omega \geqslant \omega_0 > 0$. As a result, we obtain the standard system [1, 4, 5, 46] (with proper boundary conditions)

$$
\begin{aligned}
&\frac{\mathrm{d}a}{\mathrm{d}\psi} = \frac{\varepsilon f^*(a, \psi, p, \varepsilon Q)}{\omega(a) + \varepsilon F(a, \psi, p, \varepsilon Q)}, \quad a(\psi^0) = a^0, \quad M(a)\big|_{\psi_T} = 0,\\
&\frac{\mathrm{d}p}{\mathrm{d}\psi} = -\varepsilon\left[Q\frac{\partial\omega}{\partial a} + \frac{\partial}{\partial a}(p\cdot f^*) + \varepsilon Q\frac{\partial F^*}{\partial a}\right](\omega + \varepsilon F^*)^{-1}\Big|_{q=\varepsilon Q},\\
&p(\psi_T) = \frac{\partial}{\partial a}[(\lambda \cdot M) - g]\Big|_{\psi_T}, \quad Q\big|_{\psi_T} = 0.
\end{aligned}
\tag{4.1.8}
$$

Boundary conditions (8) must be supplemented with the following relationships:

$$
\psi_T = \psi\big|_T, \qquad t - t_0 = \int_{\psi^0}^{\psi} \frac{1}{\omega + \varepsilon F^*}\Big|_{q=\varepsilon Q} \mathrm{d}\varphi. \tag{4.1.9}
$$

Thus, the original Hamiltonian system (3.2.4) has been reduced to the standard form (8). The transformations have consisted in eliminating the variable q (using integral (1) and relationship (2)) and proceeding to a new argument ψ, the phase variable adjoint with q. Such a change is known to result in the canonical system (Whittaker's equations) with the Hamiltonian function $-\varepsilon Q$ (see (2)):

$$
\begin{aligned}
&\frac{\mathrm{d}a}{\mathrm{d}\psi} = -\varepsilon\frac{\partial Q}{\partial p}, \qquad \frac{\mathrm{d}p}{\mathrm{d}\psi} = \varepsilon\frac{\partial Q}{\partial a},\\
&Q = Q(h, a, \psi, p, \varepsilon), \qquad h \in I_h, \qquad \varepsilon \in [0, \varepsilon_0].
\end{aligned}
\tag{4.1.10}
$$

Here, Q is a smooth function of the arguments a and p and parameters h and ε, 2π-periodic in ψ. For system (10), the scheme of the canonical change of variables is used which is simpler than that considered in Section 2.1.

4.1.2. CONSTRUCTION OF THE CANONICAL AVERAGED SYSTEM AND ITS INTEGRATION

For Equations (10), the method of canonical averaging with respect to the argument ψ is applicable and consists in the transformation

$$
\begin{aligned}
&\xi = a + \varepsilon\partial\sigma/\partial\eta, \qquad p = \eta + \varepsilon\partial\sigma/\partial a;\\
&(a, p) \longrightarrow (\xi, \eta); \qquad S = (a\cdot\eta) + \varepsilon\sigma(h, a, \psi, \eta, \varepsilon).
\end{aligned}
\tag{4.1.11}
$$

It is required that the function σ is 2π-periodic in ψ and that the new (averaged) Hamiltonian εR does not contain ψ, i.e.

$$\frac{d\xi}{d\psi} = \varepsilon\frac{\partial R}{\partial \eta}, \qquad \frac{d\eta}{d\psi} = -\varepsilon\frac{\partial R}{\partial \xi}, \qquad R = R(h, \xi, \eta, \varepsilon). \tag{4.1.12}$$

System (12) obviously has the integral

$$R(h, \xi, \eta, \varepsilon) = -\beta = \text{const.} \tag{4.1.13}$$

This corresponds to the constancy of the mean value of q (Sections 2.1, 3.2 and 3.3). The unknown functions σ and R can be determined with a prescribed accuracy in terms of ε if the function Q is sufficiently smooth in a and p. As in Section 2.1, we construct the expressions for σ and R in the form of the series expansions (2.1.5) [1, 4, 5, 46]

$$\begin{aligned} \sigma(h, a, \psi, \eta, \varepsilon) &= \sigma_0(h, a, \psi, \eta) + \varepsilon\sigma_1 + \cdots + \varepsilon^k\sigma_k + \cdots, \\ R(h, \xi, \eta, \varepsilon) &= R_0(h, \xi, \eta) + \varepsilon R_1 + \cdots + \varepsilon^k R_k + \cdots. \end{aligned} \tag{4.1.14}$$

The functions σ, R, and Q are connected by the nonlinear partial differential equation [1, 4, 5, 46]

$$\frac{\partial \sigma}{\partial \psi} - Q\left(h, a, \psi, \eta + \varepsilon\frac{\partial \sigma}{\partial a}, \varepsilon\right) = R\left(h, a + \varepsilon\frac{\partial \sigma}{\partial \eta}, \eta, \varepsilon\right). \tag{4.1.15}$$

Having substituted expansions (14) in (15) we obtain the sequence of linking relationships for σ_i and R_i ($i \geqslant 0$):

$$\partial\sigma_i/\partial\psi - Q_i(h, a, \psi, \eta) = R_i(h, a, \eta). \tag{4.1.16}$$

Here, the functions Q_i are known at each step of the procedure; they are determined by Q, its derivatives, and the functions $\sigma_0, R_0, \ldots, \sigma_{i-1}, R_{i-1}$ calculated in previous steps. In particular, for $i = 0, 1$, the following expressions are valid:

$$\begin{aligned} Q_0 &= Q(h, a, \psi, \eta, 0), \\ Q_1 &= \left(\frac{\partial Q}{\partial \varepsilon}\right)_0 + \left(\frac{\partial Q_0}{\partial \eta}\cdot\frac{\partial \sigma_0}{\partial a}\right) + \left(\frac{\partial R_0}{\partial a}\cdot\frac{\partial \sigma_0}{\partial \eta}\right). \end{aligned} \tag{4.1.17}$$

Under the requirement of periodicity of σ in ψ, Equations (16) imply that

$$
\begin{aligned}
&R_i(h,\xi,\eta) = -\langle Q_i(h,\xi,\psi,\eta)\rangle \equiv -Q_{i0}(h,\xi,\eta),\\
&\sigma_i(h,a,\psi,\eta) = \int \left[Q_i(h,a,\psi,\eta) - Q_{i0}(h,a,\eta)\right] \mathrm{d}\psi.
\end{aligned}
\tag{4.1.18}
$$

It should be noted that the functions σ_i, as in (2.1.8), are determined not uniquely but with an accuracy up to arbitrary functions of a, η, and h. The nonuniqueness disappears when returning to the original variables a and p according to the change formulae (11).

Let the coefficients $R_0, R_1, \ldots, R_k$ in expansion (14) be calculated according to (17) and (18). Then system (12) of the $(k+1)$th approximation is obtained by omitting the terms $O(\varepsilon^{k+2})$, which leads, generally speaking, to the error $O(\varepsilon^{k+1})$ in determining ξ and η, for $\psi - \psi^0 \sim \varepsilon^{-1}$. After having introduced the slow phase $\theta = \varepsilon\psi$, the system of the $(k+1)$th approximation is written in the form

$$
\begin{aligned}
&\frac{\mathrm{d}\xi}{\mathrm{d}\theta} = \sum_{i=0}^{k} \varepsilon^i \frac{\partial R_i}{\partial \eta}, \qquad \frac{\mathrm{d}\eta}{\mathrm{d}\theta} = -\sum_{i=0}^{k} \varepsilon^i \frac{\partial R_i}{\partial \xi},\\
&R_i = R_i(h,\xi,\eta), \qquad \varepsilon\psi^0 = \theta_0 \leqslant \theta \leqslant \theta_T = \varepsilon\psi_T, \qquad \theta_{0,T} \sim 1.
\end{aligned}
\tag{4.1.19}
$$

The general solution of system (19) on the interval $\theta \in [\theta_0, \theta_T]$, $\theta_{0,T} \sim 1$, is constructed with the error $O(\varepsilon^{k+1})$ for some range of values of the constants of integration

$$
\xi(\theta_0) = c_\xi, \quad \eta(\theta_T) = c_\eta; \qquad c_\xi \in D_\xi, \quad c_\eta \in D_\eta. \tag{4.1.20}
$$

The calculations are performed based on the solution of system (19) of the first approximation, i.e. for $k = 0$:

$$
\begin{aligned}
&\frac{\mathrm{d}\xi}{\mathrm{d}\theta} = \frac{\partial R_0}{\partial \eta}, \quad \xi(\theta_0) = c_\xi; \qquad \frac{\mathrm{d}\eta}{\mathrm{d}\theta} = -\frac{\partial R_0}{\partial \xi}, \quad \eta(\theta_T) = c_\eta;\\
&R_0(h,\xi,\eta) = -\langle Q_0\rangle = -\omega^{-1}(\xi)\left[h - (\eta \cdot f_0(\xi,\eta))\right].
\end{aligned}
\tag{4.1.21}
$$

Admissible domains D_ξ and D_η for parameters c_ξ and c_η in (20) are determined by the solution of the boundary-value problem of the first approximation; they are convex neighbourhoods of proper values, which depend on yet unknown parameters h and $\theta_T = \varepsilon\psi_T \sim 1$. In what follows, the general solution of problem (21) is considered to be known

$$\xi = \xi_0(\theta, c_\xi, c_\eta), \qquad \eta = \eta_0(\theta, c_\xi, c_\eta), \qquad \theta \in [\theta_0, \theta_T]. \tag{4.1.22}$$

Here, the dependence on other parameters (θ_0, θ_T, h) is not indicated for the sake of brevity.

The approximate solution of problem (19) and (20) is constructed in the following form:

$$\xi_{(k)} = \xi_0 + \sum_{i=1}^{k} \varepsilon^i \xi_i, \qquad \eta_{(k)} = \eta_0 + \sum_{i=1}^{k} \varepsilon^i \eta_i. \tag{4.1.23}$$

Unknown functions ξ_i and η_i $(i = 1, \dots, k)$ are found by successive solution of boundary-value problems for linear nonhomogeneous equations with zero boundary conditions:

$$\begin{aligned} \frac{d\xi_i}{d\theta} &= \left(\frac{\partial^2 R_0}{\partial\eta\partial\xi}\right)\xi_i + \left(\frac{\partial^2 R_0}{\partial\eta^2}\right)\eta_i + v_i^\xi(\theta, c_\xi, c_\eta), \qquad \xi_i(\theta_0) = 0, \\ \frac{d\eta_i}{d\theta} &= -\left(\frac{\partial^2 R_0}{\partial\xi^2}\right)\xi_i - \left(\frac{\partial^2 R_0}{\partial\eta\partial\xi}\right)\eta_i + v_i^\eta(\theta, c_\xi, c_\eta), \qquad \eta_i(\theta_T) = 0. \end{aligned} \tag{4.1.24}$$

Derivatives of the function R_0 are taken along the generating solution (22). The functions v_i^ξ and v_i^η are determined consecutively: at each step they are calculated on the basis of $R_0, R_1, \dots, R_{i-1}$ and the coefficients $\xi_1, \eta_1, \dots, \xi_{i-1}, \eta_{i-1}$ found at previous steps, e.g.,

$$v_1^\xi = (\partial R_1/\partial\eta), \qquad v_1^\eta = -(\partial R_1/\partial\xi), \qquad \xi = \xi_0, \quad \eta = \eta_0. \tag{4.1.25}$$

The general solution of a linear nonhomogeneous system is constructed by the method of variation of integration constants on the basis of the general solution of the corresponding homogeneous system.

Analogously to Section 2.2, the fundamental matrix X of the dimension $(2n \times 2n)$ is obtained by differentiating (ξ_0, η_0) with respect to the parameters c_ξ and c_η. Taking account of boundary conditions (24) leads to the desired expressions

$$\begin{aligned} \begin{pmatrix} \xi_i \\ \eta_i \end{pmatrix} &= X(\theta)\begin{pmatrix} b_i^\xi \\ b_i^\eta \end{pmatrix} + X(\theta)\int_{\theta_0}^{\theta} X^{-1}(s)\begin{pmatrix} v_i^\xi(s) \\ v_i^\eta(s) \end{pmatrix} ds \\ &\equiv X(\theta)\begin{pmatrix} b_i^\xi \\ b_i^\eta \end{pmatrix} + \begin{pmatrix} w_i^\xi(\theta) \\ w_i^\eta(\theta) \end{pmatrix}, \end{aligned} \tag{4.1.26}$$

$$X(\theta) = \begin{bmatrix} \partial\xi_0/\partial c_\xi & \partial\eta_0/\partial c_\eta \\ \partial\xi_0/\partial c_\xi & \partial\eta_0/\partial c_\eta \end{bmatrix}, \qquad \begin{pmatrix} b_i^\xi \\ b_i^\eta \end{pmatrix} = \begin{pmatrix} 0 \\ -\omega_i^\eta(\theta_T) \end{pmatrix}.$$

Here, only the dependence on the argument θ is indicated; the functions w_i^ξ and w_i^η are determined according to the identity in (26). As a result, formulae (23)–(26) give the general solution of the $(k+1)$th approximation for the boundary-value problem (19) and (20), for $c_\xi \in D_\xi$ and $c_\eta \in D_\eta$.

4.2. Construction of Higher Approximations for Solutions of the Maximum Principle Boundary-Value Problem and the Optimal Control Problem

4.2.1. APPROXIMATE CALCULATION OF ORIGINAL VARIABLES AND SOLUTION OF THE MAXIMUM PRINCIPLE BOUNDARY-VALUE PROBLEM

The solution of the $(k+1)$th approximation for the original standard system (1.10) is constructed in the form of an expansion similar to (1.23)

$$a_{(k)} = \xi_0 + \sum_{i=1}^{k} \varepsilon^i a_i, \qquad p_{(k)} = \eta_0 + \sum_{i=1}^{k} \varepsilon^i p_i. \tag{4.2.1}$$

Here, ξ_0 and η_0 are known functions of θ, θ_0, θ_T, c_ξ, c_η, and h; the terms $O(\varepsilon^{k+1})$ are not taken into account. Unknown expansion coefficients a_i and p_i, depending on the indicated parameters and (2π-periodically) on ψ, are found in the usual way as a result of substitution of (1.23) and (1) into (1.11), taking expressions (1.14), (1.17), and (1.18) for σ_i into account; in particular

$$a_1 = \xi_1 - (\partial\sigma_0/\partial\eta), \qquad p_1 = \eta_1 + (\partial\sigma_0/\partial a). \tag{4.2.2}$$

The parentheses mean that the derivatives are taken at $a = \xi_0$ and $\eta = \eta_0$. Without loss of accuracy, in the functions σ_i $(i = 0, 1, \ldots k-1)$, approximations of lower order $(k-i)$ are substituted for ξ and η. Taking account of σ_k does not make a and p more accurate and, therefore, this function is not to be calculated in (1.14) and (1.18) (Section 2.2). For the general solution of the boundary-value problem (8) and (10) we obtain the following representation

$$\begin{aligned} a_{(k)} &= \xi_0 + \varepsilon A_{(k)}(\psi, \theta, c_\xi, c_\eta, \varepsilon), & a_{(k)}\big|_{\psi^0} &= c_\xi + \varepsilon A_{(k)}\big|_{\psi^0}; \\ p_{(k)} &= \eta_0 + \varepsilon P_{(k)}(\psi, \theta, c_\xi, c_\eta, \varepsilon), & p_{(k)}\big|_{\psi_T} &= c_\eta + \varepsilon P_{(k)}\big|_{\psi_T}. \end{aligned} \tag{4.2.3}$$

Here, $A_{(k)}$ and $P_{(k)}$ are 2π-periodic in ψ; the dependence on the parameters θ_0, θ_T and h is not indicated for the sake of brevity. Now, to solve the boundary-value problem, i.e. to determine the unknowns c_ξ and c_η and other parameters (θ_T, ψ_T, h or Θ), we substitute the constructed functions $a_{(k)}$ and $p_{(k)}$ (3) into relationships (1.9) and use boundary conditions (1.8). As a result, we obtain the relationships

$$\begin{aligned}
&c_\xi + \varepsilon A_{(k)}\big|_{\psi^0} = a^0, \qquad M(\xi_0 + \varepsilon A_{(k)})\big|_{\psi_T} = 0, \\
&c_\eta + \varepsilon P_{(k)}\big|_{\psi_T} = \frac{\partial}{\partial a}\left[(\lambda \cdot M_{(k)}) - g_{(k)}\right]\Big|_{\psi_T}, \qquad Q_{(k)}\big|_{\psi_T} = 0, \\
&\Theta - \tau_0 = \int_{\theta_0}^{\theta_T} \left[\omega(a_{(k)}) + \varepsilon F_{(k)}\right]^{-1} d\theta.
\end{aligned} \tag{4.2.4}$$

It is supposed that for fixed h (or Θ when $h = 0$), boundary conditions (4) (excluding the penultimate one, which is used for determining h or Θ in the first approximation in terms of ε) admit a simple real root

$$c_\xi^0 = a^0, \quad c_\eta = c_\eta^0, \quad \theta_T = \theta_T^0, \quad \lambda = \lambda^0. \tag{4.2.5}$$

Moreover, it is supposed that this root is isolated for considered values of $a^0 \in D_a$ and $h \in I_h$ (or $\Theta \in I_\Theta$), i.e. the corresponding determinant of the Jacobian matrix of the $(2n + l + 1)$th order is not equal to zero. Certainly, λ^0, c_η^0, and θ_T^0 in (5) depend on h or Θ. The approximate solution of the mentioned Equations (4) with respect to c_ξ, c_η, and θ_T (or ψ_T) depending on h (or Θ) and other known parameters is constructed by using asymptotic expansions in terms of powers of ε. Here, one should bear in mind that the phase is determined with a larger (by one order) error in terms of ε. As follows from (4), one can firstly determine c_ξ, c_η, and λ as functions of θ_T (periodic in $\psi_T = \theta_T\varepsilon^{-1}$) and the parameter h (or Θ) from the first three vector equations, and then, using the latter equation of (4), find the quantity $\theta_T(h)$ (or $\theta_T(\Theta)$) with the error $O(\varepsilon^{k+1})$.

Indeed, the following representations take place:

$$c_\xi = c_{\xi 0} + \sum_{i=1}^{k} \varepsilon^i c_{\xi i}, \quad c_\eta = c_{\eta 0} + \sum_{i=1}^{k} \varepsilon^i c_{\eta i}, \quad \lambda = \lambda_0 + \sum_{i=1}^{k} \varepsilon^i \lambda_i. \tag{4.2.6}$$

Here, $c_{\xi 0} = c_\xi^0 = a^0$ is a known constant which is substituted into $(l + n)$ equations for determining $c_{\eta 0}$ and λ_0:

$$M(\xi_0)\big|_{\theta_T} = 0, \qquad c_{\eta 0} = \frac{\partial}{\partial a}\left[(\lambda_0 \cdot M(\xi_0)) - g(\xi_0)\right]\Big|_{\theta_T}. \tag{4.2.7}$$

It is assumed in what follows, that the root $(c_{\eta 0}, \lambda_0)$ is simple in the range of parameters considered. Then the coefficients $c_{\xi i}$, $c_{\eta i}$, and λ_i $(i = 1, \dots, k)$ are determined uniquely from linear equations in terms of functions of θ_T and h (or Θ) (periodic in ψ_T) and other known parameters of the problem. The coefficients $c_{\xi i}$ are determined on each ith step independently of $c_{\eta i}$ and λ_i and then they are substituted into the equations to find $c_{\eta i}$ and λ_i. As it turns out, the coefficients $c_{\xi i} = 0$ due to the choice of σ_i and ξ_i in the form (1.18), (1.24) and (1.26). The linear algebraic equations with a nonsingular matrix (Jacobian matrix) used for determining $c_{\eta i}$ and λ_i have known nonhomogeneous terms. In particular, the coefficients $c_{\xi 1}$, $c_{\eta 1}$, and λ_1 are determined from the equations

$$c_{\xi 1} = \left(\frac{\partial \sigma_0}{\partial \eta}\right)\Big|_{\psi^0} = 0, \qquad \left(\frac{\partial M_0}{\partial a}\right)\frac{\partial \xi_0}{\partial c_\eta} c_{\eta 1}\Big|_{\theta_T} = 0,$$
$$\left\{I - \frac{\partial^2}{\partial a^2}[(\lambda_0 \cdot M_0) - g_0]\Big|_{\theta_T}\right\} c_{\eta 1} - \frac{\partial}{\partial a}(\lambda_1 \cdot M_0)\Big|_{\theta_T} = -\left(\frac{\partial \sigma_0}{\partial a}\right)\Big|_{\psi_T}. \tag{4.2.8}$$

As a result, we obtain for c_ξ and c_η the expressions of the $(k+1)$th approximation (the expression for λ is similar to that for c_η):

$$c_\xi = a^0, \qquad c_\eta = c_{\eta 0} + \varepsilon C_{\eta(k)}, \qquad \lambda = \lambda_0 + \varepsilon \Lambda_{(k)}. \tag{4.2.9}$$

$$\begin{gathered} c_{\eta 0},\ \lambda_0\colon\ a^0,\ \theta_T - \theta_0,\ h(\Theta); \\ C_{\eta(k)},\ \Lambda_{(k)}\colon\ a^0,\ \psi^0,\ \theta_0,\ \psi_T,\ \theta_T,\ h(\Theta),\ \varepsilon. \end{gathered} \tag{4.2.10}$$

The dependence on ψ^0 and ψ_T in (10) is periodic. Having substituted (9) and (10) into (3) and omitted the terms $O(\varepsilon^{k+1})$ we obtain solutions of the $(k+1)$th approximation for a and p in the form $(\mu = h, \Theta)$:

$$\begin{aligned} a_{(k)} &= \xi_0(\theta - \theta_0, \theta_T - \theta_0, a^0, \mu) + \\ &\quad + \varepsilon A_{(k)}(\theta, \psi, \theta_0, \psi^0, a^0, \theta_T, \psi_T, \mu, \varepsilon), \\ p_{(k)} &= \eta_0(\theta - \theta_0, \theta_T - \theta_0, a^0, \mu) + \\ &\quad + \varepsilon P_{(k)}(\theta, \psi, \theta_0, \psi^0, a^0, \theta_T, \psi_T, \mu, \varepsilon). \end{aligned} \tag{4.2.11}$$

Here, $A_{(k)}$ and $P_{(k)}$ are 2π-periodic functions of the argument ψ and the parameters ψ^0 and ψ_T. Thus, the solution of the boundary-value problem (1.8) for a and p is a sum of "smooth functions" ξ_0 and η_0 depending on the slow argument θ and small, 2π-periodic in ψ, oscillations with the slowly changing amplitude whose order is ε. The aforementioned constructions can be performed in other ways convenient for implementation using a computer, for instance, by using the recursive scheme of successive approximations or the method of Newton, which gives a quadratic rate of convergence in terms of ε [49].

Thus, let the requirements of smoothness of the system and the boundary conditions for problem (1.8), as well as the assumption of the existence and "isolatedness" of the solution of the boundary-value problem of the first approximation be fulfilled. These requirements imply the existence of the solution in the form (1.21)–(1.23) and the inverse matrix $X^{-1}(\theta)$ (1.26), as well as the nonsingularity of the matrix of Equations (9) which is a Jacobian matrix for initial and boundary conditions. These requirements must be satisfied in some range of values of the argument ψ and parameters a^0, ψ^0, ψ_T, and h (or Θ) and ε:

$$a^0 \in D_a, \quad \psi \in [\psi^0, \psi_T], \quad \varepsilon \in (0, \varepsilon_0], \quad h \in I_h, \quad \Theta \in I_\Theta. \tag{4.2.12}$$

Then there exists the solution of the $(k+1)$th approximation which can be expressed in the form (11). It smoothly depends on a^0, ε, and h (or Θ) and is differentiable with respect to ψ and ψ_T. The following estimate is valid for this solution ($\psi \in [\psi^0, \psi_T]$):

$$a - a_{(k)} = O(\varepsilon^{k+1}), \qquad p - p_{(k)} = O(\varepsilon^{k+1}), \tag{4.2.13}$$

where a and p is the exact solution of the boundary-value problem under fixed values of parameters ψ_T and h (or Θ).

To determine the solution of the original boundary-value problem (1.3)–(1.6) it is necessary to find the phase ψ as a function of time t from (1.9) and to establish the connection between θ_T and Θ. Substitution of the functions $a_{(k)}$ and $p_{(k)}$ (11) into the last equation (4) enables us to rewrite it in the following form ($\mu = h, \Theta$):

$$\begin{gathered}\varepsilon(T - t_0) = \Theta - \tau_0 = \int_{\theta_0}^{\theta_T} \left[1 + \varepsilon\Psi_{(k)}\left(\frac{\theta}{\varepsilon}, \frac{\theta_T}{\varepsilon}, \theta, \theta_T, \mu, \right)\right] \frac{d\theta}{\omega(\xi_0)}, \\ \Psi_{(k)}(\psi, \psi_T, \theta, \theta_T, \mu, \varepsilon) = \sum_{i=0}^{k-1} \varepsilon^i \Psi_i(\psi, \psi_T, \theta, \theta_T, \mu).\end{gathered} \tag{4.2.14}$$

Here, $\Psi_{(k)}$ and Ψ_i are 2π-periodic functions of ψ and ψ_T; their dependence on other arguments is not indicated for the sake of brevity. Since the terms $O(\varepsilon^{k+1})$ are omitted in (14), the accuracy of determining θ_T for fixed $h \in I_h$ (or $\Theta \in I_\Theta$) will be the same as for slow variables (13) and the order of the error in determining ψ and ψ_T will be less by one. The parameter θ_T is obtained in the form of asymptotic expansion

$$\theta_T = \theta_T(\mu, \varepsilon) = \theta_T^0(\mu) + \sum_{i=1}^{k} \varepsilon^i \theta_{Ti}(\mu, \varepsilon); \qquad \mu = h, \Theta. \tag{4.2.15}$$

The quantity θ_T^0 is determined from (14) when $\varepsilon \to 0$:

$$\Theta - \tau_0 = \zeta(\theta_T^0, \mu) \equiv \int_{\theta_0}^{\theta_T^0} \omega^{-1}(\xi_0(\theta - \theta_0, \theta_T^0 - \theta_0, a^0, c_\eta^0, \mu))\mathrm{d}\theta. \qquad (4.2.16)$$

We suppose that the root $\theta_T^0(\mu)$ of Equation (16) exists and is simple, i.e. $\partial\zeta/\partial\theta_T^0 \neq 0$, $h \in I_h$; $\Theta \in I_\Theta$. Then the coefficient θ_{T1} in (15) is found, according to (14), as a root of the transcendental equation

$$\frac{\partial\zeta}{\partial\theta_T^0}\theta_{T1} = -\int_{\theta_0}^{\theta_T^0} \langle\Psi_0\rangle(\psi_{T0} + \theta_{T1}, \theta, \theta_T^0, \mu)\frac{\mathrm{d}\theta}{\omega(\xi_0)}. \qquad (4.2.17)$$

Here, the function ξ_0 in ω is taken at the determined value θ_T^0 and the averaging of ψ_0 over ψ is performed, which does not decrease the accuracy of determining the quantity θ_{T1}. The solution θ_{T1} of Equation (17) certainly exists due to the simplicity of the root θ_T^0 and the boundedness of the function $\langle\psi_0\rangle$ in θ_{T1}. Moreover, due to the periodicity of $\langle\psi_0\rangle$, this equation may have several roots as is shown in Fig. 4.1. If the chosen root θ_{T1} is simple, then subsequent coefficients θ_{Ti} ($i \geqslant 2$) of the asymptotic expansion (15) are defined uniquely from linear equations of the following form:

$$\left[\frac{\partial\zeta}{\partial\theta_T^0} + \int_{\theta_0}^{\theta_T^0} \frac{\partial\langle\Psi_0\rangle}{\partial\theta_{T1}}\frac{\mathrm{d}\theta}{\omega(\xi_0)}\right]\theta_{Ti} = \vartheta_i, \qquad i = 2, 3, \ldots, k. \qquad (4.2.18)$$

Here, ϑ_i are functions of h (or Θ), known on each ith step and depending on previous approximations. As a result, we obtain the quantities $\theta_T(\mu, \varepsilon)$ and $\psi_T(\mu, \varepsilon)$ with the error $O(\varepsilon^{k+1})$ and $O(\varepsilon^k)$, respectively.

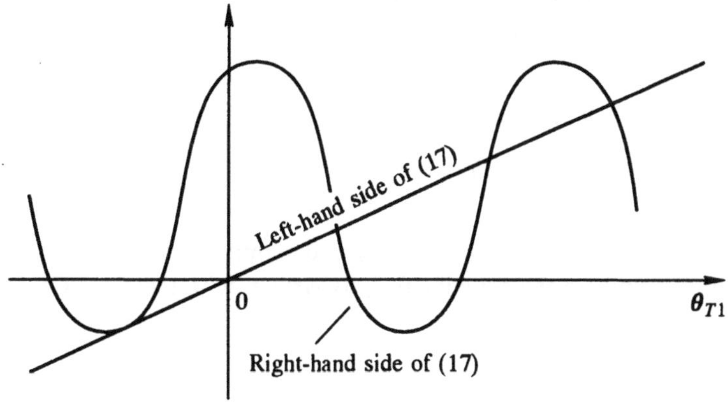

Fig. 4.1.

From the second relationship (1.9), having substituted the determined values of θ_T and ψ_T into (11), we determine, analogously to that expounded above, the functions $\theta(t,\mu,\varepsilon)$ and $\psi(t,\mu,\varepsilon)$ with the indicated error:

$$\theta = \theta(t,\mu,\varepsilon) = \theta_{(0)}(\tau,\mu) + \sum_{i=0}^{k} \varepsilon^i \theta_i(t,\mu,\varepsilon), \qquad \psi = \theta/\varepsilon. \tag{4.2.19}$$

The function $\theta_{(0)}$ in (19) is uniquely determined by the relationship

$$\tau - \tau_0 = \int_{\theta_0}^{\theta_{(0)}} \omega^{-1}\big(\xi_0(\theta - \theta_0, \theta_T^0 - \theta_0, a^0, c_\eta^0, \mu)\big)\,d\theta, \qquad \mu = h, \Theta. \tag{4.2.20}$$

Subsequent coefficients θ_i are also determined uniquely in the form

$$\theta_i(t,\mu,\varepsilon) = \omega\big(\xi_0(\theta_{(0)}(\tau,\mu))\big)\Phi_i(t,\tau,\mu,\varepsilon), \qquad i = 1,\dots,k. \tag{4.2.21}$$

Here, Φ_i are known functions of t, h (or Θ), and ε, bounded for $t \in [t_0, T]$, $T = \Theta\varepsilon^{-1}$, $\varepsilon = [0, \varepsilon_0]$, $h \in I_h$; $\Theta \in I_\Theta$.

The asymptotic expansions for determining the parameter $\theta_T = \theta_T(\Theta,\varepsilon)$ and the slow phase $\theta = \theta(t,\Theta,\varepsilon)$ in the case of free time T of the control process termination are carried out analogously.

4.2.2. CHOICE OF THE OPTIMAL SOLUTION

In the above, the functions a, p, and $q = \varepsilon Q$ to be found have been constructed with the error $O(\varepsilon^{k+1})$ and ψ has been constructed with the error $O(\varepsilon^k)$ for $t \in [t_0, T]$, $h \in I_h$ (or $\Theta \in I_\Theta$). The parameter h is not determined uniquely from (1.5). Analogously, Θ is not determined uniquely from (1.6) when $h = 0$. Despite the error of calculation of ψ and ψ_T being of the order of ε^k, the transversality condition (1.5) determines the values of the roots $\{h_\nu\}$ (analogously, condition (1.6) determines the values of $\{\Theta_\nu\}$) with sufficient accuracy (the error being $O(\varepsilon^{k+1})$):

$$\begin{aligned} &\big(p_{(k)} \cdot f_0^*(a_{(k)}, \psi_{(k-1)}, p_{(k)})\big)\big|_{\Theta\varepsilon^{-1}} = h; \\ &h \in \{h_\nu\}; \qquad \Theta \in \{\Theta_\nu\}, \qquad h = 0. \end{aligned} \tag{4.2.22}$$

The optimal value of the parameter h (or Θ) is determined by solving the minimization problem (1.7) where the functional J_ν is written with the error $O(\varepsilon^{k+1})$:

$$\begin{aligned} J(h,\varepsilon) &= J_{(k)}(h,\varepsilon) + \varepsilon^{k+1} J_{k+1}(h,\varepsilon); \qquad J_{(k)}(h,\varepsilon) \longrightarrow \min_{h\in\{h_\nu\}}, \\ J(\Theta,\varepsilon) &= J_{(k)}(\Theta,\varepsilon) + \varepsilon^{k+1} J_{k+1}(\Theta,\varepsilon); \qquad J_{(k)}(\Theta,\varepsilon) \longrightarrow \min_{\Theta\in\{\Theta_\nu\}}. \end{aligned} \tag{4.2.23}$$

If the conditions of Lemmas 3.1 and 3.2 are satisfied, then the local minima of the functions $J_{(k)}(h,\varepsilon)$ or $J_{(k)}(\Theta,\varepsilon)$ are found in close vicinities of the points h^0 and Θ^0, respectively. The values of h^0 and Θ^0 are determined from the averaged equation (1.6). The size of the vicinities has the order of ε^{d_k}, $d_k = (k+1)/(2-d)$, in the class of power-type estimates. Here, d, $d \in (0,1]$, characterizes the power-type estimate of J_{k+1} with respect to the parameter β: $J_{k+1} = O(\beta^d)$. If $d = 1$, then $d_k = k+1$ and, certainly, conditions (23) determine the minimum points h_ν or Θ_ν with the required error $O(\varepsilon^{k+1})$.

Substitution of the values obtained of the parameters h_ν or Θ_ν into the constructed functions $a_{(k)}$, $\psi_{(k-1)}$, $p_{(k)}$, and $q_{(k)}$ gives the solution of the optimal control problem with the error $O(\varepsilon^{k+1})$ relative to the functional and the boundary conditions. This approximate solution has the property that in an ε^{k+1}-vicinity of a and an ε^k-vicinity of ψ, there exists a trajectory (a,ψ) which is the exact solution of the boundary-value problem (1.2)–(1.6). If the smoothness conditions guaranteeing differentiability of the functions $J_{k+1}(h,\varepsilon)$ or $J_{k+1}(\Theta,\varepsilon)$ with respect to the parameter h or Θ are fulfilled, then the obtained values h_ν or Θ_ν give an ε^{k+1}-proximity in a and an ε^k-proximity in ψ. The approximation accuracy of the optimal control u^* in the case of a smooth system is

$$u_s^* - u_{s(k)}^* = O(\varepsilon^{k+1}), \qquad u_p^* - u_{p(k)}^* = O(\varepsilon^k), \tag{4.2.24}$$

and for nonsmooth controls, the proximity occurs with respect to the functional or in another nonuniform metric.

4.3. Two-Parameter Scheme of Canonical Averaging

As concrete examples show, reducing rather simple systems of the type (3.2.4) to the standard Hamiltonian form (1.10), averaging over ψ, solving the averaged system and the boundary-value problem, and returning to the original variables often give rise to bulky, hardly implementable calculations. Therefore, the procedure of expansion in two parameters, ε and μ (μ is a mean value for q, $\mu = O(\varepsilon)$), developed below is of interest.

We consider a univalent canonical change of variables, characterized by the generating function S (see (2.1.3)):

$$S(a,\psi,\nu,\mu,\varepsilon) = (a\cdot\nu) + \psi\mu + \varepsilon\sigma(a,\psi,\eta,\mu,\varepsilon), \tag{4.3.1}$$

such that the new (averaged) Hamiltonian K does not depend on φ:

$$K = \omega(\xi)\mu + \varepsilon k(\xi,\eta,\mu,\varepsilon). \tag{4.3.2}$$

The connection between the original a, ψ, p, q and averaged ξ, φ, η, μ variables is given by the relationships [1] (see (2.1.1))

$$\xi = a + \varepsilon\frac{\partial\sigma}{\partial\eta}, \quad \varphi = \psi + \varepsilon\frac{\partial\sigma}{\partial\mu}, \quad p = \eta + \varepsilon\frac{\partial\sigma}{\partial a}, \quad q = \mu + \varepsilon\frac{\partial\sigma}{\partial\psi}. \tag{4.3.3}$$

If the functions σ and k have been constructed, then the equations

$$\begin{aligned} &\dot{\xi} = \varepsilon\partial k/\partial\eta, \qquad \dot{\eta} = \omega'(\xi)\mu - \varepsilon\partial k/\partial\xi, \\ &\dot{\mu} = 0, \qquad \dot{\varphi} = \omega(\xi) + \varepsilon\partial k/\partial\mu, \end{aligned} \tag{4.3.4}$$

and the second condition (1.5) imply that $\mu = O(\varepsilon)$, since

$$\mu = -\varepsilon\left.\frac{\partial\sigma}{\partial\psi}\right|_T, \qquad \sigma(a,\psi+2\pi,\eta,\mu,\varepsilon) \equiv \sigma(a,\psi,\eta,\mu,\varepsilon). \tag{4.3.5}$$

These circumstances give the possibility of expanding the unknown functions σ and k in terms of powers of ε and μ and then finding (from relationships (5)) the parameter μ as a function of ε and the other parameters of the problem.

Thus, we propose to construct the functions σ and k in the form

$$\sigma = \sum_{i,j\geqslant 0} \varepsilon^i\mu^j\sigma_{ij}(a,\psi,\eta), \qquad k = \sum_{i,j\geqslant 0} \varepsilon^i\mu^j k_{ij}(\xi,\eta). \tag{4.3.6}$$

Unknown coefficients σ_{ij} and k_{ij}, $i,j \geqslant 0$, are defined as a result of substituting expansions (6) into the equation of the form (2.1.2) ($\partial S/\partial t \equiv 0$):

$$\begin{aligned} &H^*\left(a,\psi,\frac{\partial S}{\partial a},\frac{\partial S}{\partial\psi},\varepsilon\right) = K(\xi,\mu,\varepsilon), \\ &\varepsilon\omega(a)\frac{\partial\sigma}{\partial\psi} + \varepsilon h\left(a,\psi,\eta+\varepsilon\frac{\partial\sigma}{\partial a},\mu+\varepsilon\frac{\partial\sigma}{\partial\psi},\varepsilon\right) + \\ &\quad + \left[\omega(a) - \omega\left(a+\varepsilon\frac{\partial\sigma}{\partial\eta}\right)\right]\mu = \varepsilon k\left(a+\varepsilon\frac{\partial\sigma}{\partial\eta},\eta,\mu,\varepsilon\right). \end{aligned} \tag{4.3.7}$$

Expanding the functions in (7) in terms of powers $\varepsilon^i\mu^j$ and comparing the corresponding terms yields a set of linking linear partial differential equations. The successive solution of these equations in a way similar to that given in Section 2.1.1 or 4.1.2 leads to expressions for the required coefficients k_{ij} and σ_{ij}:

$$\begin{aligned} k_{ij}(\xi,\eta) &= \langle h_{ij}(\xi,\psi,\eta)\rangle \equiv h_{ij0}(\xi,\eta), \qquad i,j=0,1,\dots,r, \\ \sigma_{ij}(a,\psi,\eta) &= \frac{1}{\omega(a)}\int_{\psi^0}^{\psi}\left[h_{ij}(a,y,\eta)-h_{ij0}(a,\eta)\right]dy + e_{ij}(a,\eta). \end{aligned} \tag{4.3.8}$$

In particular, for $i,j=0,1$, the functions h_{ij} are

$$\begin{aligned} h_{00}(a,\psi,\eta) &= h(a,\psi,\eta,0,0), \\ h_{10} &= \left(\frac{\partial h}{\partial p}\right)\cdot\frac{\partial\sigma_{00}}{\partial a} + \left(\frac{\partial h}{\partial q}\right)\cdot\frac{\partial\sigma_{00}}{\partial\psi} + \left(\frac{\partial h}{\partial\varepsilon}\right) - \frac{\partial k_{00}}{\partial a}\cdot\frac{\partial\sigma_{00}}{\partial\eta}, \\ h_{01} &= \left(\frac{\partial h}{\partial p}\right) - \omega'(a)\cdot\frac{\partial\sigma_{00}}{\partial\eta}. \end{aligned} \tag{4.3.9}$$

Derivatives of the functions h, σ_{ij}, k_{ij}, and ω in (9) are taken at the point $(a,\psi,\eta,0,0)$. To construct some $(r+1)$th approximation, we calculate the coefficients σ_{ij} and k_{ij} and substitute them into (6). Then we substitute $\sigma_{(r)}(a,\psi,\eta,\mu,\varepsilon)$ into (3) and solve the obtained equation with respect to a, ψ, p, and q. To do this we have to find a and ψ as functions of the arguments ξ, φ, η, μ, and ε, using the first two Equations (3), and substitute them into the remaining equations:

$$\begin{aligned} a &= \xi + \varepsilon A_{(r)}(\xi,\varphi,\eta,\mu,\varepsilon), \\ \psi &= \varphi + \varepsilon\Psi_{(r)}(\xi,\varphi,\eta,\mu,\varepsilon), \\ p &= \eta + \varepsilon\frac{\partial}{\partial a}\sigma_{(r)}(\xi+\varepsilon A_{(r)},\varphi+\varepsilon\Psi_{(r)},\eta,\mu,\varepsilon), \\ q &= \mu + \varepsilon\frac{\partial}{\partial\psi}\sigma_{(r)}(\xi+\varepsilon A_{(r)},\varphi+\varepsilon\Psi_{(r)},\eta,\mu,\varepsilon). \end{aligned} \tag{4.3.10}$$

Substitution of (10) into the left-hand side of the first relationship (7) leads to the expression for $K_{(r)}$:

$$K_{(r)} = \omega(\xi)\mu + \varepsilon k_{(r)}(\xi,\eta,\mu,\varepsilon) + \varepsilon k_{r+1}(\xi,\varphi,\eta,\mu,\varepsilon), \tag{4.3.11}$$

where k_{r+1} is a 2π-periodic function in φ whose order with respect to small parameters ε^i, μ^j is $r+1$, i.e. $i+j=r+1$. Since $\mu=O(\varepsilon)$, $\partial k_{r+1}/\partial\varphi = O(\varepsilon^i\mu^j)$, the equation $\dot{\mu} = \varepsilon \partial k_{r+1}/\partial \varphi$ implies that $\mu = \varepsilon\beta + O(\varepsilon^{r+2})$ for $t - t_0 \sim 1/\varepsilon$, β being a constant. Having omitted the term εk_{r+1} in (11), we obtain for ξ and η the set of equations of the $(r+1)$th approximation:

$$\dot{\xi} = \frac{\partial}{\partial \eta} k_{(r)}(\xi, \eta, \varepsilon\beta, \varepsilon),$$
$$\dot{\eta} = -\omega'(\xi)\beta - \frac{\partial}{\partial \xi} k_{(r)}(\xi, \eta, \varepsilon\beta, \varepsilon). \tag{4.3.12}$$

Thus, we have obtained the canonical set of $2n$ equations of the $(r+1)$th approximation (12) which generalizes equations of the first approximation (3.2.17). When differentiating with respect to μ, the order of the term omitted is less by one. Therefore, the phase φ is determined from the equation

$$\dot{\varphi} = \varepsilon^{-1}\omega(\xi) + \frac{\partial}{\partial \mu} k_{(r)}(\xi, \eta, \varepsilon\beta, \varepsilon), \qquad (\dot{\,}) \equiv \left(\frac{\mathrm{d}}{\mathrm{d}\tau}\right), \tag{4.3.13}$$

with a larger (by one order) error $O(\varepsilon^r)$. Further constructions in solving the boundary-value problem and separation of the optimal solution are carried out analogously according to the technique developed in Section 4.2.

4.4. Analysis of Optimal Control Problems on a Relatively Short Time Interval

4.4.1. STATEMENT OF THE PROBLEM

As has been mentioned in Section 3.2.1, equations of the boundary-value problem of the maximum principle for an essentially nonlinear system (3.2.4) are similar to equations of multi-frequency nonlinear oscillations in the neighbourhood of a resonance point, see [1, 4, 5, 38, 46, 60, 69, 77, 94, 115, 120, 123, 124, 130, 133, 150]. The analysis of motion of a nonlinear resonant system on an asymptotically large time interval $t - t_0 \sim 1/\varepsilon$ causes significant difficulties [1, 46, 123, 124, 150]. If a highly complicated general situation takes place, some researchers confine themselves to the time interval $t - t_0 \sim 1/\sqrt{\varepsilon}$ during which the system performs $\sim 1/\sqrt{\varepsilon}$ oscillations, and slow variables change by an amount $O(\sqrt{\varepsilon})$.

One can consider the optimal control problem in a similar way, which leads to the analysis of equations and boundary conditions (3.2.4) on a relatively short time interval $t \in [t_0, T]$, $T = \Theta\varepsilon^{-1/2}$. As in the cases considered above, the statement must be consistent, i.e. it must lead to control functions of the order unity and the variation of the slow variables a and p by an amount of $O(\varepsilon^{1/2})$. The terminal set given by a relationship of the type (3.2.1) can have the form

$$M(a-a^*)\big|_T = \delta N(a)\big|_T, \qquad a, a^* \in D_a, \quad \delta = \sqrt{\varepsilon},$$
$$M = (M_1, \ldots, M_l), \quad N = (N_1, \ldots, N_l), \tag{4.4.1}$$

where $M(0) = 0$, rank $M'(0) = l$, the point a^* belongs to a δ-neighbourhood of the initial point a^0 and N is a smooth function. We suppose that $T \sim \delta^{-1}$ in the problem with a fixed time of termination or solution of the control problem leads to such an estimate when T is not fixed.

One can establish that under some general conditions, the solution of the boundary-value problem is unique and can be constructed with any accuracy in terms of δ determined by the smoothness of the functions included in relationships (3.2.4). It should be noted that an approximate solution of the set of equations requires asymptotic methods of separating the fast and slow motions, since direct application of the perturbation technique (expansion in terms of powers of δ) on a time interval $t - t_0 \sim \delta^{-1}$ leads to difficulties.

4.4.2. CONSTRUCTION OF THE SOLUTION OF THE BOUNDARY-VALUE PROBLEM WITH AN ARBITRARY ACCURACY

Let the averaged system of the $(r+1)$th approximation in terms of ε for a and p and of the rth approximation for ψ be constructed in the form (3.12) and (3.13). Its solution is constructed on the time interval $t - t_0 \sim \delta^{-1}$ with the error $O(\varepsilon^{r+1})$ for ξ and η and with the error $O(\varepsilon^{r+1/2})$ for φ. As follows from the equations, taking into account the estimates of the omitted terms ($O(\varepsilon^{r+2})$ and $O(\varepsilon^{r+1})$, respectively), the estimates for ξ, η, and φ will be valid if it is possible to determine the parameter β with an accuracy $O(\varepsilon^{r+1/2})$ from the boundary conditions (1) and transversality conditions of the type (3.2.4), where a, ψ, p, and q are specified according to (3.10). This is proved by direct construction of the general solution of system (3.12). Indeed, having introduced the slow time $s = \delta t$, $s \in [s_0, \Theta]$, $\Theta \sim 1$, we obtain the equations

$$\dot{\xi} = \delta \frac{\partial}{\partial \eta} k_{(r)}(\xi, \eta, \delta^2\beta, \delta^2), \qquad \xi(s_0) = \xi^0, \qquad (\dot{\ }) \equiv \left(\frac{\mathrm{d}}{\mathrm{d}s}\right),$$
$$\dot{\eta} = -\delta\beta\omega'(\xi) - \delta \frac{\partial}{\partial \xi} k_{(r)}(\xi, \eta, \delta^2\beta, \delta^2), \qquad \eta(s_0) = \eta^0. \tag{4.4.2}$$

The solution of Equations (2) is constructed by using the perturbation technique [1, 49, 122]:

$$\begin{aligned} \xi &= \xi_0 + \delta\xi_1 + \delta^2\xi_2 + \cdots + \delta^{2r+1}\xi_{2r+1} + O(\varepsilon^{r+1}), \\ \eta &= \eta_0 + \delta\eta_1 + \delta^2\eta_2 + \cdots + \delta^{2r+1}\eta_{2r+1} + O(\varepsilon^{r+1}). \end{aligned} \tag{4.4.3}$$

Substitution of (3) into (2) leads to the solution in the form of polynomials

$$\begin{aligned} &\xi_0 = \xi^0, \qquad \eta_0 = \eta^0; \\ &\xi_1 = \frac{\partial k_0}{\partial \eta}(s - s_0), \qquad \eta_1 = -\left[\omega'(\xi^0)\beta + \frac{\partial k_0}{\partial \xi}\right](s - s_0); \\ &\xi_2 = \frac{1}{2}\frac{\partial^2 k_0}{\partial\eta\partial\xi}\frac{\partial k_0}{\partial\eta}(s - s_0)^2 - \frac{1}{2}\frac{\partial^2 k_0}{\partial\eta^2}\left[\omega'(\xi^0)\beta + \frac{\partial k_0}{\partial\xi}\right](s - s_0)^2, \\ &\eta_2 = -\frac{1}{2}\beta\omega''(\xi^0)\frac{\partial k_0}{\partial\eta}(s - s_0)^2 - \frac{1}{2}\frac{\partial^2 k_0}{\partial\xi^2}\frac{\partial k_0}{\partial\eta}(s - s_0)^2 + \\ &\qquad + \frac{1}{2}\frac{\partial^2 k_0}{\partial\eta\partial\xi}\left[\omega'(\xi^0)\beta + \frac{\partial k_0}{\partial\xi}\right](s - s_0)^2. \end{aligned} \tag{4.4.4}$$

Subsequent coefficients (3) are determined analogously. The written form of the general solution to be found is significantly simplified if one represents system (2) as one vector equation and seeks its solution as an expansion in terms of powers of the small quantity $\delta(s - s_0)$ with unknown coefficients in the form

$$x_{(r)} = x^0 + \sum_{i=1}^{2r+1} c_i\delta^i(s - s_0)^i \qquad (\dot{x} = \delta X(x),\ x(0) = x^0). \tag{4.4.5}$$

Substitution of $x_{(r)}$ results in the equation for c_i:

$$\sum_{i=1}^{2r+1} ic_i\delta^{i-1}(s - s_0)^{i-1} = X\left(x^0 + \sum_{i=1}^{2r+1} c_i\delta^i(s - s_0)^i\right). \tag{4.4.6}$$

Suppose the coefficients c_i $(i = 1, 2, \ldots, 2r + 1)$ have been determined and, from this, unknown variables ξ and η have been constructed with an error $O(\varepsilon^{r+1})$, i.e. with the accuracy $O(\delta^{2r+1})$, in the form of polynomials of the order $(2r + 1)$. Then the phase φ is determined from Equation (3.13) with an accuracy $O(\varepsilon^r)$, i.e. with the error $O(\delta^{2r+1})$. Substitution of the functions $\xi_{(2r+1)}$, $\eta_{(2r+1)}$, and $\varphi_{(2r)}$ into (3.10) allows determining the variables a and p with the error $O(\varepsilon^{r+1})$, the phase ψ with the error $O(\delta^{2r+1})$, and the variable q with the required accuracy (with the error $O(\delta^{2r+3})$). Then substituting expression (3.10) into boundary conditions (3.2.4) and (1) one can obtain the relationships for determining unknown parameters ξ^0, η^0, φ^0, λ, and β for fixed $T \sim \delta^{-1}$. If T is not fixed, then the transversality conditions are to be supplemented with the relationship (1.6) when $h = 0$. This relationship determines the quantity T with the error $O(\delta^{2r+1})$. The aforenamed parameters are obtained in the form of expansions in terms of powers of δ from the resulting equations

$$
\begin{aligned}
&a^0 = \xi^0 + \varepsilon A_{(r)}(\delta^0, \varphi^0, \eta^0, \delta^2\beta, \delta^2) + O(\delta^{2r+2}),\\
&M\left(a_{(r)}(\Theta) - a^*\right) = \delta N\left(a_{(r)}(\Theta)\right) + O(\delta^{2r+2}),\\
&\psi^0 = \varphi^0 + \varepsilon \Psi_{(r)}(\xi^0, \varphi^0, \eta^0, \delta^2\beta, \delta^2) + O(\delta^{2r+1}),\\
&\eta_{(r)}(\Theta) + \varepsilon P_{(r)}\left(a_{(r)}, \psi_{(r)}, p_{(r)}, \delta^2\beta, \delta^2\right)\big|_\Theta \qquad\qquad (4.4.7)\\
&\quad = p_{(r)}(\Theta) = \frac{\partial}{\partial a}\left[(\lambda \cdot (M_{(r)} - \delta N_{(r)})) - g_{(r)}\right]\Big|_\Theta + O(\delta^{2r+2}),\\
&\beta + Q_{(r)}\left(a_{(r)}, \psi_{(r)}, p_{(r)}, \delta^2\beta, \delta^2\right)\big|_\Theta = O(\delta^{2r+1}),\\
&\left((\eta_{(r)} + \varepsilon P_{(r)}) \cdot f_0^*(a_{(r)}, \psi_{(r)}, \eta_{(r)} + \varepsilon P_{(r)})\right)\big|_\Theta = O(\delta^{2r+1}).
\end{aligned}
$$

In particular, when $M, N \equiv 0$, the solution of the boundary-value problem (7) to the first approximation for fixed $T = \Theta\delta^{-1}$ has the form

$$
\begin{aligned}
&a_{(1)} = a^0 + \delta f_0(a^0, -g'(a^0))(s - s_0) + O(\delta^2),\\
&\psi_{(1)} = \psi^0 + \delta^{-1}\omega(a^0)(s - s_0) +\\
&\qquad + (1/2)\omega'(a^0) f_0(a_0, -g'(a^0))(s - s_0)^2 + O(\delta),\\
&p_{(1)} = -g'(a^0) +\\
&\qquad + \delta\left[\omega'(a^0)\beta_{(1)} - \left(g'(a^0) \cdot \frac{\partial}{\partial a} f_0(a^0, -g'(a^0))\right)\right](\Theta - s) - \qquad (4.4.8)\\
&\qquad - g''_{a^2}(a^0) f_0(a^0) f_0(a^0, -g'(a^0))(\Theta - s) + O(\delta^2),\\
&\beta_{(1)} = \omega^{-1}(a^0)\Big(g'(a^0) \times\\
&\qquad \times \left[f_0^*(a^0, \psi_{(1)}(\Theta, \delta), -g'(a^0)) - f_0(a^0, -g'(a^0))\right]\Big) + O(\delta).
\end{aligned}
$$

Substitution of expressions (8) into the function u^* (3.2.3) allows determining the optimal control, both in the open-loop and feedback ($a^0 \to a$, $s_0 \to s$) forms. Here, one should bear in mind a singularity of the feedback control at $t = T$ ($s = \Theta$) if T is fixed and the terminal manifold (1) is given. The value of the functional $J[u]$ (3.2.2) is given by

$$\begin{aligned} J_{(1)} = g(a^0) + \delta\big(g'(a^0)\cdot f_0(a^0, -g'(a^0))\big)(\Theta - s_0) + \\ + \delta G_0(a^0, -g'(a^0))(\Theta - s_0) + O(\delta^2). \end{aligned} \tag{4.4.9}$$

Thus, in the considered case of the optimal control problem leading to a relatively short time interval $(T - t_0) \sim \varepsilon^{-1/2}$, construction of the approximate solution is substantially simplified. Theoretically, the problem does not seem to be of interest. However, the application of the developed approach may turn out to be acceptable when investigating concrete optimal control problems for nonlinear oscillatory systems, for instance when constructing a simplified "quasi-optimal" feedback control.

Solution of concrete optimal control problems for non- linear oscillatory and rotatory systems admitting mechanical interpretations is given in Chapters 5 and 6.

4.4.3. ANALYSIS OF MULTI-FREQUENCY SYSTEMS USING THE AVERAGING METHOD ON A RELATIVELY SHORT TIME INTERVAL

Let ψ be a vector of any dimension and q be the variable adjoint with it. Then the boundary-value problem of the maximum principle is a generalization of the problem (3.2.4) and (1):

$$\begin{aligned} &\dot a = \varepsilon f^*(a, \psi, p, q), \qquad a(t_0) = a^0, \\ &\dot\psi = \omega(a) + \varepsilon F^*(a, \psi, p, q), \qquad \psi(t_0) = \psi^0, \\ &M(a(T), \delta) = 0; \qquad t \in [t_0, T], \quad T = \Theta\delta^{-1}, \quad \delta = \varepsilon^{1/2}, \\ &\dot p = \frac{\partial}{\partial a}(q\cdot\omega) - \varepsilon\frac{\partial}{\partial a}\big[(p\cdot f^*) + (q\cdot F^*)\big], \\ &p(T) = -\frac{\partial g}{\partial a}\Big|_T + \frac{\partial}{\partial a}(\lambda\cdot M)\Big|_T, \qquad q(T) = 0. \end{aligned} \tag{4.4.10}$$

Here, as in Section 4.1, the condition $M(a^0, 0) = 0$ is fulfilled. Without loss of generality, we suppose that the functional $J[u] = g(a(T))$, and the function under the integral sign is $G \equiv 0$. It should be noted that, in the general case, the average of the right-hand side of the equation for q is not equal to zero, i.e. it is not possible to perform independent averaging over components of the phase ψ. By using the change

$$a = a^0 + \delta\alpha, \quad \psi = \omega(a^0)(t - t_0) + \psi^0 + \varphi, \quad p = p, \quad q = \delta\rho, \tag{4.4.11}$$

where α, φ, p, and ρ are unknown variables of the order of unity, the system is reduced to the standard form:

$$\begin{aligned}
&\dot{\alpha} = \delta f_*\left(\omega\left(a^0\right)t, \alpha, \varphi, p, \rho, \delta\right), \qquad \alpha(t_0) = 0,\\
&\dot{\varphi} = \delta\omega'\left(a^0\right)\alpha + \delta^2 F_*\left(\omega\left(a^0\right)t, \alpha, \varphi, p, \rho, \delta\right), \qquad \varphi(t_0) = 0,\\
&\dot{p} = -\delta\frac{\partial}{\partial a}\left(\rho\cdot\omega\left(a^0\right)\right) + \delta^2 P_*\left(\omega\left(a^0\right)t, \alpha, \varphi, p, \rho, \delta\right),\\
&\left(\frac{\partial M}{\partial a}\right)_{0T}\alpha(T) + \left(\frac{\partial M}{\partial \delta}\right)_{0T} + \delta N(\alpha(T), \delta) = 0,\\
&p(T) = -\left(\frac{\partial q}{\partial a}\right)_{0T} + \frac{\partial}{\partial a}(\lambda\cdot M)_{0T} + \delta L\left(\alpha(T), \lambda, \delta\right), \qquad \rho(T) = 0.
\end{aligned} \tag{4.4.12}$$

The right-hand side of the standard set of differential equations (12) is a sophisticated quasiperiodic function of t with a constant frequency basis $\{\omega(a^0)\}$, see [39, 115]. Further asymptotic analysis of the boundary-value problem (12) on a time interval $t - t_0 \sim \delta^{-1}$ is performed analogously to Section 2.5 [1, 10]. The choice of the method of averaging essentially depends on the arithmetical properties of the constant vector $\omega(a^0)$ and the properties of the Fourier series in ψ for the functions f^* and F^*.

Thus, when corresponding requirements of the smoothness are satisfied, the solution of the boundary-value problem (12), and, by this, the solution of the optimal control problem for a multi-frequency, essentially nonlinear, system, can be constructed with a prescribed accuracy in terms of $\delta = \varepsilon^{1/2}$ on a relatively short time interval $t - t_0 \sim \delta^{-1}$ on which small (of the order of δ) change of slow variables occurs. The described approach may appear to be usable in some applied problems when constructing quasioptimal feedback control. Problems of time-optimal type (with free time of the process termination) are treated analogously. In this case, terminal conditions must lead, as was the case above, to small (of the order of $\delta = \varepsilon^{1/2}$) change of slow state variables of the system.

4.5. Remarks

4.5.1. MULTI-FREQUENCY SYSTEMS

Analysis of essentially nonlinear oscillatory systems with many phases on a time interval $t - t_0 \sim 1/\varepsilon$ leads to substantial difficulties when constructing general procedures for asymptotic solution of control problems. The basic difficulty is that the vector variable q, adjoint with the phase ψ, is not small in the general case, since $\langle\dot{g}\rangle$ does not equal zero. Then the variable p, adjoint with a, will not be slow: $\dot{p} = O(q) = O(1)$, $p \sim \varepsilon^{-1}$ for $t - t_0 \sim \varepsilon^{-1}$. Since the right-hand sides of equations of the boundary-value problem are not uniformly bounded in p, the application of asymptotic methods of the separation of motions on a time interval $t - t_0 \sim \varepsilon^{-1}$ is inconsistent. It is required to take into account specific features of control problems for nonlinear oscillations ("passing through resonances" [1, 15, 123] "sticking at resonances" [1, 115, 150] and so on) and to make more delicate estimates which enable one to reduce the system of the type (4.10) to the standard form, i.e. to show that $q = O(\varepsilon)$ for $t - t_0 \sim \varepsilon^{-1}$ [1]. For example, in some problems, the ratio of frequencies appears to be constant; then, using the appropriate change of phases, one can reduce the system to the single-frequency form. Such situations arise in problems of control of motions of a symmetrical rigid body with a fixed point, close to the regular precession; the ratio of frequencies is defined by the principal moments of inertia. Furthermore, if one succeeds in establishing that $p = O(\varepsilon)$, then, in the case of the two-frequency problem, one can obtain unimprovable error estimates, $O(\sqrt{\varepsilon})$ or $O(\sqrt{\varepsilon}\ln\varepsilon^{-1})$, with respect to slow variables [123], depending on the features of "passing through the resonance"; the smoothness conditions and some general requirements supposed to be fulfilled.

4.5.2. AVERAGING FOR SMALL FREQUENCIES AND PASSING THE SEPARATRIX

In some applications, it is important to estimate the error of the averaging method for problems leading to small frequencies, when the rate of the phase change is $\dot{\psi} \sim \omega(a) = O(\varepsilon)$ and can pass through the zero value. The estimate of the error of the averaging method essentially depends on the features of passing the zero value.

In problems connected with passing "separatrices", a logarithmic singularity takes place: $\omega \sim 1/\ln|a - a^*|^{-1}$, $a \to a^*$. Such situations arise when a pendulum proceeds from an oscillatory mode to a rotatory one or vice versa, when rotations of a nonsymmetrical rigid body proceed from one domain to another (in these domains, polhodes [138] surround the axes with extremal values of principal moments of inertia at the mass centre), and in other problems (see Chapters 5 and 6). In reference [123], it is established that the application of the averaging method gives the error $O(\varepsilon\ln\varepsilon^{-1})$ with respect to the slow variable a for $t - t_0 \sim \varepsilon^{-1}$.

The estimate of the averaging method in the problem of passing the zero frequency value with restricted velocity $O(\varepsilon)$ is similar to the case of passing the resonance. We can formally introduce an additional "fictitious" phase which is not included in the right-hand sides of the system. Then the original phase has the sense of the amount off-resonance for the problem of passing the resonance. Provided the smoothness conditions mentioned above, as well as some others (which are more easily verified for the single-frequency system in question than for the two-frequency one), are satisfied, we obtain the estimates $O(\varepsilon^{1/2})$ or $O(\varepsilon^{1/2}\ln\varepsilon^{-1})$ for $t - t_0 \sim \varepsilon^{-1}$, depending on the features of passing the zero frequency value. These estimates are also unimprovable [123, 124].

The cases when passing through the resonance or through the zero value of the frequency has a more general character (such cases are considered in Sections 2.5 and 2.6 for quasilinear oscillatory systems) require additional investigation. In applied optimal control problems, the aforenamed conditions are often satisfied, which makes it possible to simplify substantially the equations of motion and to provide a sufficiently high accuracy for the averaging method.

4.5.3. THE AVERAGING METHOD IN MORE GENERAL CONTROLLED SYSTEMS

Consider a controlled system of the form (3.1.1) under more general, assumptions, than in Chapter 3, concerning the generating controlled system. Analogously to Section 2.5, we suppose that the family of unperturbed rotatory-oscillatory motions has an incomplete dimension ($n_c + n_\psi < n_y$, where $n_c \geqslant 0$, $n_\psi \geqslant 1$, $n_y \geqslant 2$) and is asymptotically stable. Unlike the quasilinear case considered above, we suppose here that the frequency vector $\omega = \omega_*(c, x) \equiv \omega(a)$. Using the theory of Lyapunov and Floquet [38, 45, 47, 74, 75, 94, 115, 120, 137] we construct the nonsingular change of variables $(x, y) \rightarrow (a, \psi, z)$, where z is a vector of the dimension $n_y - (n_c + n_\psi) \geqslant 1$. The equations of controlled motion are reduced to the form

$$\begin{aligned}
&\dot a = A(a, \psi, z, u, \varepsilon), && a(t^0) = a_0, \quad a \in D_a,\\
&\dot\psi = \omega(a) + \Psi(a, \psi, z, u, \varepsilon), && \psi(t^0) = \psi^0, \quad \psi \in R^{n_\psi},\\
&\dot z = \Lambda(a)z + Z(a, \psi, z, u, \varepsilon), && z(t^0) = z^0, \quad z, 0 \in D_z;\\
&A, \Psi, Z = O(\varepsilon + |z|^2), && \varepsilon \in [0, \varepsilon_0].
\end{aligned} \tag{4.5.1}$$

Here, $\Lambda(a)$ is a stable matrix, i.e. its eigenvalues $\lambda(a)$ have negative real parts for the considered range of changing the "slow" variable a. The motion of system (1) will occur in an ε-vicinity of the stable integral manifold $z = 0$ of the generating system, i.e. $z = O(\varepsilon)$ for $t \geqslant t_0$, $t_0 - t^0 \sim \ln \varepsilon^{-1}$.

Then we state the optimal control problem for a weakly controlled system (1) with respect to the slow variable a on an asymptotically large time interval $t \in [t_0, T]$, $T = \Theta\varepsilon^{-1}$:

$$M\big(a(T), \varepsilon\big) = 0, \qquad J_0[u] = g\big(a(T), \varepsilon\big) \longrightarrow \min_u, \qquad u \in U. \tag{4.5.2}$$

Here, the value of T can be either specified beforehand or defined in the course of solving problem (1) and (2). Since, for the variable z, the estimate $z(t) = O(\varepsilon)$, $t \in [t_0, \Theta\varepsilon^{-1}]$, takes place, the influence of this variable on the variable a will be of the order of ε. Therefore, to a first approximation, we can confine ourselves to the asymptotic solution of problem (2) for the system of the form

$$\dot a = \varepsilon A_0^{(1)}(a, \psi, u), \qquad \dot\psi = \omega(a) + \varepsilon\Psi_0^{(1)}(a, \psi, u). \tag{4.5.3}$$

The functions $A_0^{(1)}$ and $\Psi_0^{(1)}$ in (3) are obtained when analysing the structure of the functions A, Ψ (and Z). The function A has the form

$$A = \varepsilon A^{(1)}(a, \psi, u, z, \varepsilon) + A^{(2)}(a, \psi, z, \varepsilon),$$

$$A_0^{(1)} = A^{(1)}(a, \psi, u, 0, 0), \qquad A^{(2)} = O(|z|^2).$$

The functions Ψ and Z are expressed analogously. It should be noted that, as in Section 2.5, the variable $r(t)$, adjoint with $z(t)$, is $r(t) = O(\varepsilon)$. Therefore, the influence of the variables z and r on the remaining state and adjoint variables will be $O(\varepsilon)$ for $t - t_0 \sim \varepsilon^{-1}$. These circumstances enable carrying out asymptotic expansions of the solution of the maximum principle boundary-value problem with a prescribed accuracy in terms of ε using the general scheme of the averaging method developed by Volosov [38, 149, 150]. One succeeds in carrying out constructive manipulations for single-frequency systems or when the resonance does not take place in the required order (in terms of ε) of the averaging method. The construction of solutions of higher approximations essentially relies on the solution of the problem of the first approximation (2) and (3) (see Sections 3.2, 3.3, 4.1 and 4.2).

CHAPTER 5

Control of Motions of "Pendulum-Type" Systems

Using the methods developed in Chapters 1–4, we analyse here optimal control problems for quasilinear and nonlinear pendulum type systems. The acceleration or velocity of the suspension point (the equilibrium position) is taken as a control action. This is of practical importance, for instance when controlling loading machines, taking into account load oscillations, as well as manipulation systems, considering the elastic compliance of their parts, etc.

5.1. Spatial Motions of a Pendulum Having a Controlled Position of the Suspension Point

5.1.1. EQUATIONS OF MOTION

Consider oscillations of a spatial (spherical) pendulum in the neighbourhood of the lower equilibrium position ($z < 0$), see Fig. 5.1. We suppose that the constraint is not violated during the motion, i.e. the constraint force is positive. To describe the motion we introduce the fixed (inertial) reference frame $OXYZ$ and the moving (König's) reference frame $O'xyz$:

$$\mathbf{R} = \mathbf{r}_0 + \rho, \quad \rho = (x, y, z); \quad z = -\left(l^2 - r^2\right)^{1/2}, \quad r^2 = x^2 + y^2. \tag{5.1.1}$$

Here, $\mathbf{r}_0(t)$ is a specified motion of the point O' with respect to the reference frame $OXYZ$; the pendulum length l being assumed constant. We introduce new scales for length (l) and time ($1/\nu$):

$$\rho_* = \frac{\rho}{l}, \quad \mathbf{r}_{0*} = \frac{\mathbf{r}_0}{l}, \quad t_* = \nu t, \quad \nu^2 = \frac{g}{l}. \tag{5.1.2}$$

Then the exact equations of motion are reduced to the following compact form:

$$\begin{aligned} &\frac{\ddot{r}}{1 - r^2} - r\dot{\varphi}^2 + \frac{r}{(1 - r^2)^{1/2}} = -\ddot{x}_0 \cos\varphi - \ddot{y}_0 \sin\varphi - \ddot{z}_0 \frac{r}{(1 - r^2)^{1/2}}, \\ &\ddot{\varphi} r + 2\dot{r}\dot{\varphi} = \ddot{x}_0 \sin\varphi - \ddot{y}_0 \cos\varphi \qquad (r > 0). \end{aligned} \tag{5.1.3}$$

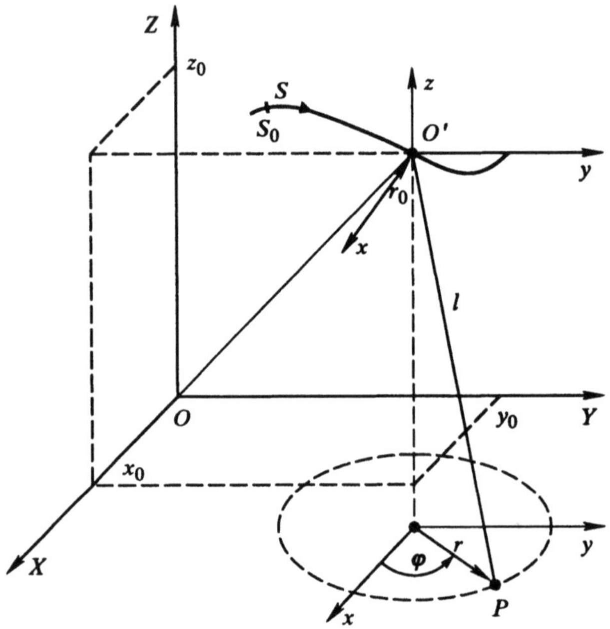

Fig. 5.1.

Here, r and φ are polar coordinates characterizing the motion of the point P in the plane $O'xy$, i.e.

$$x = r\cos\varphi, \qquad y = r\sin\varphi \qquad \left(z = -\left(1 - r^2\right)^{1/2}\right). \tag{5.1.4}$$

For small r, system (3) is similar to Equations (1.4.18) governing the oscillations of a "plane" oscillator.

5.1.2. STATEMENTS OF CONTROL PROBLEMS

We consider the acceleration $\ddot{\mathbf{r}}_0$ of the suspension point O' to be a control function taking values from a three-dimensional domain U:

$$\ddot{\mathbf{r}}_0 = \mathbf{u}, \qquad \mathbf{u}(t) \in U \quad (\mathbf{u} = 0 \in U). \tag{5.1.5}$$

Using the methods of Chapters 1–4, it is possible to investigate some optimal control problems for system (3) and (5), for both quasilinear and nonlinear cases. Here are basic statements which are of definite interest for applications.

1) Bringing the suspension point O' to a specified position and quenching relative oscillations

$$t = t_0: \qquad r^0, \dot{r}^0; \quad \varphi^0, \dot{\varphi}^0; \quad \mathbf{r}_0^0, \dot{\mathbf{r}}_0^0; \tag{5.1.6}$$

$$t = T: \qquad r^* = \dot{r}^* = 0; \quad \mathbf{r}_0^*, \dot{\mathbf{r}}_0^* = 0. \tag{5.1.7}$$

If for $t > T$ the control $\mathbf{u}(t) \equiv 0$, then system (3) remains at the specified position (7).

2) Bringing the system from an arbitrary initial state (6) to a state of translatory motion as a whole (without relative oscillations) with a specified constant velocity $\dot{\mathbf{r}}_0^*$, i.e.

$$t = T: \qquad r^* = \dot{r}^* = 0, \quad \dot{\mathbf{r}}_0^*. \tag{5.1.8}$$

In both cases (7) and (8), the values of the angle φ and the angular velocity $\dot{\varphi}$ are not fixed because they are not defined for $r = 0$ (and are not substantial). Moreover, under conditions (8), the terminal value $\mathbf{r}_0^*$ is not fixed.

Other statements are possible as well which take into account additional state constraints on the velocity or trajectory of the suspension point O', on the trajectory of the particle P, etc. The control problems can also be extended to a pendulum with variable length l (see Sections 5.4 and 1.4), a pendulum with a massive suspension point O' acted on by a control force, the presence of disturbances, and so on. The control process can be considered both on fixed and nonfixed time intervals for different performance criteria (the time of motion, energy consumption, and others). For example, it seems interesting, from a fundamental point of view, to consider whether it is possible to transfer a spatial pendulum in a limited time to a specified point and quench its oscillations, provided the suspension point can move only along a prescribed curve, in particular along a circle. This problem was proposed by Ishlinsky, see below.

5.1.3. CONTROL OF ENERGY OF RELATIVE OSCILLATIONS FOR A SYSTEM OF QUASILINEAR APPROXIMATION

We assume the oscillations to be small and introduce a small parameter ε, characterizing the amplitude of these oscillations and the magnitude of the control action. Different orders of these quantities in terms of ε lead, analogously to Section 1.1, to diffèrent expressions for the perturbing functions, see (1.1.20 a, b), To take nonlinear effects into account, we assume (the ways of introducing a small parameter for the problem of transferring a pendulum are discussed in details in Section 5.3)

$$r = \varepsilon^{1/2} r_*, \quad \ddot{x}_0 = \varepsilon^{3/2} u_x, \quad \ddot{y}_0 = \varepsilon^{3/2} u_y \quad \left(\ddot{z}_0 = O(\varepsilon^2)\right). \tag{5.1.9}$$

Substitution of (9) into (3) leads to a system of the type (1.4.18) ($r_* \to r$):

$$\ddot{r} - r\dot{\varphi}^2 + r = \varepsilon r^3(1/2 - \dot{\varphi}^2) + \varepsilon u_r + \varepsilon^2 + \cdots,$$
$$\ddot{\varphi} r + 2\dot{r}\dot{\varphi} = \varepsilon u_\varphi + \varepsilon^2 + \cdots, \tag{5.1.10}$$
$$u_r = -(u_x \cos\varphi + u_y \sin\varphi), \qquad u_\varphi = (u_x \sin\varphi - u_y \cos\varphi).$$

System (10) differs from (1.4.8) by the presence of the additional term $(-\varepsilon r^3 \dot{\varphi}^2)$. When $\varepsilon = 0$, system (10) is completely integrated in the form (1.4.19).

We pose the problem of optimal (in the sense of a quadratic functional of the type (1.4.23)) change of the energy E of small relative oscillations. Since $u_x^2 + u_y^2 = u_r^2 + u_\varphi^2$, the solution of this problem for system (10) and (11) has the form (1.4.24)–(1.4.26). The only difference is in the expression for the averaged phase of the first approximation, more exactly, the frequency is

$$\Omega(\tau,\varepsilon) = 1 - \varepsilon\left[\frac{3}{8}\,E(\tau) - \frac{1}{2}\frac{K^2(\tau)}{dE(\tau)}\left(1 - \frac{|K(\tau)|}{E(\tau)}\right)\right]. \tag{5.1.11}$$

Here, $E(\tau)$, $K(\tau)$, and d have the former sense and are calculated according to (1.4.20), (1.4.24).

5.1.4. TRANSFER OF A PENDULUM ALONG A CURVILINEAR TRAJECTORY

Consider the motion of system (10), having assumed that the suspension point O' moves along the prescribed curve $\mathbf{r}_0(s)$ so that conditions (9) are satisfied, Fig. 5.1. Let s be a parameter of the curve, for instance, its length counted from some point S_0, and the velocity of the point O' be controlled. Then, supposing that the ratio $\dot{s}/L\nu$ is small where L is a characteristic linear dimension and $\nu = (g/l)^{1/2}$, the set of equations of motion (10) in dimensionless Cartesian coordinates OXY takes the form (the terms $O(\varepsilon^2)$ are omitted)

$$\ddot{\mathbf{R}} + [\mathbf{R} - \mathbf{r}_0(s)] = \varepsilon \mathbf{r} r^2(1/2 - \dot{\varphi}^2) = \varepsilon \mathbf{f},$$
$$\dot{s} = v, \qquad v \in [v_1, v_2], \quad v = 0 \in [v_1, v_2]. \tag{5.1.12}$$

Here, the denotations

$$\mathbf{R} = (x, Y), \qquad \mathbf{r}_0 = (x_0, y_0), \qquad \mathbf{r} = \mathbf{R} - \mathbf{r}_0,$$
$$r^2 = x^2 + y^2, \qquad \dot{\varphi} = r^{-1}(x\dot{y} - y\dot{x}), \tag{5.1.13}$$

are introduced. When $\varepsilon = 0$, system (12) performs oscillations about a fixed equilibrium position: $\mathbf{r}_0 = \text{const}$. Using the change similar to (1.1.13) (**a**, **b** are Van der Pohl's variables)

$$\mathbf{R} = \mathbf{r}_0(s) + \mathbf{a}\sin t + \mathbf{b}\cos t, \qquad \dot{\mathbf{R}} = \mathbf{a}\cos t - \mathbf{b}\sin t, \tag{5.1.14}$$

we can reduce the controlled system (12) to the standard form (1.1.14):

$$\begin{aligned} \dot{\mathbf{a}} &= -\varepsilon\left[\mathbf{r}'_{0s}(s)v + \mathbf{f}\right]\sin t, \qquad \mathbf{a}(0) = \mathbf{a}^0, \\ \dot{\mathbf{b}} &= -\varepsilon\left[\mathbf{r}'_{0s}(s)v + \mathbf{f}\right]\cos t, \qquad \mathbf{b}(0) = \mathbf{b}^0, \\ \dot{s} &= \varepsilon v, \qquad v \in [v_1, v_2], \quad s(0) = s^0. \end{aligned} \tag{5.1.15}$$

Here, the expressions for r and φ which are calculated on the basis of the change (14):

$$\begin{aligned} x &= a_x \sin t + b_x \cos t, \qquad y = a_y \sin t + b_y \cos t, \\ \dot{x} &= a_x \cos t - b_x \sin t, \qquad \dot{y} = a_y \cos t - b_y \sin t, \end{aligned} \tag{5.1.16}$$

are substituted into the function **f**. According to (12), (13), and (16) we have a cubic estimate for **f**:

$$|\mathbf{f}| = O\left(r^3\right) = O\left(|\mathbf{a}|^3 + |\mathbf{b}|^3\right). \tag{5.1.17}$$

The following statement of the control problem is of applied interest. It is required to bring system (15) to the state

$$t = T: \qquad \mathbf{a}(T) = \mathbf{b}(T) = 0, \quad s(T) = s^* \neq s^0 \tag{5.1.18}$$

in the finite time T.

For the optimality criterion we can take an integral quadratic functional of the form B from (1.2.20) or the time of the control process. The optimality of the control v does not seem to be very essential. More important is the issue of the existence of the solution, since the controllability of the nonlinear system (15) is not evident. In particular, if $\mathbf{r}_0(s)$ is a straight line, then it is impossible to quench the oscillations in the orthogonal plane. Below, we consider the case when the suspension point O' is moving along the circle of radius $\mathbf{r}_0$. Equations of motion of the point O' are

$$\begin{aligned} x_0 &= r_0 \cos\alpha, \qquad y_0 = r_0 \sin\alpha \qquad \left(s = s^0 + r_0\left(\alpha - \alpha^0\right)\right), \\ \dot{\alpha} &= \varepsilon\omega, \qquad \omega \in [\omega_1, \omega_2], \quad \alpha(0) = \alpha^0 \ (= 0), \quad \alpha(t) = \alpha^*. \end{aligned} \tag{5.1.19}$$

Consider firstly the control problem (15)–(18) with fixed instant T of the process termination and the integral performance index

$$J[\omega] = \frac{\varepsilon}{2}\int_0^T \omega^2\,dt \longrightarrow \min_{\omega}, \qquad |\omega(t)| < \infty. \tag{5.1.20}$$

The variables **p**, **n**, and q, adjoint with **a**, **b**, and α, are introduced, and the Hamiltonian function of the problem is composed according to (1.2.3). The conditions of maximality of H with respect to ω imply the expression for the optimal control ω^*:

$$\begin{gathered}\omega^* = q + r_0(P\sin t + N\cos t),\\ P = (\mathbf{p}\cdot\mathbf{d}), \qquad N = (\mathbf{n}\cdot\mathbf{d}), \qquad \mathbf{d} = (\sin\alpha, -\cos\alpha).\end{gathered} \tag{5.1.21}$$

Substituting (21) into the Hamiltonian function H and averaging over t we obtain the Hamiltonian of the first approximation (see (1.2.2)):

$$\begin{gathered}\varepsilon^{-1}\langle H^*\rangle = \frac{1}{2}q^2 + \frac{1}{4}r_0^2(P^2+N^2) + \langle f\rangle = h = \text{const},\\ \langle f\rangle = (\mathbf{p}\cdot\mathbf{f}_s) + (\mathbf{n}\cdot\mathbf{f}_c), \qquad \mathbf{f}_s = \langle \mathbf{f}\sin t\rangle, \quad \mathbf{f}_c = \langle \mathbf{f}\cos t\rangle.\end{gathered} \tag{5.1.22}$$

Having retained the former notations for the averaged variables, we obtain the following Hamiltonian system in the slow time τ:

$$\begin{gathered}\dot{\mathbf{a}} = \frac{1}{2}r_0^2P\mathbf{d} + \mathbf{f}_s, \qquad \mathbf{a}(0) = \mathbf{a}^0, \quad \mathbf{a}(\Theta) = 0,\\ \dot{\mathbf{b}} = \frac{1}{2}r_0^2N\mathbf{d} + \mathbf{f}_c, \qquad \mathbf{b}(0) = \mathbf{b}^0, \quad \mathbf{b}(\Theta) = 0,\\ \dot\alpha = q, \qquad \alpha(0) = \alpha^0, \quad \alpha(\Theta) = \alpha^*,\\ \dot q = -\frac{\partial}{\partial a}\left[\frac{r_0^2}{4}(P^2+N^2) + \langle f\rangle\right],\\ \dot{\mathbf{p}} = -\partial\langle f\rangle/\partial\mathbf{a}, \qquad \dot{\mathbf{n}} = -\partial\langle f\rangle/\partial\mathbf{b}, \qquad (\dot{\ }) \equiv (d/d\tau).\end{gathered} \tag{5.1.23}$$

From Equations (21) we determine the variable q as a function of α and other slow variables, as well as of the parameter h:

$$q = \operatorname{sign}(\alpha^* - \alpha^0)\left[2h - \frac{1}{2}r_0^2(P^2+N^2) - \langle f\rangle\right]^{1/2}. \tag{5.1.24}$$

In what follows, we consider the variables **a**, **b**, **p**, and **n** to be functions of α. Equations for them are derived by dividing (12) by $\dot{\alpha} = q$:

$$\mathbf{a}' = \frac{1}{2} r_0^2 q^{-1} P \mathbf{d} + q^{-1} \mathbf{f}_s, \qquad \mathbf{b}' = \frac{1}{2} r_0^2 q^{-1} N \mathbf{d} + q^{-1} \mathbf{f}_c,$$
$$\mathbf{p}' = -q^{-1} \partial \langle f \rangle / \partial \mathbf{a}, \qquad \mathbf{n}' = -q^{-1} \partial \langle f \rangle / \partial \mathbf{b}, \tag{5.1.25}$$
$$\big(\mathbf{a} = \mathbf{a}(\alpha),\ \mathbf{b} = \mathbf{b}(\alpha),\ \mathbf{p} = \mathbf{p}(\alpha),\ \mathbf{n} = \mathbf{n}(\alpha),\ (') \equiv (\mathrm{d}/\mathrm{d}\alpha)\big).$$

Here, the prime denotes the derivative with respect to α. The initial and terminal conditions (15) and (18) take the form

$$\mathbf{a}(\alpha^0) = \mathbf{a}^0, \qquad \mathbf{b}(\alpha^0) = \mathbf{b}^0, \qquad \mathbf{a}(\alpha^*) = \mathbf{b}(\alpha^*) = 0. \tag{5.1.26}$$

The connection between τ and α, defining the parameter h when $\tau = \Theta$ and $\alpha = \alpha^*$, is found from (23) and (24) after having solved the problem (25) and (26), i.e.

$$\tau - \tau_0 = \int_{\alpha^0}^{\alpha} q^{-1}\big(\beta, h(\mathbf{a}^0, \mathbf{b}^0), \mathbf{a}^0, \mathbf{b}^0\big)\, \mathrm{d}\beta, \qquad \Theta = \tau\big|_{\alpha^*}. \tag{5.1.27}$$

In the general case, the solution of the boundary-value problem can be constructed numerically. However, if initial displacements are small and one can consider **f**, according to (17), to be a perturbation ($|\mathbf{f}| = O(r^3)$), then the solution is found approximately by successive approximations in terms of powers of some parameter characterizing the magnitude of the perturbation in the considered range of variables. In particular, if the perturbation is neglected ($\mathbf{f} \equiv 0$, see (1.1.20)), then **p** and **n** are constant, and integration of Equations (25) for **a** and **b** leads to elliptic integrals whose arguments and moduli depend on the unknown slow variables. They can be determined numerically from transcendental equations.

We will consider a simplified boundary-value problem (25)–(27) where $q = \text{const}$. Then, certainly, the control is not optimal. However, one can prove that the boundary-value problem has a solution. It arises from (27) that $q = (\alpha^* - \alpha^0)(\Theta - \tau_0)^{-1}$, and the solution of the unperturbed problem (when $\mathbf{f} = \alpha^0 = \tau_0 = 0$) immediately gives the expressions for the adjoint variables **p** and **n**:

$$\mathbf{p}_0 = -2r_0^{-2} q W^{-1} \mathbf{a}^0, \qquad \mathbf{n}_0 = -2r_0^{-2} q W^{-1} \mathbf{b}^0,$$
$$W = \begin{bmatrix} \alpha^*/2 - \frac{1}{4}\sin\alpha^* & -\frac{1}{2}\sin^2\alpha^* \\ -\frac{1}{2}\sin^2\alpha^* & \alpha^*/2 + \frac{1}{4}\sin 2\alpha^* \end{bmatrix}, \tag{5.1.28}$$
$$\det W = (1/4)\big(\alpha^{*2} - \sin^2\alpha^*\big) \approx \alpha^{*4}/12 \qquad \big(|\alpha^*| \ll 1\big).$$

The solution of the boundary-value problem (25) for small $|\mathbf{a}^0|$ and $|\mathbf{b}^0|$ is constructed using a perturbation technique. The adjoint variables $\mathbf{p}$, $\mathbf{n}$ are considered to be constant, and to a first approximation one takes the quantities (28). For convenience of the calculations, the change $\mathbf{f} \to \mu\mathbf{f}$ ($\mu = 1$) is made. The approximate solution is constructed using the change $\mathbf{a} = \mathbf{a}_0 + \mu\Delta\mathbf{a}$, $\mathbf{b} = \mathbf{b}_0 + \mu\Delta\mathbf{b}$, where $\mathbf{a}_0$ and $\mathbf{b}_0$ are obtained by elementary integration of (25), $\mathbf{f} = 0$; the variable q being specified and $\mathbf{p}^0$ and $\mathbf{n}^0$ being calculated according to (28). The unknowns $\mathbf{p}$ and $\mathbf{n}$ are constructed analogously: $\mathbf{p} = \mathbf{p}_0 + \mu\Delta\mathbf{p}$, $\mathbf{n} = \mathbf{n}_0 + \mu\Delta\mathbf{n}$. The constants $\Delta\mathbf{p}$ and $\Delta\mathbf{n}$ are determined from the boundary condition $\Delta\mathbf{a}(\alpha^*) = \Delta\mathbf{b}(\alpha^*) = 0$ and calculated by successive approximations in terms of powers of μ. In the first approximation with respect to μ we have

$$\Delta\mathbf{p} = -\frac{2}{r_0^2}W^{-1}\int_0^{\alpha^*} \mathbf{f}_s^0\, d\alpha, \qquad \Delta\mathbf{n} = -\frac{2}{r_0^2}W^{-1}\int_0^{\alpha^*} \mathbf{f}_c^0\, d\alpha.$$

Here, $\mathbf{f}^0$ is the function $\mathbf{f}$, where $\mathbf{a} = \mathbf{a}_0(\alpha)$ and $\mathbf{b} = \mathbf{b}_0(\alpha)$ have been substituted. Thus, the solution of the boundary-value problem (23) and, with this, the solution of the control problem (15)–(19) exist when the initial oscillations are sufficiently small. The control ω^* (21) is a function periodic in t, with slowly changing phase addition and amplitude.

The time-optimal control of system (15) and (18) under the constraint (19) in the first approximation is given by the expressions for ω^* and $\langle H^*\rangle$ obtained on the basis of the technique of Section 1.3:

$$\begin{aligned}
\omega^* &= \frac{1}{2}(\omega_1 + \omega_2) + \frac{1}{2}(\omega_2 - \omega_1)\,\mathrm{sign}(q/r_0 + P\sin t + N\cos t),\\
\langle H^*\rangle &= \frac{1}{2}\,q(\omega_1 + \omega_2) + \frac{1}{2}r_0(\omega_2 - \omega_1)\left(P^2 + N^2\right)^{1/2} g(k) + \langle f\rangle,\\
g(k) &= \begin{cases} k, & |k| > 1 \quad \left(k \equiv (q/r_0)\left(P^2 + N^2\right)^{-1/2}\right),\\ (2/\pi)\left[\left(1 - k^2\right)^{1/2} + k\arcsin k\right], & |k| < 1. \end{cases}
\end{aligned} \qquad (5.1.29)$$

The averaged boundary-value problem of the first approximation is obtained on the basis of $\langle H^*\rangle$ (29) (see Section 1.3, 2.1, and 2.2), the function $\langle H^*\rangle$ being constant: $\langle H^*\rangle = h = \text{const}$. It follows from (29) that the control process can contain parts of two types: 1) the motion of the suspension point O' with extremal velocity $\omega_{1,2}$, which corresponds to $|k| \geqslant 1$ in (29); 2) alternating motion, corresponding to $|k| < 1$. In the first case, active quenching of the oscillations is not performed. In the second case, the control is such that the process of quenching the oscillations accompanies the travel of the suspension point towards the desired position. This situation is to a large extent similar to that investigated below, in Sections 5.3 and 5.4, for a "plane" pendulum. Considering the function f as a perturbation when initial displacements are small, one can construct the solution of the boundary-value problem in the first approximation using numerical methods.

5.2. The Control of Oscillations and Rotations of a Plane Pendulum with Regulated Acceleration of the Suspension Point

5.2.1. STATEMENT OF THE CONTROL PROBLEM

Consider oscillations and rotations of a physical pendulum in the vertical plane OXY (see Fig. 5.2). The equation of motion has the form

$$A\ddot{\varphi} + mgl\sin\varphi = -ml\big(\ddot{x}_0\cos\varphi + \ddot{y}_0\sin\varphi\big),$$
$$\varphi(0) = \varphi^0, \qquad \dot{\varphi}(0) = \dot{\varphi}^0. \tag{5.2.1}$$

Here, A is a moment inertia with respect to the axis O', m is a mass, l is an "arm"; and x_0, y_0 are displacements of the point O'. System (1) is reduced to the dimensionless form ($t_* = \nu t$, $\nu^2 = mgl/A$). The acceleration of the point O' is assumed to be small, i.e. $\varepsilon\mathbf{u} = \ddot{\mathbf{r}}_0/g$, $\varepsilon \ll 1$, $u \sim 1$. Using new variables we obtain the equation ($t_* \to t$):

$$\ddot{\varphi} + \sin\varphi = -\varepsilon(u_x\cos\varphi + u_y\sin\varphi). \tag{5.2.2}$$

On the basis of the integrals, the total energy E and the phase ψ, known for $\varepsilon = 0$

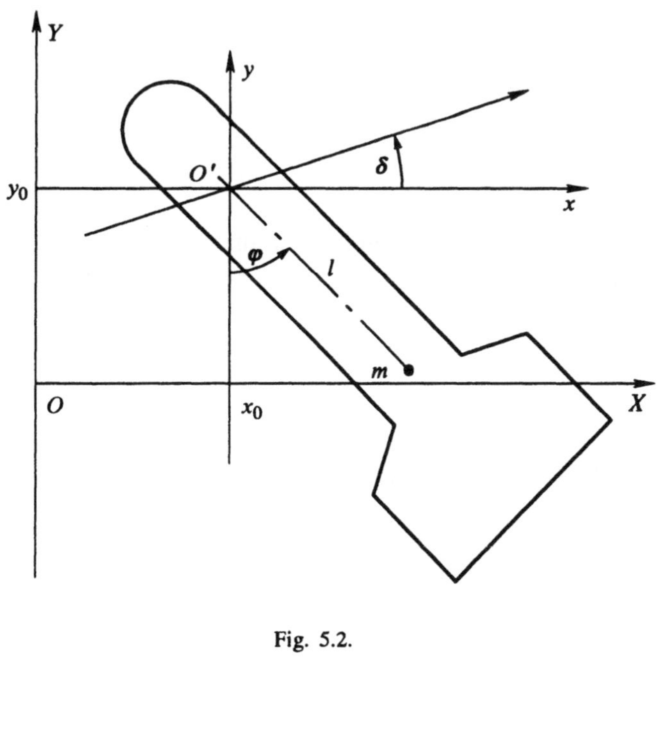

Fig. 5.2.

(see Section 3.1), Equation (2) is reduced to the standard form with rotating phase (3.1.13)

$$
\begin{aligned}
&\dot{E} = -\varepsilon(u_x \cos\varphi + u_y \sin\varphi)\dot{\varphi}, \qquad E(0) = E^0,\\
&\dot{\psi} = \omega(E) + E\int_{\varphi^0}^{\varphi}\left[\frac{\omega'(E)}{\dot{\varphi}(y,E)} - \frac{\omega(E)}{\dot{\varphi}^2(y,E)}\frac{\partial\dot{\varphi}}{\partial E}\right]dy, \qquad \psi(0) = \psi^0,\\
&\dot{\varphi}(\varphi, E) = \pm\sqrt{2}[E - (1-\cos\varphi)]^{1/2}, \qquad E = \frac{1}{2}\dot{\varphi}^2 + (1-\cos\varphi),\\
&\psi = \omega(E)\int_{\varphi^0}^{\varphi}\frac{dy}{\dot{\varphi}(y,E)} + \psi^0, \qquad T_0(E) = \oint\frac{d\varphi}{\dot{\varphi}(\varphi,E)}, \quad \omega = \frac{2\pi}{T_0}.
\end{aligned}
\tag{5.2.3}
$$

We consider two statements of the optimal control problem. The first one is the minimization or maximization of the total energy of oscillations and rotations E over a fixed time interval $t \in [0, T]$:

$$
J_E[u] = \pm E(T) \longrightarrow \min_u, \qquad u \in U, \quad T = \Theta\varepsilon^{-1}. \tag{5.2.4}
$$

The second problem is the minimization of the process time, given the terminal value of energy E_*:

$$
\begin{aligned}
&M(E)\big|_T = E(T) - E^* = 0, \qquad E^* \geqslant 0;\\
&J_T[u] = \varepsilon T \longrightarrow \min_u, \qquad u \in U.
\end{aligned}
\tag{5.2.5}
$$

These problems appear to be dual if the quantity T in the statement (4) is chosen in a reasonable way. Using the averaging method developed in Sections 3.2 and 3.3, below we analyse, optimal control problems (3)–(5) in the first approximation with respect to ε under different assumptions concerning the admissible set U for values of the vector function u.

5.2.2. THE TRAVEL OF THE SUSPENSION POINT ALONG THE STRAIGHT LINE

If the point O' can travel only along the straight guide inclined at a constant angle δ with respect to the horizon, then U is a segment (see Fig. 1.8)

$$
U = \{u\colon\ u_x = u\cos\delta,\ u_y = u\sin\delta,\ |u| \leqslant 1,\ \delta = \text{const}\}. \tag{5.2.6}
$$

After substitution of (6) into (3), the optimal control u^* is determined from the maximality conditions of the Hamiltonian function, provided $p(t) \not\equiv 0$:

$$u^* = -\operatorname{sign}\left[p\dot{\varphi}\cos(\varphi - \delta)\right]. \tag{5.2.7}$$

Here, p is the variable adjoint with E. The variable adjoint with the phase ψ has the order ε and is not taken into account when considering the first approximation. The averaged equation for E (3) has the form

$$\begin{aligned} &\dot{\xi} = I(\xi, \delta)v, \qquad v = \operatorname{sign}\eta, \quad |v(\tau)| \leqslant 1, \quad \xi(0) = E^0, \\ &\tau = \varepsilon t \in [0, \Theta], \quad (\dot{\ }) = (\mathrm{d}/\mathrm{d}\tau), \quad I = \left\langle\left|\dot{\varphi}\cos(\varphi - \delta)\right|\right\rangle \geqslant 0. \end{aligned} \tag{5.2.8}$$

Here, ξ and η are the averaged values of E and p. The function $v(\tau)$ is easily determined on the basis of the analysis of Equation (8) and the corresponding equation for η, where one assumes $\beta = 0$ (see Section 3.2). For problem (4) ($T = \Theta\varepsilon^{-1}$ is fixed), we have

$$\int_{E_*}^{E_0} I^{-1}(\xi, \delta)\,\mathrm{d}\xi = \int_0^{\Theta} v(\tau)\,\mathrm{d}\tau, \qquad |v| \leqslant 1, \quad E_* = \xi(\Theta). \tag{5.2.9}$$

Here, $v = \operatorname{sign}\eta = 1$ for the problem of maximization of energy ($-E(T) \to \min$). In the case of minimization of $E(\tau)$, the function $v(\tau)$, $|v| \leqslant 1$, is determined, according to (9), in a more complicated way, since in this statement singular controls are possible. If $\Theta \leqslant \vartheta(E^0, 0, \delta)$, where $\vartheta(E_*, E^0, \delta)$ is the left-hand side of relationship (9), then $v(\tau) = -1$, $\tau \in [0, \Theta]$. If $\Theta > \vartheta(E^0, 0, \delta)$, then $v(\tau)$ can assume arbitrary values such that the integral should be equal to $\vartheta(E^0, 0, \delta)$. To construct the solution of the control problem in the feedback form, the relationships defining the current value of v, $|v| \leqslant 1$, are to be considered for $E^0 = E$, $\tau_0 = \tau$, i.e.

$$\int_E^{E_*} I^{-1}(\xi, \rho)\,\mathrm{d}\xi = \int_\tau^{\Theta} v(\sigma)\,\mathrm{d}\sigma. \tag{5.2.10}$$

One should bear in mind a singularity of the feedback control as $\tau \to \Theta$.

For the optimal control problem (5), the function $v(\tau)$, $|v| \leqslant 1$, as follows from Section 3.3, is given by

$$v = \operatorname{sign}\left(E^* - E\right), \quad u_s^* = \operatorname{sign}\left[\left(E^* - E\right)\dot{\varphi}\cos(\varphi - \delta)\right]. \tag{5.2.11}$$

If $v(\tau)$ is determined uniquely, then the approximate optimal value

$$E = \xi(\tau) + \varepsilon + \cdots$$

is found by the quadrature

$$\int_{E^0}^{\xi} I^{-1}(\zeta,\delta)\,d\zeta = \tau \operatorname{sign}(E^* - E^0). \tag{5.2.12}$$

For the rotatory mode ($\xi > 2$), calculation of the function $I(\xi,\delta)$ from (8) yields

$$\begin{aligned} I(\xi,\delta) = I_2(\xi,\delta) &= \frac{1}{2\pi}\int_0^{2\pi} \left|\dot\varphi\cos(\varphi-\delta)\right| d\psi \\ &= \frac{1}{T_2(\xi)}\int_0^{2\pi}\left|\cos(\varphi-\delta)\right| d\varphi = \frac{2}{k_2\mathbf{K}(k_2)}, \\ k_2^2 = \frac{2}{\xi} < 1, &\qquad T_2 = 2k_2\mathbf{K}(k_2). \end{aligned} \tag{5.2.13}$$

Here, $\mathbf{K}$ is the full elliptic integral of the first kind with the modulus k_2; the index 2 refers to rotations; T_2 is a period of rotation. The function I_2 (13) does not depend on δ.

In the case of oscillations, when $O \leqslant \xi < 2$, the function I (8) is given by

$$\begin{aligned} I(\xi,\delta) = I_1(\xi,\delta) &= \frac{2}{T_1(\xi)}\int_{-\varphi_0}^{\varphi_0}\left|\cos(\varphi-\delta)\right| d\varphi \\ &= \frac{2}{T_1(\xi)}\int_{-\varphi_0-\delta}^{\varphi_0-\delta}\left|\cos x\right| dx, \\ \varphi_0(\xi) = \cos^{-1}(1-\xi), &\quad k_1^2 = \xi/2 < 1, \quad T_1(\xi) = 4\mathbf{K}(k_1). \end{aligned} \tag{5.2.14}$$

Here, φ_0 is the amplitude of oscillations, k_1 is the modulus, and T_1 is the period of oscillations; the index 1 refers to oscillations. The quadrature in (14) is calculated in terms of elementary functions of the arguments $(\varphi_0 - \delta)$ and $-(\varphi_0 + \delta)$. However, the resultant expression for I_1 is rather lengthy. Expanding the function $|\cos x|$ into a Fourier series we obtain the following expression for I_1:

$$I_1(\xi,\delta) = \frac{8}{\pi T_1}\left[\varphi_0 + \sum_{j=1}^{\infty}\frac{(-1)^j}{j(4j^2-1)}\sin(2j\varphi_0)\cos(2j\delta)\right]. \tag{5.2.15}$$

The series in (15) converges rather quickly: the difference of the sum of the first j_* terms from the exact value has the order $(1/8)(j_* + 1)^{-2}$.

If the motion occurs in a neighbourhood of the separatrix $k_1 \uparrow 1$ or $k_2 \uparrow 1$, which takes place when $\xi \uparrow 2$ or $\xi \downarrow 2$ respectively, then $T_{1,2}(\xi) \to \infty$ $(\omega \downarrow 0)$ and the phase ψ ceases to be rotating (fast variable). This circumstance breaks the applicability conditions for the averaging method. As has been established in [124], application of the averaging method in the case of passing a separatrix leads to somewhat greater error (of the order of $\varepsilon \ln \varepsilon^{-1}$) on an interval $t - t_0 \sim \varepsilon^{-1}$.

It is interesting to analyse the dependence of the solution $E \approx \xi(\tau, E^0, \delta)$ on the parameter δ, $\delta \in [0, \pi/2]$. In the rotatory mode ($\xi > 2$), the controlled motion does not depend on $\delta = \text{const}$, as was established in (13).

For the oscillatory mode ($0 < \xi < 2$), differentiating the dependence I_1 (14) with respect to δ, which is admissible according to (15), one can obtain

$$\frac{\partial I_1}{\partial \delta} = \frac{2}{T_1}\left[\left|\cos(\varphi_0 + \delta)\right| - \left|\cos(\varphi_0 - \delta)\right|\right]. \tag{5.2.16}$$

It follows from (16) that if $\varphi_0 = 0,\ \pi/2,\ \pi$, i.e. $\xi = 0, 1, 2$, then $\partial I_1/\partial\delta \equiv 0$, $\delta \in [0, \pi/2]$. In other cases, $\partial I_1/\partial\delta$ vanishes only if $\delta = 0$ or $\delta = \pi/2$. Hence, I_1 monotonically depends on δ and reaches the minimum over δ at one endpoint of the interval $\delta \in [0, \pi/2]$ and the maximum at the other endpoint. Calculation of I_1 for $\delta = 0$ and $\delta = \pi/2$ on the basis of formula (14) leads to the expressions

$$\begin{aligned} I_1(\xi, 0) &= \frac{4}{T_1(\xi)}\begin{cases} \sin\varphi_0(\xi), & 0 \leqslant \xi \leqslant 0 \quad (0 \leqslant \varphi_0 \leqslant \pi/2), \\ 2 - \sin\varphi_0(\xi), & 1 < \xi \leqslant 2 \quad (\pi/2 > \varphi_0 \leqslant \pi), \end{cases} \\ I_1(\xi, \pi/2) &= (4/T_1(\xi))\left[1 - \cos\varphi_0(\xi)\right], \quad 0 \leqslant \xi \leqslant 2 \quad (0 \leqslant \varphi_0 \leqslant \pi). \end{aligned} \tag{5.2.17}$$

Having compared formulae (17), we establish that the maximum of $I_1(\xi, \delta)$ over δ is reached at $\delta = 0$ if the amplitude of the oscillations is "small": $\varphi_0(\xi) < \pi/2$ $(0 < \xi < 1)$, and at $\delta = \pi/2$ if the amplitude is "large": $\pi/2 < \varphi_0(\xi) < \pi$ $(1 < \xi < 2)$. In the rotatory mode, when $\xi > 2$, as well as when $\varphi_0 = 0$ $(\xi = 0)$, $\varphi_0 = \pi/2$ $(\xi = 1)$ and $\varphi_0 = \pi$ $(\xi = 2)$, the function I_1 does not depend on δ.

Thus, the rate of changing ξ depends on the constant angle δ of the guide inclination as follows. For oscillations of "small" amplitude ($\varphi_0 < \pi/2$), the most efficient is a horizontal travel of the suspension point O', i.e. $\delta = 0$. For oscillations of "large" amplitude, $\pi/2 < \varphi_0 < \pi$, vertical motions ($\delta = \pi/2$) produce the most effect. In the case of rotations ($\xi > 2$) all values $\delta \in [0, \pi/2]$ are equivalent. Thus, if E^0, $E^* < 1$, then the most preferable is the value $\delta = 0$. Vice versa, if $2 > E^0$, $E^* > 1$, then $\delta = \pi/2$. Figure 5.3 shows the dependences $\xi(\tau, E^0, \delta)$ for $E^0 = 0$ and the different values $\delta_i = (i/12)\pi$, $i = 0, 1, \ldots, 6$. The curve for $i = 1$ is not drawn, since

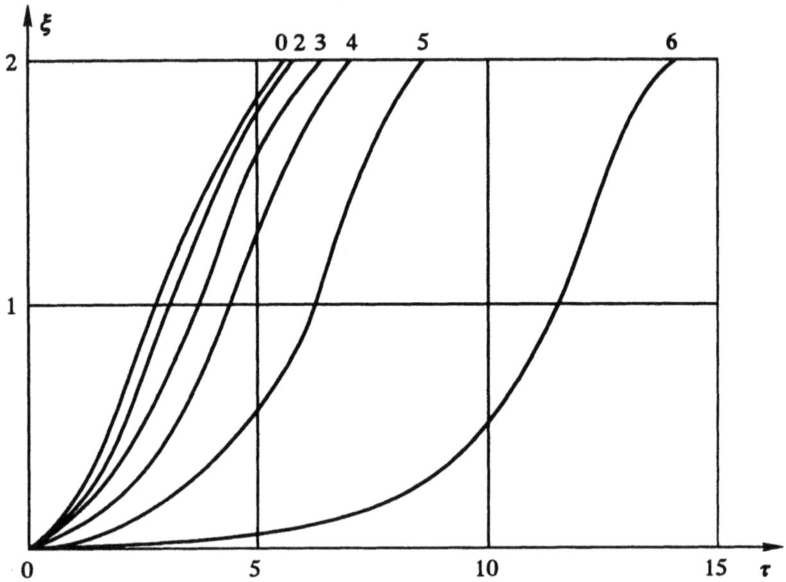

Fig. 5.3.

it virtually coincides with the curve for $i = 0$. These graphs confirm the analysis of the influence of δ on the control, which has been carried out.

It should be noted that for the lower equilibrium position ($\xi = 0$, $\varphi_0 = 0$) the function $I_1(0, \delta) \equiv 0$ (see (14) or (15)), i.e. $\xi = 0$ is an equilibrium point of Equation (8). However, for $0 \leqslant \delta < \pi/2$, there exists an unique solution, not being constant, which hits this point or leaves it in a finite time, since for small φ_0 (and ξ) the asymptotic behaviour of the right-hand side of (8) has the form $I_1 \sim \varphi_0 \sim \xi^{1/2}$. But if $\delta = \pi/2$, the asymptotic behaviour will be different: $I_1 \sim \varphi_0^2 \sim \xi$. The time of travel to the point $\xi = 0$ ($\varphi_0 = 0$) or leaving it will be infinite for $\delta = \pi/2$. This follows from physical considerations. Therefore, it is assumed $\xi(0) = 0.001$ for the curve corresponding to $i = 6$ ($\delta = \pi/2$) in Fig. 5.3.

Using the family of curves given in Fig. 5.3, one can determine the minimal time $\Theta(\delta, E^0, E^*)$ for the motion from the point $\xi = E^0$ to the point $\xi = E^*$ as a function of δ. The curves corresponding to different E^0 and E^* are given in Fig. 5.4; the curve 1 corresponds to $E^0 = 0$, $E^* = 2$, i.e. to the transfer of the pendulum from the lower equilibrium position to the upper one ($\varphi^* = \pi$); the curve 2 is presented for $E^0 = 0$, $E^* = 1$; the curve 3 corresponds to $E^0 = 1$, $E^* = 2$. It is evident that the dependence 1 can be obtained by adding the dependences 2 and 3. The curve 3 weakly monotonically decreases as $\delta \to \pi/2$, which is in agreement with the analysis carried out.

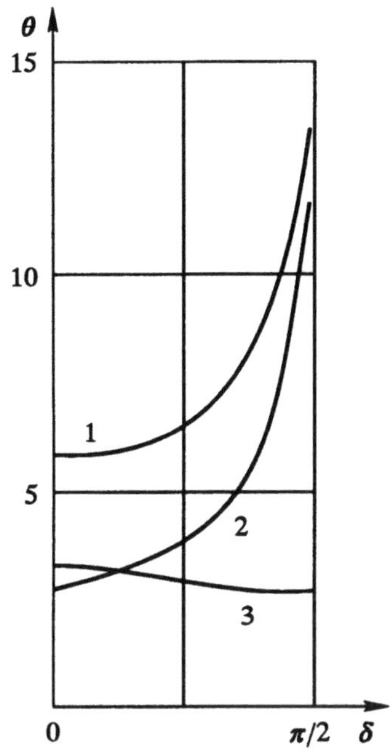

Fig. 5.4.

In the case of rotations, integration of Equations (8) and (13) with $v = \operatorname{sign}\eta = +1$ and $E^0 = 2$ leads to a dependence close to the quadratic parabola $\xi(\tau) \approx 2[1 + (\tau - \tau_*)/\pi]^2$, see Fig. 5.5. Let $\tau_*(\delta)$ be the time of reaching the value $\xi(\tau_*, \delta) = 2$ by the curve $\xi(\tau, \delta)$, see Fig. 5.3. Then, based on the analysis, one can construct the complete family of trajectories for both statements (4) and (5) of the control problem for oscillations and rotations of system (3). Indeed, by choosing the curve from Fig. 5.3 or Fig. 5.5 corresponding to a given $\delta \in [0, \pi/2]$ one can find $\tau = \tau_0$ for which $\xi = E^0$. For the statement (4), one has to measure the segment $\tau - \tau_0 = \Theta$ from the point $\tau = \tau_0$ along the abscissa axis τ in the direction opposite to the sign in the functional (4). For the time-optimal problem (5), the endpoint of the optimal trajectory is determined by the value Θ from the equation

$$\xi\left(\tau_0 + \Theta \operatorname{sign}\left(E^* - E^0\right)\right) = E^*.$$

Thus, the solution of the problem of optimal control of the energy of the "plane" pendulum in oscillatory and rotatory modes has been constructed. During the control process one can change the inclination angle δ, for instance, when passing the value $\varphi_0(\xi) = \pi/2$, without loss of accuracy of the solution. The cases when δ can be regarded as a control are analysed below.

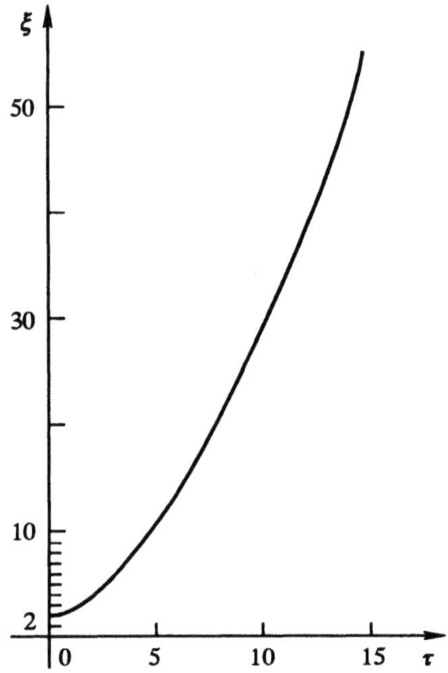

Fig. 5.5.

5.2.3. GENERALIZATION OF THE CONTROL PROBLEM

Consider the statements (4) and (5) for wider admissible ranges for the vector u of the suspension point acceleration.

(1) *The constraint in the form of a rectangle.* Let the domain U have the form of a rectangle (see Fig. 1.8):

$$U = \left\{u\colon\ |u_{x,y}| \leqslant u^0_{x,y};\ u^0_{x,y} = \text{const} > 0\right\}. \tag{5.2.18}$$

For these constraints, analogously to Section 5.2.2, we find the approximate optimal control u^*:

$$\begin{gathered} u^*_{x,y} = -u^0_{x,y}\, v \operatorname{sign}\left(\dot{\varphi} n_{x,y}\right), \\ n_x = \cos\varphi, \qquad n_y = \sin\varphi, \qquad |v(\tau)| \leqslant 1. \end{gathered} \tag{5.2.19}$$

Substituting (19) into Equation (3) for E and averaging over ψ one can obtain the equation to be found. For the rotatory mode, i.e. for $\xi > 2$, we have

$$\dot{\xi} = \frac{4}{T_2(\xi)}(u_x^0 + u_y^0)v = 2\frac{u_x^0 + u_y^0}{k_2\mathbf{K}(k_2)}v. \tag{5.2.20}$$

For oscillations ($0 \leqslant \xi < 2$)

$$\dot{\xi} = \left[u_x^0 I_1(\xi, 0) + u_y^0 I_1(\xi, \pi/2)\right]v, \qquad |v(\tau)| \leqslant 1. \tag{5.2.21}$$

The right-hand sides of Equations (20) and (21) are defined according to Section 5.2.2. The function $v(\tau)$, $|v| \leqslant 1$, is defined analogously to (9) and (10). It follows from (20) and (21) that when $u_x^0 = u_y^0 = 1$, the control (19) is more efficient, since for the same ξ, absolute values of the right-hand sides of these equations are greater than those of Equations (8). This is natural, since the constraint (19) is "wider" than (6). Integration and analysis of Equations (20) and (21) are conducted analogously to Section 5.2.2. Having divided Equations (20) and (21) by the quantity $u^0 = u_x^0 + u_y^0$ and introduced the angular variable γ using the relationships $\cos^2\gamma = u_x^0/u_0$ and $\sin^2\gamma = u_y^0/u_0$ for $\gamma \in [0, \pi/2]$, we obtain families of curves $\xi(\theta, E^0, \gamma)$, $\theta = u^0\tau$, similar to those presented in Figs. 5.3, 5.4, and 5.5.

(2) *Constraints in the form of an ellipse*. In the case when the domain U is bounded by an ellipse (see Fig. 1.8) we have

$$U = \left\{u\colon\ a^2u_x^2 + b^2u_y^2 \leqslant 1;\ a, b = \text{const} > 0\right\}. \tag{5.2.22}$$

Analogously to Section 5.2.2, we find the approximate optimal control u^*:

$$u_{x,y}^* = -n_{x,y}\left(a^2n_x^2 + b^2n_y^2\right)^{-1/2}v, \qquad |v(\tau)| \leqslant 1. \tag{5.2.23}$$

The averaged energy $\xi(\tau)$ changes according to the equation

$$\dot{\xi} = L(\xi, a, b)v, \qquad L \equiv \left\langle|\dot{\varphi}|\left(a^2n_x^2 + b^2n_y^2\right)^{-1/2}\right\rangle. \tag{5.2.24}$$

For the rotatory mode ($\xi > 2$), the function L has the form

$$L = L_2(\xi, a, b) = \frac{4\mathbf{K}(\varkappa(a, b))}{aT_2(\xi)}, \qquad \varkappa(a, b) = \left(1 - \frac{b^2}{a^2}\right)^{1/2}, \quad a \geqslant b. \tag{5.2.25}$$

If $a \leqslant b$, then one has to permute a and b in expression (25). For the oscillatory mode, the function L in (24) is expressed in terms of complete and incomplete elliptic integrals of the first kind. In the case of "small" amplitudes of oscillations $0 < \varphi_0(\xi) \leqslant \pi/2$

$$L = L_1(\xi, a, b) = \begin{cases} (4/aT_1)F(\varphi_0, \varkappa(a,b)), & a \geqslant b, \\ (4/bT_1)\left[\mathbf{K}(\varkappa(b,a)) - F(\pi/2 - \varphi_0, \varkappa(b,a))\right], & a \leqslant b. \end{cases} \tag{5.2.26}$$

For "large" amplitudes $\pi/2 \leqslant \varphi_0(\xi) < \pi$, when $1 \leqslant \xi < 2$,

$$L_1(\xi, a, b) = \begin{cases} (4/aT_1)\left[2\mathbf{K}(\varkappa(a,b)) - F(\pi/2 - \varphi_0, \varkappa(a,b))\right], & a \leqslant b, \\ (4/bT_1)\left[\mathbf{K}(\varkappa(b,a)) + F(\varphi_0 - \pi/2, \varkappa(b,a))\right], & a \leqslant b. \end{cases} \tag{5.2.27}$$

For $a = b$, expressions (26) and (27) are simplified:

$$L_1(\xi, a, a) = 4a^{-1}\varphi_0(\xi)T_1^{-1}(\xi), \qquad 0 \leqslant \varphi_0 \leqslant \pi. \tag{5.2.28}$$

Determination of the function $v(\tau)$, $|v| \leqslant 1$, and construction and analysis of the phase trajectory $\xi(\tau, E^0)$ are carried out following the scheme of Section 5.2.2.

5.3. Oscillatory Systems with Velocity-Controlled Equilibrium Positions

Here, we investigate time-optimal control problems for oscillatory systems governed by equations of the type (3.1.15) which are a special case of Equations (3.1.11).

5.3.1. SOME MECHANICAL MODELS

Consider, as an illustration, some models of mechanical systems and formulate the conditions which enable reducing the equations of motion to the form (3.1.15).

(1) *A pendulum whose suspension point travels along the horizontal line.* The motion of a pendulum of constant length l with controlled velocity v of travel of the suspension point O' is governed by the equations (see Fig. 5.6)

$$l\ddot{\varphi} + g\sin\varphi = -\ddot{x}\cos\varphi, \qquad \dot{x} = v, \qquad v \in [v_1, v_2]. \tag{5.3.1}$$

The constraint force is supposed to be positive and the velocity v can quickly (virtually instantaneously) change within prescribed limits. If the suspension is implemented by means of an inextensible perfectly flexible thread, then an additional

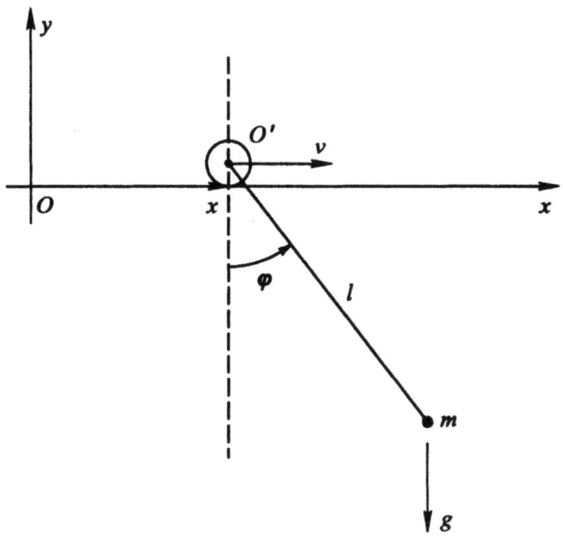

Fig. 5.6.

analysis of the motion is required in connection with the possibility of "leaving the constraint" [46]. Assuming $\varphi \sim \varepsilon \ll 1$, where $\varepsilon = |v|_{\max}(\nu l)^{-1}$, $\nu = (g/l)^{1/2}$, one can ignore this effect.

We introduce in system (1) a small parameter ε and dimensionless variables $t_* = \nu t$, $x_* = x/L$ and $y_* = (x + l\varphi)/L$ where L is a characteristic length. Then nonlinear terms are dropped in the expansions $\sin\varphi = \varphi + O(\varphi^3)$ and $\cos\varphi = 1 + O(\varphi^2)$. This is admissible, since the integral of $\ddot{x}\varphi^2$, where $\ddot{x}$ can be a generalized impulse function of the intensity ε, will lead, in general, to the absolute error $O(\varepsilon^2)$ on a time interval $t - t_0 \sim \varepsilon^{-1}$; the relative error will be $O(\varepsilon)$. Omitting the index in the dimensionless variables one can arrive at the equations of motion of the form (3.1.15) (to the system with "controlled equilibrium position"):

$$\ddot{y} + (y - x) = 0, \qquad \dot{x} = \varepsilon u, \qquad u \in [u_1, u_2]. \tag{5.3.2}$$

The approach expounded above has an essential drawback: it enables one to treat the problem of transferring the pendulum from the state of rest to an arbitrary point $x(T) = x^*$, $|x^* - x| < \infty$, on a sufficiently large time interval $T - t_0 < \infty$ ($T = t_0 + (x^* - x^0)(\varepsilon\bar{u})^{-1}$, $\bar{u} \in [u_1, u_2]$; for instance, $\bar{u} = u_1$ or $\bar{u} = u_2$), but the amplitude of relative oscillations will be of the order of ε. More accurate bringing the pendulum to the state of rest, for instance with the error $O(\varepsilon^2)$, requires the solution of the synthesis problem (the problem of constructing the feedback control) for the system (1) linearized with respect to φ. To obtain the feedback control and to implement it in practice is rather difficult [1, 46, 156]. It appears to be useful for applications if one supplements the aforenamed simple mode of transferring the suspension point to the given position x^* with a correction mode, i.e. with a control having a higher order in terms of ε, for example, $\dot{x} = \varepsilon^2 u$ or $\dot{x} = \varepsilon^{3/2} u$, $u \sim 1$. Then in the first case, the state point is brought into the ε^2-vicinity of the prescribed position in a time $t - t_0 \sim \varepsilon^{-1}$; in the second case, it is brought, in a shorter time $t - t_0 \sim \varepsilon^{-1/2}$, into an $\varepsilon^{3/2}$-vicinity. The idea of combining the modes of transfer and correction is rather fruitful and is often used in practice (when operating loading machines, manipulation robots, etc). In what follows, we mostly investigate the correction problem for a "pendulum-type" system using asymptotic methods. At the same time, we also consider more general models with variable length and controlled length.

It is worth noting that the set of equations in the canonical "angle-action" variables is more convenient for analysis. These equations do not include impulse functions, and they make it possible to derive easily the set of equations of the type (2) governing the correction mode and to use the averaging technique expounded above. This approach is connected with the introduction of Routh's variables [1, 127, 138] which are more natural and convenient for analysis. These variables are widely used when investigating different problems of dynamics of a rigid body, gyroscopes, vibration technological machines, and others [1, 84, 91, 97].

(2) *Derivation of equations of motion and statements of control problems.* In the problem being considered we are to proceed from the equation for the angular variable in the Lagrangian form with the Lagrangian function $\mathcal{L}$ to the equations in the Hamiltonian form with the Hamiltonian function $\mathcal{H}$ by introducing the momentum (the action) p, i.e.

$$\begin{gathered}
\mathcal{L} = \mathcal{K} - \mathcal{U}, \qquad \mathcal{H} = p\widehat{\varphi} - \mathcal{L}, \qquad \mathcal{U} = mgl(1 - \cos\varphi), \\
\mathcal{K} = \frac{1}{2} m\left(\dot{x}^2 + \dot{l}^2 + l^2\dot{\varphi}^2 + 2l\dot{x}\dot{\varphi}\cos\varphi + 2\dot{l}\dot{x}\sin\varphi\right), \\
p = \frac{\partial \mathcal{L}}{\partial \dot{\varphi}} = ml\left(l\dot{\varphi} + \dot{x}\cos\varphi\right), \qquad \widehat{\varphi} = \frac{p}{ml^2} - \frac{\dot{x}}{l}\cos\varphi, \\
\mathcal{H} = \frac{p^2}{2ml^2} - p\frac{\dot{x}}{l}\cos\varphi - \frac{m}{2}\dot{x}^2\sin^2\varphi - m\dot{x}\dot{l}\sin\varphi + mgl(1 - \cos\varphi).
\end{gathered} \tag{5.3.3}$$

(The variables $x = x(t)$ and $l = l(t)$ are specified kinematically.)

Using the expressions (3) for the Hamiltonian function $\mathcal{H}$ the equations of pendulum motion (1) are reduced to the form which does not include impulse functions $\ddot{x}$

(provided $\dot{x}$ is a piecewise smooth function of t)

$$\begin{aligned} \dot{\varphi} &= \frac{\partial \mathcal{H}}{\partial p} = \frac{p}{ml^2} - \frac{v}{l}\cos\varphi, \qquad \dot{x} = v, \qquad v \in [v_1, v_2], \\ \dot{p} &= -\frac{\partial \mathcal{H}}{\partial \varphi} = -p\frac{v}{l}\sin\varphi + mv^2\sin\varphi\cos\varphi + m\dot{l}v\cos\varphi - mgl\sin\varphi. \end{aligned} \tag{5.3.4}$$

Henceforth throughout this section, we consider only a pendulum of constant length ($l = \text{const}$).

In the dimensionless variables (see above), Equations (4) are reduced to the form with $m = l = g = 1$; $t_* = \nu t$, $\nu^2 = g/l$, $v_* = v/(\nu l)$, $x_* = x/l$, and $p_* = p/(m\nu l)$. The set of equations (4) is suitable for the asymptotic analysis of the optimal control problem, both in the nonlinear and quasilinear case, under different assumptions concerning orders of characteristic quantities in terms of a small parameter. For $v = \text{const}$, the system for φ and p has a rotatory-oscillatory solution expressed in terms of elliptic functions. In particular it admits the solution $\varphi = 0$, $p = v = \text{const}$ which corresponds to the motion of the pendulum with constant velocity without relative oscillations ($\varphi = \dot{\varphi} = 0$). It follows from (4) that if $\varphi(T) = p(T) = 0$ and $x(T) = x^*$, then, provided $v \equiv 0$ for $t > T$, the system remains at the state $\varphi = \dot{\varphi}(t) = p(t) \equiv 0$ and $x(t) = x^*$.

In what follows, we assume $v = \varepsilon^d u$, $u \sim 1$, $d > 0$, and consider the following statements of time-optimal control problems for the motion of system (4) in the correction mode.

1) Nonlinear rotatory-oscillatory system; the change of either the total energy or the energy and the position (E, or E and x) is given.

$$\begin{aligned} &E = \frac{1}{2}p^2 + (1 - \cos\varphi), \qquad E = O(1), \qquad \dot{E} = O(\varepsilon), \\ &\varphi = O(\varepsilon^{-1}) \ (\varphi = O(1)), \qquad p, \dot{\varphi} = O(1), \\ &x^* - x^0 = O(1), \\ &\dot{x} = \varepsilon u \ (d = 1, u \sim 1), \qquad t = t_0 = O(\varepsilon^{-1}). \end{aligned} \tag{5.3.5}$$

As has been noted, during the mode of transfer the state point of system (4) is brought to the state (5) by using the control $u = \text{const}$, $u \in [u_1, u_2]$. At the end of the process the state point finds itself in an ε-neighbourhood of the terminal manifold.

2) Quasilinear oscillatory system; the change of either the amplitude or the amplitude and the position is given. It is assumed that, having used the small control $v = \varepsilon u$, system (4) has been brought into an ε-neighbourhood of the specified point

$\varphi, \dot{\varphi}, p = O(\varepsilon)$, $x = x^* = O(\varepsilon)$. Then by the changes $\varphi = \varepsilon\Phi$, $p = \varepsilon P$, $x = \varepsilon X$, and $v = \varepsilon^2 u$, where $\Phi, P, X, u = O(1)$, we obtain the control problem which can be reduced to the form (2) or (3.1.15) ($m = g = l = 1$):

$$\begin{aligned} &\dot{\Phi} = P - \varepsilon u + O(\varepsilon^2), \qquad \Phi(0) = \Phi^0, \quad \Phi(T) = 0, \\ &\dot{P} = -\Phi + O(\varepsilon^2), \qquad P(0) = P^0, \quad P(T) = 0, \\ &\dot{X} = \varepsilon u, \qquad u \in [u_1, u_2], \quad X(0) = X^0, \quad X(T) = 0. \end{aligned} \tag{5.3.6}$$

Control problems for system (6) are solved with the error $O(\varepsilon)$ on a time interval $t - t_0 \sim \varepsilon^{-1}$ by using the methods of Chapters 1 and 2. Since $P(T) = O(\varepsilon)$, assuming $u(t) \equiv 0$ for $t > T$ one can bring the system to the state $p(t) = O(\varepsilon^2)$, $\varphi(t) = O(\varepsilon^2)$, $x(t) - x^* = O(\varepsilon^2)$, $\dot{\varphi}(t) = O(\varepsilon^2)$; $T = \Theta\varepsilon^{-1}$.

If the correction control is taken in the form $v = \varepsilon^{3/2}u$, $u \in [u_1, u_2]$, then, in the time $t - t_0 \sim \varepsilon^{-1/2}$, system (4) arrives at an $\varepsilon^{3/2}$-neighbourhood of the specified terminal state.

Suppose, having used the control $v = \varepsilon^{1/2}u$ in the mode of transfer, the system has been brought into an $\varepsilon^{1/2}$-neighbourhood of the specified state point. Then, having made the changes $\varphi = \varepsilon^{1/2}\Phi$, $p = \varepsilon^{1/2}P$, and $x = \varepsilon^{1/2}X$ and using the control $v = \varepsilon^{3/2}u$ in the correction mode we obtain the system which can be reduced to the form (3.1.15) ($m = g = l = 1$):

$$\begin{aligned} &\dot{\Phi} = P - \varepsilon u + O(\varepsilon^2), \qquad \dot{X} = \varepsilon u, \qquad u \in [u_1, u_2], \\ &\dot{P} = -\Phi + (1/6)\varepsilon\Phi^3 + O(\varepsilon^2) \end{aligned} \tag{5.3.7}$$

with initial and boundary conditions of the type (6). System (7) allows to solve the control problem with the error $O(\varepsilon^{3/2})$.

It should be noted that if the original system (4) is at the state t_0, $\varphi = O(\varepsilon^{1/2})$, $p = O(\varepsilon^{1/2})$, $x = O(1)$, then, by using the control $v = \varepsilon u$, it is brought into an ε-neighbourhood of the specified terminal point $\varphi = p = x = 0$ in a time $t - t_0 \sim \varepsilon^{-1}$. In dimensionless Lagrangian variables, the equations of motion have the form (3.1.15)

$$\ddot{y} + (y - x) = (\varepsilon/6)(y - x)^3 + O(\varepsilon^2), \qquad \dot{x} = \varepsilon u. \tag{5.3.8}$$

By the change $Y = \Phi + X$, system (7) can be also reduced to the form (8). Thus, time-optimal control problems described by equations of the form (2) and (8) or, which is the same, by sets of equations (6) and (7) are the subject of further analysis when solving control problems for the pendulum having a velocity-controlled equilibrium position.

(3) *Controlled torsional oscillations.* Consider the motion of a symmetrical rigid body G connected to the turning unit D by means of a bifilar suspension of the length l (see Fig. 5.7). The unit D rotates about the vertical axis of symmetry OZ with the controlled angular velocity $\dot{\gamma} = \omega$. When a sufficiently small angular difference appears, the body G begins to perform torsional oscillations described by the equations

$$\begin{aligned}
&J\ddot{\varphi} - \frac{1}{2}mS'(\varphi-\gamma)(\dot{\varphi}-\dot{\gamma})^2 + mS(\varphi-\gamma)(\ddot{\varphi}-\ddot{\gamma}) + mg\partial z/\partial\varphi = 0,\\
&s(\varphi-\gamma) = h^4 z^{-2}\sin^2(\varphi-\gamma),\\
&z = -\left\{l^2 - 2h^2[1-\cos(\varphi-\gamma)]\right\}^{-1/2}, \qquad \dot{\gamma} = \omega, \quad \omega \in [\omega_1, \omega_2].
\end{aligned} \tag{5.3.9}$$

Here, J is the moment of inertia of the body G with respect to the axis OZ, m is the mass of the body, and h is the distance from the axis of rotation to the suspension points. In what follows, the quantities $|\varphi-\gamma|$, $|\dot{\varphi}-\dot{\gamma}|$, and ω are assumed to be small in the sense mentioned above. Having introduced the small parameter ε: $\varepsilon^d = |\omega|_{\max}/(\nu\Gamma)$, $d > 0$, where $\nu^2 = mgh^2/(Jl)$ and Γ is a characteristic rotation angle of the system, and having ignored small nonlinear terms, we obtain the equations of the form (2), (6)–(8). For Equation (9), the remarks made in Sections 5.3.1, item (1),

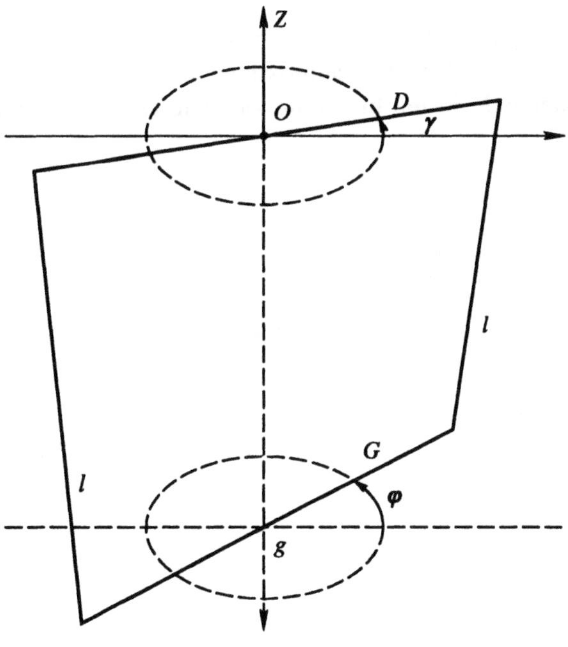

Fig. 5.7.

and 5.3.1, item (2), are valid. Equations (1), (9) give a simplified description of the basic motions of loading machines with ideal drives. They can be used for an estimation of the limiting performance characteristics of such machines transferring large loads (containers) and the facilities of control of their angular position.

5.3.2. SOLUTION OF THE PROBLEM OF TRANSFERRING A QUASILINEAR SYSTEM

For systems of the type (2) and (8), we make the change

$$y = x + a\sin\psi, \qquad \dot{y} = a\cos\psi, \qquad x = x,$$

where a is the amplitude of relative oscillations. As a result, the problem of time-optimal transferring the pendulum to the specified point accompanied by quenching relative oscillations is reduced to the form (the equation for ψ is not essential)

$$\begin{aligned} &\dot{a} = -\varepsilon u\sin\psi + \varepsilon\alpha a^3\sin^3\psi\cos\psi, \qquad \dot{x} = \varepsilon u, \\ &a(0) = a^0 = \left[(y^0 - x^0)^2 + \dot{y}^{02}\right]^{1/2}, \quad x(0) = x^0, \\ &a(T) = 0, \quad x(T) = x^* \ (= 0), \quad \Theta = \varepsilon T \longrightarrow \min, \quad u \in [u_1, u_2]. \end{aligned} \tag{5.3.10}$$

The particular case of the quasilinear system (10), which is considered below, admits analytical integration of the problem of the first approximation. The optimal control u^* and the averaged Hamiltonian K in the first approximation are (the average of $\sin^3\psi\cos\psi$ is zero)

$$\begin{aligned} u^* &= \frac{1}{2}(u_1 + u_2) + \frac{1}{2}(u_2 - u_1)\operatorname{sign}(r - p\sin\psi), \\ \varepsilon^{-1}K &= \frac{1}{2}(u_1 + u_2)r + \frac{1}{2}(u_2 - u_1)\langle |r - p\sin\psi|\rangle + \beta. \end{aligned} \tag{5.3.11}$$

Here, p, q, and r are the variables adjoint with a, ψ, and x; the average in (11) is found analytically:

$$\langle |r - p\sin\psi|\rangle = \begin{cases} |r|, & |k| \geqslant 1 \quad (k \equiv r/p), \\ (2/\pi)|p|\left[(1 - k^2)^{1/2} + k\sin^{-1}k\right], & |k| \leqslant 1. \end{cases} \tag{5.3.12}$$

It follows from (11), (12) that $p, r = \text{const}$, since k does not depend on a and x. The canonical system has the form

$$\dot{a} = 0, \quad \dot{x} = \frac{1}{2}(u_1 + u_2) + \frac{1}{2}(u_2 - u_1)\,\mathrm{sign}\, r, \qquad |k| \geqslant 1;$$

$$\begin{cases} \dot{a}(1/\pi)(u_2 - u_1)(1 - k^2)^{1/2}\,\mathrm{sign}\, p, & |k| \leqslant 1, \\ \dot{x} = \tfrac{1}{2}(u_1 + u_2) + (1/\pi)(u_2 - u_1)\sin^{-1} k\ \mathrm{sign}\, p. \end{cases} \tag{5.3.13}$$

If $|k| \geqslant 1$ and $a^0 \neq 0$, then the fulfillment of the terminal condition for the amplitude a is impossible, according to (13).

Let $|k| < 1$. Then integration of the problem (13) gives

$$a(\tau) = a^0\left(1 - \frac{\tau}{\Theta}\right), \qquad x(\tau) = x^0 + \left(x^* - x^0\right)\frac{\tau}{\Theta}, \qquad \tau = \varepsilon t. \tag{5.3.14}$$

The quantities k and Θ are determined from (13) by solving the set of transcendental equations

$$\begin{aligned} &(1/\pi)(u_2 - u_1)\left(1 - k^2\right)^{1/2}\Theta = a^0, \\ &\left[\frac{1}{2}(u_1 + u_2) - (1/\pi)(u_2 - u_1)\sin^{-1} k\right]\Theta = x^* - x^0. \end{aligned} \tag{5.3.15}$$

Eliminating the variable $\Theta > 0$ from (15) we obtain the equation for k, $|k| < 1$:

$$\begin{aligned} &\left[\sin^{-1} k - (\pi/2)(u_1 + u_2)/(u_2 - u_1)\right]\left(1 - k^2\right)^{-1/2} = \lambda, \\ &\lambda = \left(x^0 - x^*\right)/a^0, \qquad |\lambda| < \infty. \end{aligned} \tag{5.3.16}$$

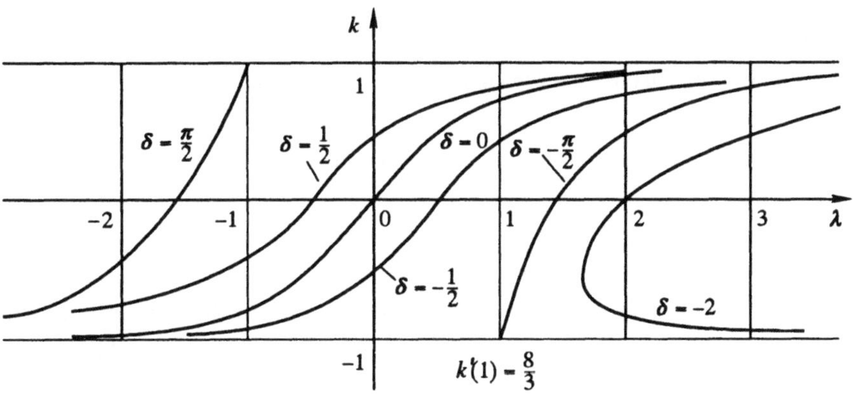

Fig. 5.8.

Having determined $k(\lambda)$ from (16), one can find the quantity $\Theta(\lambda)$ using any equation in (15). The solution of Equation (16) is found numerically in the form of a family of curves $k = k(\lambda, \sigma)$ (see Fig. 5.8), where $\sigma = (\pi/2)(u_1 + u_2)/(u_2 - u_1)$, $|\sigma| < \infty$. When $|\sigma| < \pi/2$, the root $k(\lambda, \sigma)$ to be found exists and is unique for any λ. For values $|\sigma| \geqslant \pi/2$, the solution of the control problem exists not for all values of λ but only for sufficiently large $|\lambda|$. It should be noted that in this case ($|\sigma| > \pi/2$), system (10) cannot remain at the required state when $\tau > \Theta$. The two obtained values of k correspond to two values of Θ and x, a.

5.3.3. TRANSFER OF A LINEAR OSCILLATORY SYSTEM WITH SMALL LINEAR FRICTION

Consider the problem of the time-optimal transfer of a linear oscillatory system of the form (2) or (8), taking small linear dissipative forces into account:

$$\ddot{y} + 2\varepsilon\delta\dot{y} + (y - x) = 0, \qquad \dot{x} = \varepsilon u, \qquad \delta > 0. \tag{5.3.17}$$

The presence of the cubic term $\varepsilon\alpha(y - x)^3$ in (17) does not change the scheme and the results of the analysis, since the influence of this term on the amplitude in the first approximation is equal to zero.

As was done above, we pose here the problem of transferring the system from an arbitrary state $(y^0, \dot{y}^0, x^0)$ to the specified state in the minimal time

$$y(T) = x(T) = x^*, \quad \dot{y}(T) = 0, \qquad \Theta \longrightarrow \min, \quad |u| \leqslant 1. \tag{5.3.18}$$

For $t > T$, system (17) will remain in the position (18) if $u \equiv 0$. The reduction to the standard form is implemented by the change $y = x + a\sin\psi$, $\dot{y} = a\cos\psi$, and $x = x$:

$$\begin{aligned}
&\dot{a} = -\varepsilon u \sin\psi - 2\varepsilon\delta a\cos^2\psi, \qquad a(0) = a^0,\\
&\dot{x} = \varepsilon u, \quad |u(t)| \leqslant 1, \qquad x(0) = x^0,\\
&\dot{\psi} = 1 - (\varepsilon/a)u\cos\psi + 2\varepsilon\delta\sin\psi\cos\psi, \qquad \psi(0) = \psi^0.
\end{aligned} \tag{5.3.19}$$

Terminal conditions (18) for a and x are rewritten as follows:

$$a(T) = 0, \qquad x(T) = x^*. \tag{5.3.20}$$

To solve the optimal control problem (19) and (20) in the first approximation in terms of ε we use the technique of Section 3.1. Based on (1.3.4) when $q = 0$, we find the optimal control u^* in the first approximation (see (11))

$$u^* = \operatorname{sign}(r - p\sin\psi), \qquad r = \text{const}. \tag{5.3.21}$$

For the averaged Hamiltonian function K, the following expression takes place:

$$\varepsilon^{-1}K = \langle |r - p\sin\psi| \rangle - \delta pa + \beta = \text{const}. \tag{5.3.22}$$

Here, the average $\langle |r - p\sin\psi| \rangle$ is found analogously to (12). As a result, we obtain the boundary-value problem of the first approximation of the type (13):

$$\begin{aligned}
&\dot a = -\delta a, \qquad \dot x = \operatorname{sign} r, \qquad |k| \geqslant 1;\\
&\begin{cases} \dot a = (2/\pi)\left(1-k^2\right)^{1/2}\operatorname{sign} p - \delta a,\\ \dot x = (2/\pi)\sin^{-1} k \operatorname{sign} p, \qquad |k| \leqslant 1;\end{cases}\\
&\dot p = \delta p, \quad p = p^0 e^{\delta\tau}, \qquad k = k^0 e^{-\delta\tau}, \quad k^0 = r/p^0,\\
&a(0) = a^0, \quad x(0) = x^0, \quad a(\Theta) = 0, \quad x(\Theta) = x^*.
\end{aligned} \tag{5.3.23}$$

The solution of problem (23) leads to the following three control modes.

1) The motion with $|k| \geqslant 1$ for $\tau \in [0, \Theta]$ takes place only when $a^0 = 0$.

2) The motion when $|k| < 1$ for $\tau \in [0, \Theta]$ includes quenching the oscillations (decreasing a) and transferring (if $k \neq 0$). This mode occurs for sufficiently large a^0 and small $|x^* - x^0|$.

3) The general case includes modes 1) and 2) on the intervals $\tau \in [0, \tau^*]$ and $\tau \in [\tau^*, \Theta]$, respectively. The quantity τ^* is determined by the conditions

$$|k| = 1, \qquad \tau^* = \delta^{-1}\ln\left|k^0\right|, \qquad \left|k^0\right| > 1. \tag{5.3.24}$$

The motion of system (23) is governed by the equations

$$\begin{aligned}
&\dot a = (2/\pi)\left(1-k^2\right)^{1/2}\chi - \delta a(\operatorname{sign} p = -1),\\
&\dot x = (1-\chi)\operatorname{sign} r + (2/\pi)\chi\sin^{-1}|k|\operatorname{sign} r.
\end{aligned} \tag{5.3.25}$$

Here, $\chi = 1\ (\tau - \tau^*)$ is the unit function (the Heaviside function). System (25) is completely integrated, since $\operatorname{sign} r = \operatorname{sign}(x^* - x^0)$.

As a result, we obtain the set of transcendental equations for determining two unknown parameters Θ and τ^* (or k^0):

$$
\begin{aligned}
&a^0 e^{-\delta\Theta} = \frac{2}{\pi}\int_{\tau^*}^{\Theta} e^{-\delta(\tau^*-\tau)}\left[1-e^{-2\delta(\tau^*-\tau)}\right]^{1/2} \mathrm{d}\tau,\\
&x^* - \left[x^0 + \tau^* \operatorname{sign}\left(x^* - x^0\right)\right]\\
&\quad = \frac{2}{\pi}\operatorname{sign}\left(x^* - x^0\right)\int_{\tau^*}^{\Theta} \sin^{-1} e^{\delta(\tau^*-\tau)}\,\mathrm{d}\tau,\\
&\Theta > \tau^* > 0, \quad a^0 > 0, \quad x^* \neq x^0.
\end{aligned}
\tag{5.3.26}
$$

The solution $\Theta(a^0, x^0)$ and $\tau^*(a^0, x^0)$ (or $k^0(a^0, x^0)$) can be constructed numerically. Substituting k^0 into the expression (23) for k and then $k = k(\tau, a^0, x^0)$ into (21), one can obtain, having calculated the averaged phase ψ of the first approximation, the optimal control u_p^*. Characteristic behaviour of the function $u_p^*(t, \varepsilon)$ and its connection with the function χ (25) are shown in Fig. 5.9. Assuming $\tau = 0$, $a^0 = a$, and $x^0 = x$ ($k^0 = k$) we can obtain the optimal control to be found in the feedback form.

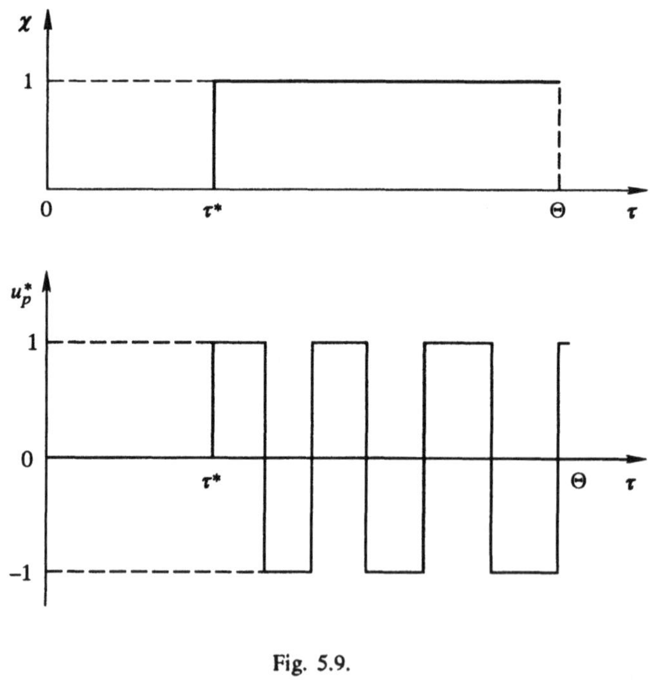

Fig. 5.9.

A similar boundary-value problem of the maximum principle and control modes arise when solving the problem of transferring a pendulum with variable length. This problem is considered below.

5.4. The Time-Optimal Feedback Control of Transferring a Pendulum of Variable Length

5.4.1. STATEMENT OF THE CONTROL PROBLEM

Consider a mechanical model of small oscillations of a plane pendulum with a variable length l. The suspension point of this pendulum can be driven with the limited velocity v, $v_1 \leqslant v(t) \leqslant v_2$, along the horizontal guide (see Figs. 1.2 and 5.6). The amplitude of oscillations is supposed to be a small quantity of the order of ε, $\varepsilon \ll 1$. The rates of changing the length and the velocity of the suspension point have the same order as small quantities. As to the intensity of accelerations, we assume that it is $O(\varepsilon)$ for x and $O(\varepsilon^2)$ for l. Making the change $y = x + l\varphi$, dropping the small terms of the third order with respect to ε, and dividing by $\varepsilon \neq 0$, we obtain the following equations of motion for the correction mode (see Section 5.3):

$$\begin{gathered} \ddot{y} + gl^{-1}(y - x) = 0, \qquad \dot{x} = v, \qquad v \in [v_1, v_2], \\ y(0) = y^0, \quad \dot{y}(0) = \dot{y}^0, \quad x(0) = x^0. \end{gathered} \tag{5.4.1}$$

By their form, these equations coincide with those for the case $l = \text{const} > 0$. Indeed, the equations of motion in the Hamiltonian form, which does not include impulse forces

$$\begin{aligned} &\dot{\varphi} = l^{-2}(p_\varphi - lv\cos\varphi) \qquad (m = 1), \\ &\dot{p}_\varphi = -p_\varphi l^{-1} v \sin\varphi + \dot{l}\varphi\cos\varphi + v^2\cos\varphi\sin\varphi - lg\sin\varphi, \\ &\mathcal{H} = \frac{1}{2} l^{-2}(p_\varphi - lv\cos\varphi)^2 + \frac{1}{2}\dot{l}^2 - \dot{l}v\sin\varphi + lg(1 - \cos\varphi), \end{aligned} \tag{5.4.2}$$

are reduced to the form (1), on the basis of the adopted assumptions and the problem statement given above. If l is the velocity-controlled suspension length, i.e. $\dot{l} = \varepsilon w$, $w \sim 1$, then Equations (2) are more convenient. These equations are reduced to the form (3.6) or (3.7) under different assumptions as to orders of the quantities $(y - x)$, x, T and others (see Section 5.3).

Then we use the dimensionless variables and the small parameter

$$\begin{gathered} t_* = \nu t, \quad \nu^2 = g/l^*, \quad \sigma = l/l^* \sim 1, \quad y_* = y/L, \quad \dot{y}_* = \dot{y}/(L\nu), \\ x_* = x/L, \quad v = \varepsilon u, \quad \varepsilon = |v|_{\max}/(L\nu) \ll 1. \end{gathered} \tag{5.4.3}$$

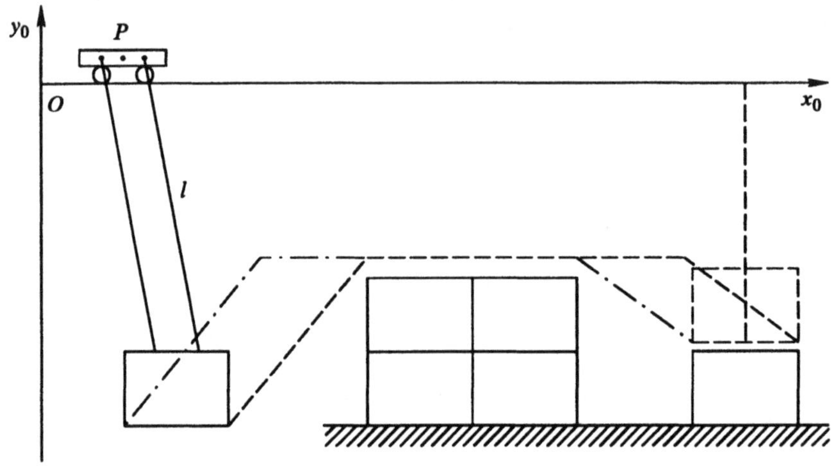

Fig. 5.10.

Here, l^* is a characteristic length of the pendulum, L is a characteristic distance of transferring the system, and $|v|_{\max}$ is the maximum of the numbers $|v_1|$ and $|v_2|$. Having dropped the indices after the substitution of (3) into (1), one can obtain the equation of motion for the correction mode:

$$\ddot{y} + \sigma^{-1}(y - x) = 0, \qquad \dot{x} = \varepsilon u, \qquad u \in [u_1, u_2]. \tag{5.4.4}$$

We study the problem of the time-optimal control of system (4). The terminal conditions have the form $y(T) = x(T) = x^*$, $\dot{y}(T) = 0$. The change reducing the system to the standard form is $((y, \dot{y}, x) \to (a, \psi, x))$:

$$y = x + a\sin\psi, \qquad \dot{y} = a\sigma^{-1/2}\cos\psi, \qquad x = x. \tag{5.4.5}$$

We assume that the relative length σ of the pendulum changes slowly and smoothly according to some law, with the rate defined by the values of the slow variables a, x, σ. As a result, the problem of the minimum-time transfer of the pendulum of variable length is specified by the relationships

$$\begin{aligned}
&\dot{a} = -\varepsilon u\sin\psi + \frac{1}{2}\varepsilon a\sigma^{-1}w\cos^2\psi,\\
&\dot{x} = \varepsilon u, \qquad \dot{\sigma} = \varepsilon w(a, x, \sigma),\\
&\dot{\psi} = \sigma^{-1/2} - \varepsilon u a^{-1}\cos\psi - \frac{1}{4}\varepsilon\sigma^{-1}w\sin 2\psi,
\end{aligned} \tag{5.4.6}$$

$$a(0) = a^0 > 0, \quad x(0) = x^0, \quad \sigma(0) = \sigma^0 > 0, \quad \psi(0) = \psi^0,$$
$$a(T) = 0, \quad x(T) = x^*, \quad \Theta = \varepsilon T \longrightarrow \min, \quad u \in [u_1, u_2].$$

The law of changing σ in (6) is supposed to be defined regarding the state constraints on the motion of the pendulum. For example, when transferring a suspended load using loading machines, it is necessary to change the suspension length because of the presence of obstacles (Fig. 5.10). The controlled system (6) will be essentially nonlinear if the function w depends on the controlled variables a and x.

5.4.2. APPROXIMATE BOUNDARY-VALUE PROBLEM OF THE MAXIMUM PRINCIPLE

Denote by p, q, r, and s the variables adjoint with a, ψ, x, and σ, respectively. According to (3.2.3) we can write down the Hamiltonian function and determine the optimal control:

$$H = \varepsilon p\Big(-u\sin\psi + \frac{1}{2}a\sigma^{-1}w\cos^2\psi\Big) + \varepsilon(ru + sw) + q\sigma^{-1/2} -$$
$$- \varepsilon q\Big[ua^{-1}\cos\psi + \frac{1}{4}\sigma^{-1}w\sin 2\psi\Big] \longrightarrow \max_u, \qquad u \in [u_1, u_2], \tag{5.4.7}$$
$$u^* = \frac{1}{2}(u_1 + u_2) + \frac{1}{2}(u_2 - u_1)\operatorname{sign}(r - p\sin\psi - qa^{-1}\cos\psi).$$

Using the technique of Section 3.3 we write down the averaged boundary-value problem:

$$\dot a = \begin{cases} \frac{1}{4}a\sigma^{-1}w(a, x, \sigma), & |k| \geqslant 1; \quad a(0) = a^0, \\ \frac{1}{\pi}(u_2 - u_1)(1 - k^2)^{1/2}\operatorname{sign} p + \frac{1}{4}a\sigma^{-1}w, & |k| \leqslant 1; \end{cases}$$
$$\dot x = \begin{cases} \frac{1}{2}(u_1 + u_2) + \frac{1}{2}(u_2 - u_1)\operatorname{sign} r, & |k| \geqslant 1; \quad x(0) = x^0, \\ \frac{1}{\pi}(u_2 - u_1)\sin^{-1}|k|\operatorname{sign} r + \frac{1}{2}(u_1 + u_2), & |k| \leqslant 1; \end{cases}$$
$$\dot\sigma = w(a, x, \sigma), \qquad \sigma(0) = \sigma^0, \tag{5.4.8}$$
$$\dot p = -(1/4)p\sigma^{-1}\partial(aw)/\partial a - s\partial w/\partial a,$$
$$\dot r = -\big((1/4)a\sigma^{-1}p + s\big)\partial w/\partial x, \qquad k = r/p,$$
$$\dot s = -\frac{1}{4}ap\partial\big(\sigma^{-1}w\big)/\partial a + \frac{1}{2}\beta\sigma^{-3/2} - s\partial w/\partial\sigma,$$
$$a(\Theta) = 0, \quad x(\Theta) = x^*, \quad s(\Theta) = 0; \qquad \Theta \longrightarrow \min_\beta, \quad \beta \in I_\beta.$$

In the general case, for an arbitrary function w, the solution of problem (8) can be constructed numerically for each specified law $w = w(a, x, \sigma)$.

5.4.3. QUASILINEAR TREATMENT OF THE CONTROL PROBLEM

In the case of a quasilinear system, when $w = w(\sigma)$ (see (1.1.7)), system (8) admits complete integration in quadratures. For $|k| \geqslant 1$,

$$a = a^0\left(\frac{\sigma}{\sigma^0}\right)^{1/4}, \quad p = p^0\left(\frac{\sigma^0}{\sigma}\right)^{1/4}, \quad \tau = \int_{\sigma^0}^{\sigma} \frac{ds}{w(s)}, \quad u_p^* = u_{1,2},$$
$$x = x^0 + \frac{1}{2}\left[(u_1+u_2) + (u_2-u_1)\,\mathrm{sign}\,(x^* - x^0)\right]\tau. \tag{5.4.9}$$

The feedback control and the minimal time are

$$u_s^* = \frac{1}{2}(u_1+u_2) + \frac{1}{2}(u_2-u_1)\,\mathrm{sign}\,(x^* - x), \qquad \Theta = \frac{x^* - x^0}{u_p^*}. \tag{5.4.10}$$

The averaged phase ψ, according to (6), (10), is expressed by the quadrature

$$\psi = \psi^0 + \frac{1}{\varepsilon}\int_{\sigma^0}^{\sigma} \frac{ds}{s^{1/2}w(s)} - \left(\frac{a^0}{4\sigma^0}\right)^2 \int_{\sigma_0}^{\sigma} \frac{ds}{s^3 w(s)}.$$

The case $|k| \geqslant 1$ corresponds to the travel of the pendulum with the extremal velocity without active quenching oscillations; the amplitude changing according to (9).

If $|k| < 1$, then using expressions for p, r, and σ, coinciding with (9), and having integrated the equations for a and x, and taken the boundary conditions (8) into account, one can obtain the transcendental equations for σ^*, γ:

$$a^0 = \frac{u_2 - u_1}{\pi}\int_{\sigma^0}^{\sigma^*} \frac{(1-\gamma^2\sigma^{1/2})^{1/2}}{\sigma^{1/4}w(\sigma)}\,d\sigma,$$
$$x^* - x^0 = -\frac{u_2-u_1}{\pi}\int_{\sigma^0}^{\sigma^*} \frac{\sin^{-1}(\gamma\sigma^{1/4})}{w(\sigma)}\,d\sigma + \frac{u_1+u_2}{2}\int_{\sigma^0}^{\sigma^*} \frac{d\sigma}{w(\sigma)}. \tag{5.4.11}$$

Here, σ^* is an unknown terminal value of σ, $\sigma^* = \sigma(\Theta)$; under real conditions, σ and σ^* are subjected to the constraints: $0 < \sigma_{\min} < \sigma,\ \sigma^* \leqslant \sigma_{\max} < \infty$; the parameter γ in (11) is $\gamma = (r/p^0)(\sigma^0)^{-1/4}$. While deriving the relationships (11), we took into account that $\mathrm{sign}\, p = 1$ and we assumed the function $\omega(\sigma)$ to have been sign-definite; besides, we considered that $|w| \geqslant w^0 > 0$ and $\sigma(\tau) \in [\sigma_{\min}, \sigma_{\max}]$. If σ^* and γ are not uniquely determined from (11), then the magnitude and the sign of γ, i.e. $|k^0|$ and $\mathrm{sign}\, r$, are found from the condition $\Theta \to \min$, where $\Theta = \tau|_{\sigma^*}$

(see (9)). When $u_2 = -u_1$ or $|u_1 + u_2|$ is sufficiently small, it uniquely follows from (11) that $\operatorname{sign} r = \operatorname{sign}(x^* - x^0)$.

It is necessary to note that due to the constraints imposed on σ and σ^*, Equations (11) are not always solvable. For example, if the rate of change of σ, i.e. $|w(\sigma)|$, is high, then in the case of decreasing the suspension length, the right-hand sides of the equations become small irrespective of γ. Therefore, for finite a^0 and $|x^* - x^0|$, the equalities become impossible. On the contrary, if σ changes relatively slowly, i.e. $|w(\sigma)|$ is small, then system (11) is solvable. In this case, the situation occurs which is close to that considered in Section 5.3, where the limit case $w \equiv 0$, $\sigma = \text{const}$ is treated. For arbitrary relationships between the quantities in question, an additional analysis of Equation (11) is required. In practice, they usually control the rate of lifting a swinging load or lowering it. The combination of operations of transferring, quenching oscillations, lifting and lowering leads to reducing the total operation time of the system. Therefore, the analysis of the control problem for the pendulum with a monotonically changing length may be of interest as part of a more general problem for the model with controlled velocity of changing the suspension length.

5.4.4. ANALYSIS OF CONTROLLED MOTION

The set of motion modes obtained by using the averaging method can be divided into the following cases.

1) As has been mentioned in Section 5.4.3 (see also Section 5.3), when $k^2 \geqslant 1$, the control corresponds to transferring the pendulum with the maximal velocity $u = u_{1,2}$ without active quenching of the oscillations. In this case, the monotone increase or decrease of their amplitudes is possible due to the change of the suspension length according to (9). For $a^0 = 0$, we have $a \equiv 0$, i.e. the motion is optimal in the first approximation in terms of ε.

2) The optimal motion mode is possible when $0 < k^2 \leqslant 1$. This takes place if system (11) is solvable.

3) Similarly to the case 1), the control mode with $k = 0$ is possible, which corresponds to the optimal quenching oscillations regardless transferring. In this case, the drift of the suspension point O' occurs with the velocity $\dot{x} = (1/2)(u_1+u_2)$. For the symmetrical constraint ($u_1 = -u_2$), the velocity $\dot{x} = 0$. This mode, as in the case 1), is incomplete, since it corresponds to special initial conditions.

4) The control mode is possible when $k^2 \geqslant 1$ on the initial stage of the motion and then it becomes less than unity due to the decrease of σ: $k^2 = k^{02}\sigma^{1/2}$. On the second stage, as it has been mentioned, transferring and the active quenching of the oscillations occur simultaneously.

5) The opposite situation is possible as well, when $k^2 \leqslant 1$ on the initial stage. This corresponds to simultaneous transferring and quenching of the oscillations (until the amplitude a becomes zero). Then transferring with the extremal velocity $u_{1,2}$

occurs, $k^2 > 1$. Such a control mode may take place when the suspension length is increasing.

To construct the feedback control and to determine the motion of the system for $k^2 \leqslant 1$ it is necessary to solve system (11) for arbitrary admissible initial values of the variables and then to treat them as current ones.

5.4.5. FEEDBACK CONTROL UNDER A CONSTANT RATE OF LIFTING AND LOWERING A LOAD

The case $w(\sigma) = w_0 = \text{const}$ [1, 24, 46] is important in applications. The dependence $\sigma(\tau) = \sigma^0 + w_0\tau$ leads to simple explicit expressions for formulae (9). In what follows, we consider for simplicity that the constraints on u (7) are symmetrical; without loss of generality we assume $u_2 = -u_1 = 1$. The nonsymmetrical case is more complicated for complete analysis.

The control mode 2) with $0 < k^2 < 1$ leads to the following set of transcendental

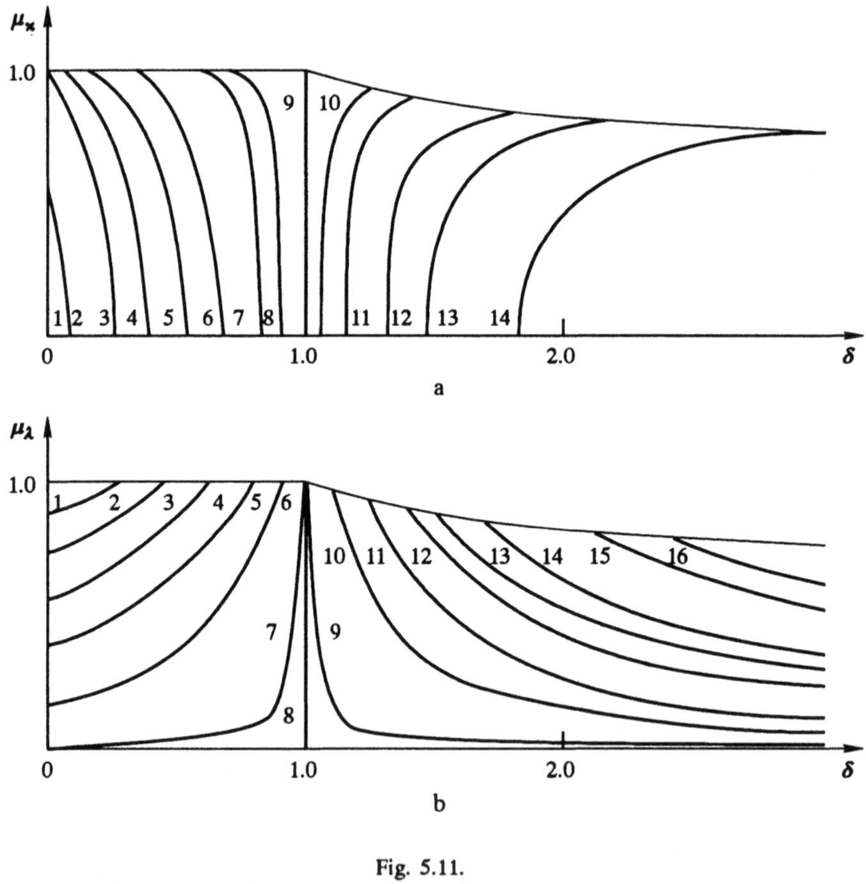

Fig. 5.11.

equations of the type (11) relative to $\mu = r/p^0$ and $\delta = \sigma^*/\sigma^0$:

$$\varkappa = \frac{2}{\pi}\int_1^\delta \left(1-\mu^2 z^{1/2}\right)^{1/2} z^{-1/4}\,dz, \qquad \varkappa = \frac{a^0}{\sigma^0} w_0, \quad |\varkappa| < \infty,$$
$$\lambda = \frac{2}{\pi}\int_1^\delta \sin^{-1}\left(\mu z^{1/4}\right)dz, \qquad \lambda = \frac{x^0 - x^*}{\sigma^0} w_0, \quad |\lambda| < \infty. \tag{5.4.12}$$

For given (or measured) values a^0, x^0, and σ^0 defining the quantities $\varkappa$ and λ, it is required to determine the unknowns δ and μ as functions of initial or current values of state variables. Quadratures in (12) are expressed in terms of transcendental functions [87]. But, as was the case for system (3.26), we cannot analytically solve this system similarly to (3.16). The solution to be found is constructed numerically. To do this we determine, using Equations (12), the following one-parametric families of curves

$$\mu_\varkappa = \pm\varphi_\varkappa(\delta), \qquad \mu_\lambda = \varphi_\lambda(\delta). \tag{5.4.13}$$

Here, δ, $0 < \delta < \infty$, is an argument and $\varkappa$ and λ are parameters of the families. These parameters can change unlimitedly according to (12). The families of curves $\mu_\varkappa$ and μ_λ are shown in Fig. 5.11, a and b.

The numbers from 1 to 14 in Fig. 5.11, a, indicate curves of the family $\mu_\varkappa$ corresponding to the following values of parameter $\varkappa$:

No	1	2	3	4	5	6	7
$\varkappa$	$-8/(3\pi)$	−0.7	−0.5	−0.4	−0.3	−0.2	−0.1
No	8	9	10	11	12	13	14
$\varkappa$	−0.05	0	0.05	0.1	0.2	0.3	0.5

Figure 5.11, a, presents a "half" of the family of curves corresponding to the plus sign in (13): $\mu_\varkappa = +\varphi_\varkappa$. The family $\mu_\varkappa = -\varphi_\varkappa$ is obtained by reflecting with respect to the δ-axis.

In Fig. 5.11, b, the numbers from 1 to 16 correspond to the following values of parameter λ of the family μ_λ:

No	1	2	3	4	5	6	7	8
λ	$-5/8$	−0.5	−0.4	−0.3	−0.2	−0.1	−0.01	0
No	9	10	11	12	13	14	15	16
λ	0.01	0.1	0.2	0.3	0.4	0.5	0.8	1.0

Here, also only a "half" of the family, with $\mu_\lambda > 0$, is presented. It corresponds to the motion in the "negative" direction: $\operatorname{sign} r = -1$. For $\mu_\lambda < 0$, corresponding to $w_0 > 0$, the family of curves is obtained by reflecting the family shown in Fig. 5.11, b, with respect to the δ-axis and changing the sign of the parameter λ: $\mu_\lambda = -\varphi_{-\lambda}(\delta)$.

Below, we give the basic properties of families of curves shown in Fig. 5.11, a and b. For the family in Fig. 5.11, a, the values $\varkappa > 0$ correspond to the values $\delta > 1$, i.e. $w_0 > 0$ (the length of the pendulum increases); the values $\mu_\varkappa$ being bounded: $|\mu_\varkappa| \leqslant \delta^- 1/4$. If sufficiently large values of δ are admitted, then there exists a root of the first equation (11) for any $\varkappa \geqslant 0$. The values $\varkappa < 0$ correspond to δ: $0 < \delta < 1$; the constraint on $\mu_\varkappa$ having the form $|\mu_\varkappa| \leqslant 1$. Taking the natural constraint $\delta > 0$ into account, one can establish from an estimate of the integral that the first equation (11) is satisfied for sufficiently small $|\varkappa|$: $0 > \varkappa > -8/(3\pi)$. Thus, if the length of the pendulum is permanently decreasing, we have enough time to quench oscillations of a finite amplitude: $a^0 \leqslant (8/3\pi)\sigma^0|w_0|^{-1}$. This amplitude can be of any large magnitude for sufficiently large values of the ratio $\sigma^0|w_0|^{-1}$. In dimensional variables, this ratio has a clear physical sense. It is the time interval during which the suspension length decreases from its initial value to zero.

When considering the families of curves in Fig. 5.11, b, one can establish, similarly to that expounded above, that the function μ_λ is bounded for $\delta \geqslant 0$: $0 < \mu_\lambda \leqslant \delta^{-1/4}$, and roots of the second equation (11) exist for any $\lambda > 0$ if admissible values of δ are sufficiently large. In the case of decreasing length, which corresponds to $\lambda < 0$, the constraint on μ_λ for $0 < \delta \leqslant 1$ has the form $0 \leqslant \mu_\lambda \leqslant 1$. In this domain, as above, due to the natural restriction $\delta > 0$, the second equation (11) can be satisfied if the quantity $|\lambda|$ is sufficiently small, i.e. $0 > \lambda > -5/8$. It means that for the model with a permanently decreasing suspension length, it is possible to transfer the pendulum by a distance which does not exceed the quantity $|x^* - x^0| = (5/8)\sigma^0|w_0|^{-1}$. This quantity can be of any large amount if the time $\sigma^0|w_0|^{-1}$ is sufficiently large.

The problem of determining the unknown parameters μ and δ using the families of curves constructed in Fig. 5.11, a and b, is solved as follows. We choose the curve from the family in Fig. 5.11, a, for given $\varkappa$, and from the family in Fig. 5.11, b, for given λ. The unique intersection point (μ^*, δ^*) of the curves $\mu_\varkappa(\delta)$ and $\mu_\lambda(\delta)$, if it exists, determines the optimal control mode in the case $k^2 \leqslant 1$. Substituting the obtained value $k_0 = \mu^*$ into the expression for k, $k = \mu^*\sigma^{1/4}$, and then into expression (7) we obtain the feedback optimal control:

$$u^* = -\operatorname{sign}\left[\mu^*(\varkappa, \lambda)\sigma^{1/4} - (y - x)\right],$$
$$\varkappa = (a/\sigma)w_0, \qquad \lambda = \big((x - x^*)/\sigma\big)w_0. \tag{5.4.14}$$

The minimal time is found after substitution of $\sigma^* = \delta^*(\varkappa^0, \lambda^0)\sigma^0$ into (9):

$$\Theta = \Theta(a^0, x^0, \sigma^0) = (\delta^* - 1)\sigma^0/w_0. \tag{5.4.15}$$

The optimal state trajectory in the first approximation is obtained by elementary integration of linear Equations (8) for a and x when

$$\begin{aligned} &u_2 = -u_1 = 1, \qquad w = w_0, \qquad \sigma = \sigma^0 + w_0\tau, \\ &k = \mu^*\sigma^{1/4}, \qquad \operatorname{sign} p = -1, \qquad \operatorname{sign} r = -1. \end{aligned} \tag{5.4.16}$$

Thus, the modes 1)–3) from Section 5.4.4 are completely investigated.

Controlled motions are also possible when the chosen curves from the families in Fig. 5.11, a, b do not intersect; this can correspond to modes 4) and 5). Let this take place for an increasing pendulum length, i.e. when $w_0 > 0$ ($\delta > 1$). Then the optimal mode is mode 5) consisting of two parts. The first part is mode 2) with $0 < k^2 \leqslant 1$ and the second is mode 1) with $k^2 > 1$. The unknown values of δ_1^* and μ_1^* for the first part are found from the condition of intersection of the curve from the family in Fig. 5.11, a with the bounding curve $\mu_* = \delta^{-1/4}$. Thus, at the first stage of mode 5) there occur a quenching of the oscillations until the amplitude becomes zero, and simultaneous travel of the suspension point in the required direction; then, since the time instant corresponding to $\sigma_1^* = \delta_1^*\sigma^0$, the pendulum travels at the extremal velocity without oscillations ($a = 0$). Determination of the optimal control, both open-loop and feedback, the minimal time and the state trajectory is performed on the basis of the modes 1) and 2) analyzed above. The solution of the control problem exists if the ratio $(\sigma_{\max} - \sigma^0)/w_0$ is sufficiently large.

Consider now a case different from that mentioned above, when the pendulum length is decreasing: $w_0 < 0$ ($\delta < 1$). Let the curves from the families in Fig. 5.11, a and b corresponding to given values $\varkappa$ and λ not intersect. In this case, mode 4) will be optimal if the problem has a solution. It consists of two parts: for the first part, $k^2 \geqslant 1$, and mode 1) is implemented; for the second part, $0 < k^2 < 1$, which corresponds to mode 2) considered above. This means that it is advantageous to quench the amplitude of oscillations on the first stage by decreasing the suspension length (see (9)) and to travel with the extremal velocity $u_1 = -1$. Then, beginning from some time instant corresponding to the suspension length $\sigma_2^* = \delta_2^*\sigma^0$, one has to transfer the system to the given position $x = x^*$ with a complete quenching of the oscillations by the end of the process. Here, we suppose that the final value $\sigma^* = \delta^*\sigma^0 > 0$. The unknown quantities δ^* and δ_2^* are determined from equations similar to (12). These equations are obtained as a result of the integration of the system (8) for $k^2 \geqslant 1$ (when $\sigma^0 \geqslant \sigma \geqslant \sigma_2^*$) and for $k^2 \leqslant 1$ (when $\sigma^* \leqslant \sigma < \sigma_2^*$). They have the form

$$\begin{aligned} &\varkappa\delta_2^{1/4} = \frac{2}{\pi}\int_{\delta_2}^{\delta} \left(1 - \mu_*^2 z^{1/2}\right)^{1/2} \frac{dz}{z^{1/4}}, \qquad \mu_* = \delta_2^{-1/4} > 1, \\ &\lambda + (1 - \delta_2) = \frac{2}{\pi}\int_{\delta_2}^{\delta} \sin^{-1}\left(\mu_* z^{1/4}\right) dz. \end{aligned} \tag{5.4.17}$$

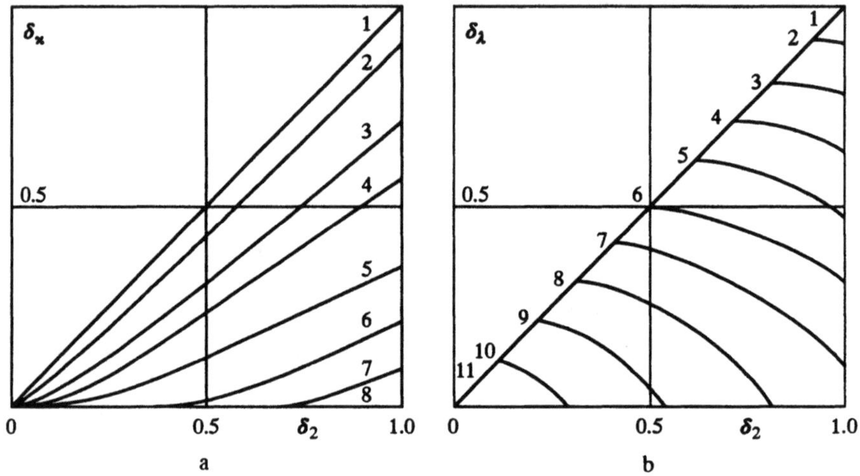

Fig. 5.12.

Given $\varkappa$ and λ, Equations (17) are solved with respect to δ_2 and δ by a numerical procedure similar to that used for Equations (12). We construct numerically two families of curves corresponding to the first and second equations (17) (Fig. 5.12, a and b). Here, $\delta_\varkappa$ and δ_λ are functions of the argument $\delta_2 \in [0, 1]$, and $\varkappa$ and λ are parameters of the families. The curves $\delta_\varkappa$ in Fig. 5.12, a, having numbers from 1 to 8, correspond to the following values of parameter $\varkappa$.

No	1	2	3	4	5	6	7	8
$\varkappa$	0	−0.01	−0.05	−0.1	−0.2	−0.3	−0.4	−0.5

Analogously, the curves δ_λ in Fig. 5.12, b having the numbers 1 to 11 correspond to the following values of parameter λ.

No	1	2	3	4	5	6	7	8	9	10	11
λ	0	−0.1	−0.2	−0.3	−0.4	−0.5	−0.6	−0.7	−0.8	−0.9	−1

Unknown values of δ^* and δ_2^* are determined by the intersection point of the respective curves from the families in Fig. 5.12, a and b. As follows from the equations and the constructed curves, control mode 4) can exist if $\varkappa > -1/2$ and

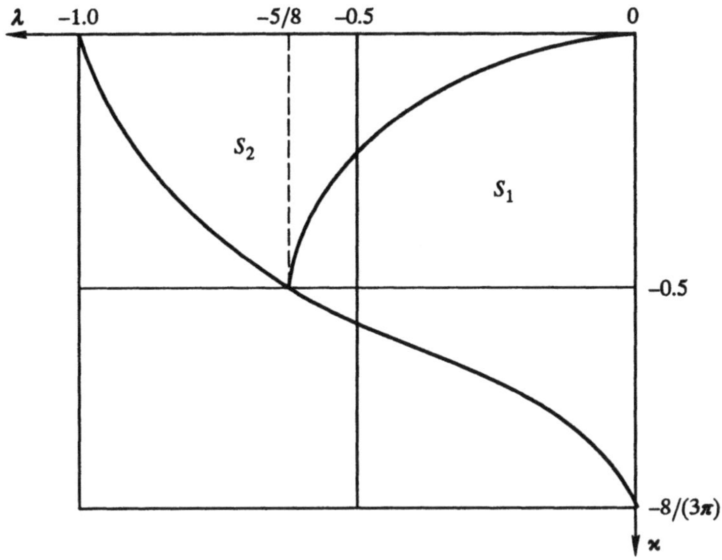

Fig. 5.13.

$\lambda \geqslant -1$. Moreover, it is evident that the point of intersection (δ^*, δ_2^*) of the curves $\delta_\varkappa$ and δ_λ does not exists not for each pair of curves satisfying these restrictions.

Below, we determine the range of parameters $\varkappa$ and λ for which mode 2) or 4) takes place under the condition of decreasing the suspension length: $w_0 < 0$ $(0 < \delta < 1)$. We find numerically the range of parameters $\varkappa$ and λ for which there exist points of intersection of curves from the families in Fig. 5.11, a and b, and Fig. 5.12, a and b. The results of the calculations are given in Fig. 5.13. Here, the domain S_1 defines the values of parameters $\varkappa$ and λ for which mode 2) with $k^2 \in (0, 1)$ is implemented. The domain S_2 corresponds to those values of $\varkappa$ and λ for which the solution of the optimal control problem is given by mode 4). Unification of the domains S_1 and S_2 for each time instant defines the range of the initial or current values of parameters $\varkappa$ and λ, connected with a^0, x^0, and σ^0 (or a, x, and σ), for which the solution of the optimal control problem is possible within the framework of the model having a permanently decreasing suspension length. Outside this domain, we may define the solution as a controlled motion of the pendulum with constant suspension length. As is shown in Section 5.3, this problem always has a solution.

5.4.6. REMARKS

It should be noted that the mechanical model of the pendulum-type system with controlled suspension length and other parameters appears to be rather useful and is often encountered in practice. From the viewpoint of applications, different generalizations of the problem considered above are of significant interest: construction of optimal and simple suboptimal control laws for spatial motions of a pendulum with a controlled suspension point and length; taking into account the state constraints on a load; taking into account the dynamics of drives implementing force action on the system; taking into account perturbations of various physical nature (the environment, imperfections of the mechanical model, control systems, etc.). It seems an important generalization to consider a two-mass model, i.e. a system of two bodies connected by a flexible rope: a carrying body, acted on by control forces, and a carried body. These are objects similar to a helicopter – a load, a spacecraft – a satellite (probe), etc. It is also interesting, both in theoretical and applied aspects, to consider parametric control of oscillations and rotations of pendulum-type systems (such as a swing).

CHAPTER 6

Optimal Control of Orbital Motions and Rotations of Spacecrafts Using "Low Thrust"

Using the methods of Chapters 1–4, we study to a first approximation problems of the optimal control of plane orbital motions of spacecrafts. We consider also problems of the optimal and suboptimal control of spacecraft rotation relative to its centre of mass, in particular, stabilization problems.

6.1. Time-Optimal Correction of an Orbit Close to a Circular One

Consider the time-optimal control problem for the motion of a particle in a central gravitational field on an orbit close to a circular one (see Fig. 1.4) using a low thrust with a restricted modulus (see Example (4) in Section 1.1.3). It is convenient to write down the equations of the first approximation (1.1.23) with respect to ε for a perturbed motion using osculating variables a, b, c, and γ:

$$\alpha = a\sin\gamma + b\cos\gamma, \qquad \beta = -a\cos\gamma + b\sin\gamma. \tag{6.1.1}$$

Assuming external perturbations (f_r, f_φ) in (1.1.23) to be small amounts of higher order in terms of ε, we can obtain, using the standard procedure of Chapter 1, the system

$$\begin{aligned}
&\dot a = \varepsilon u_r \sin\varphi + 2\varepsilon u_\varphi \cos\varphi + \varepsilon F_a, \qquad a(t_0) = a^0,\\
&\dot b = \varepsilon u_r \cos\varphi - 2\varepsilon u_\varphi \sin\varphi + \varepsilon F_b, \qquad b(t_0) = b^0,\\
&\dot c = \varepsilon u_\varphi + \varepsilon F_c, \qquad c(t_0) = c^0,\\
&\dot\gamma = -3\varepsilon c + \varepsilon F_\gamma, \qquad \gamma(t_0) = \gamma^0, \qquad t \in [t_0, T].
\end{aligned} \tag{6.1.2}$$

Here, $\varphi = t - t_0 + \gamma$ and the functions $F_{a,b,c,\gamma}$ have zero averages over φ; hence, their form is insignificant for what follows. We state the time-optimal control problem with a given goal:

$$a(T) = a^*, \quad b(T) = b^*, \quad c(T) = c^*, \quad \gamma(T) = \gamma^*; \qquad (6.1.3)$$
$$T = \Theta\varepsilon^{-1}, \quad T \longrightarrow \min, \quad u_r^2 + u_\varphi^2 \leqslant 1.$$

According to the technique of Section 1.3, we compile the Hamiltonian function H and find the optimal control u^* (see (1.3.4))

$$H = \varepsilon\left[p_a F_a + p_b F_b + p_c F_c + p_\gamma(F_\gamma - 3c)\right] +$$
$$+ \varepsilon u \Pi \sin(\varphi + \mu) + \varepsilon u_\varphi \Pi\left[2\cos(\varphi + \mu) + \delta\right],$$
$$\Pi = \left(p_a^2 + p_b^2\right)^{1/2}, \quad \sin\mu = \frac{p_b}{\Pi}, \quad \delta = \frac{p_c}{\Pi}, \qquad (6.1.4)$$
$$u_r^* = \frac{\sin(\varphi + \mu)}{\Phi}, \quad u_\varphi^* = \frac{\delta + 2\cos(\varphi + \mu)}{\Phi},$$
$$\Phi = \Phi(\varphi + \mu, \delta) = \left\{\sin^2(\varphi + \mu) + [\delta + 2\cos(\varphi + \mu)]^2\right\}^{1/2}.$$

Carrying out the averaging over φ, according to (2.1.7), and using the former notations for averaged variables we obtain the Hamiltonian of the first approximation

$$k = -3cp_\gamma + \Pi\langle\Phi\rangle = \text{const}. \qquad (6.1.5)$$

The expression for the average $\langle\Phi\rangle$ in (5) generalizes the elliptic integral of the second kind $\mathbf{E}$ [87]

$$\langle\Phi\rangle = f(\delta) = \frac{1}{2\pi}\int_0^{2\pi}\left(1 - \delta^2 + 4\delta\cos\psi + 3\cos^2\psi\right)^{1/2} d\psi, \qquad (6.1.6)$$
$$f(0) = (4/\pi)\mathbf{E}(\sqrt{3}/2).$$

The analytical properties of the function $f(\delta)$ are studied below. Using the fact that the variables a, b, and γ are cyclic in the averaged system and expressions (5), (6), and (2.1.10), we obtain the equations of motion in the slow time $\tau = \varepsilon t \in [\tau, \Theta]$:

$$\dot{a} = \cos\mu\left[f(\delta) - \delta f'(\delta)\right], \qquad \dot{b} = \sin\mu\left[f(d) - \delta f'(\delta)\right], \qquad (6.1.7)$$
$$\dot{c} = f'(\delta), \quad \dot{\gamma} = -3c; \qquad p_{a,b,\gamma} = \text{const}, \quad \dot{p}_c = 3p_\gamma.$$

System (7) is integrated in quadratures, since the right-hand sides of the equations for a, b, and c become known functions after the substitution of the expression for $\delta = \delta^0 + 3\lambda(\tau - \tau_0)$ where δ^0, $\lambda = p_\gamma/\Pi$, and Π are constants. Putting the state coordinates a, b, c, and γ at some instant $\tau = \Theta$ equal to specified values (3) leads to the set of transcendental equations relative to unknown parameters of the boundary-value problem μ, δ^0, λ, and Θ (or δ^*):

$$\begin{gathered}
\frac{\cos\mu}{3\lambda}\int_{\delta^0}^{\delta^*}[f(\delta) - \delta f'(\delta)]\,d\delta = a^* - a^0 \equiv \Delta a, \\
\frac{\sin\mu}{3\lambda}\int_{\delta^0}^{\delta^*}[f(\delta) - \delta f'(\delta)]\,d\delta = b^* - b^0 \equiv \Delta b, \\
\frac{1}{3\lambda}\int_{\delta^0}^{\delta^*} f'(\delta)\,d\delta = c^* - c^0 \equiv \Delta c, \qquad \delta^* = \delta^0 + 3\lambda(\Theta - \tau_0), \\
-3c^0(\Theta - \tau_0) + f(\delta^0)\frac{\Theta - \tau_0}{\lambda} - \frac{1}{3\lambda^2}\int_{\delta^0}^{\delta^*} f(\delta)\,d\delta = \gamma^* - \gamma^0 \equiv \Delta\gamma.
\end{gathered} \tag{6.1.8}$$

By simple algebraic transformations system (8) is reduced to two transcendental equations relative to δ^0 and δ^*, which include three essential parameters of the problem, $\Delta a^2 + \Delta b^2$, Δc, and $\Delta\gamma$. In the general case, their solution is obtained numerically.

The solution of the problem of the control of the orbit elements a, b, and c is significantly simplified if the position of the particle on the orbit, i.e. the quantity γ, at the instant T is not essential, since in this case $p_\gamma \equiv 0$, $\delta, p_c = \text{const}$. The set of equations (8) is reduced to the form

$$\begin{gathered}
\cos\mu\,[f(\delta) - \delta f'(\delta)](\Theta - \tau_0) = \Delta a, \\
\sin\mu\,[f(\delta) - \delta f'(\delta)](\Theta - \tau_0) = \Delta b, \qquad f'(\delta)(\Theta - \tau_0) = \Delta c.
\end{gathered} \tag{6.1.9}$$

The graph of the function $f(\delta)$ is given in Fig. 6.1. It is essential that the function f is even, $f > 0$, it has a minimum (6) at $\delta = 0$ and unlimitedly increases as $|\delta| \to \infty$, approaching the asymptotes $\pm\delta$ as $\delta \to \pm\infty$. The derivative $f'(0) = 0$ and, evidently, $f'(\delta) \to \pm 1$ as $\delta \to \pm\infty$.

For $\Delta c \neq 0$, system (9) is reduced to one equation relative to the parameter δ:

$$\begin{gathered}
d = D(\delta), \qquad D(-\delta) = -D(\delta), \\
d \equiv \left(\Delta a^2 + \Delta b^2\right)^{1/2}/(\Delta c), \qquad D(\delta) \equiv [f(\delta) - \delta f'(\delta)]/f'(\delta).
\end{gathered} \tag{6.1.10}$$

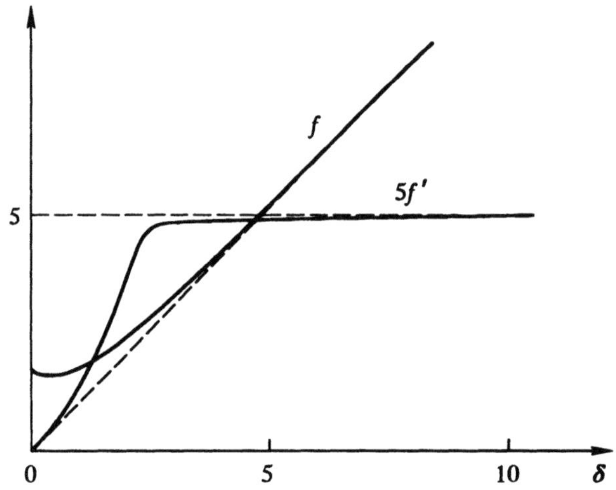

Fig. 6.1.

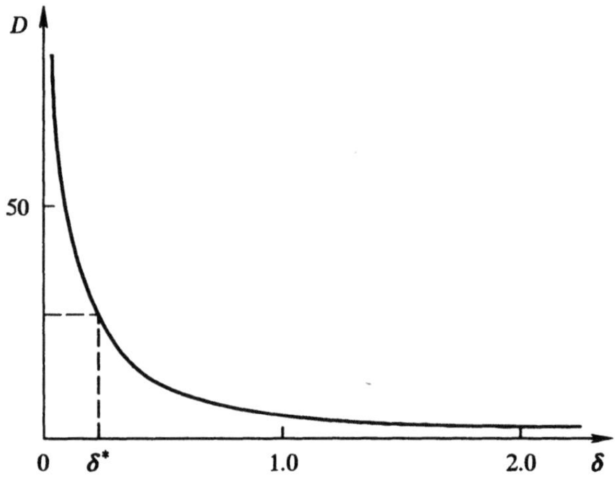

Fig. 6.2.

The solution of the transcendental equation (10) can be obtained graphically using Fig. 6.2 where the right-hand side of this equation is presented. For a given or measured value of d, the quantity δ^* is determined with an accuracy up to the sign which is found from the third equation (9) and, evidently, coincides with the sign of Δc. It follows from (9) and (10), see Fig. 6.2, that $\delta = 0$ when $\Delta c = 0$. After having determined the parameter δ^*, the quantity $\Theta - \tau_0$ is found easily:

$$\Theta - \tau_0 = \begin{cases} \Delta c / f'(\delta^*), & \Delta c \neq 0, \\ \dfrac{(\Delta a^2 + \Delta b^2)^{1/2}}{|f(\delta^*) - \delta^* f'(\delta^*)|}, & \Delta a^2 + \Delta b^2 \neq 0. \end{cases} \tag{6.1.11}$$

The quantities $\cos\mu$ and $\sin\mu$ are determined uniquely from (9), provided $\Delta a^2 + \Delta b^2 \neq 0$. Otherwise, the square brackets contain an expression equal to zero, which corresponds to $|\delta| = \infty$, and the control, according to (4), is given by $u_r^* = 0$, $u_\varphi^* = \operatorname{sign}\delta$ and does not depend on $\mu = \text{const}$ which can take any value. For different aspects of these problems, see in [29, 56–59, 71, 72, 76, 86, 88, 99, 107, 108, 116, 134, 153] and others.

6.2. Optimal Evolution of Elements of a Plane Elliptic Orbit

6.2.1. STATEMENT OF THE OPTIMAL CONTROL PROBLEM FOR THE MOTION ON AN ELLIPTIC ORBIT IN A GRAVITATIONAL FIELD BY LOW THRUST

Equations of motion in polar coordinates (1.1.23) without taking perturbations into account have the form [1, 72, 108] (Fig. 6.3)

$$\begin{aligned} &\dot r = v_r, \; r(0) = r^0, \qquad \dot\varphi = v_\varphi r^{-1}, \qquad \varphi(0) = \varphi^0, \\ &\dot v_r = v_\varphi^2 r^{-1} - r^{-2} + \varepsilon u_r, \qquad v_r(0) = v_r^0, \\ &\dot v_\varphi = -v_r v_\varphi r^{-1} + \varepsilon u_\varphi, \qquad v_\varphi(0) = v_\varphi^0. \end{aligned} \tag{6.2.1}$$

A thrust is considered to be small compared with the local gravitational force. Unlike the quasilinear treatment in Sections 6.1 and 1.1, we suppose here that the eccentricity e of the orbit is not small: $0 < e_1 \leqslant e \leqslant e_2 < 1$. To derive the standard system of equations for controlled motion, we use the following complete set of integrals [72, 108, 127]:

$$\begin{aligned} &\frac{1}{2}(v_r^2 + v_\varphi^2) - r^{-1} = E < 0, \qquad r v_\varphi = N \neq 0, \\ &r = p(1 + e\cos x)^{-1}, \qquad r = \frac{1}{2}(-E)^{-1}(1 - e\cos\xi), \\ &t + \delta = (-2E)^{-3/2}(\xi - e\sin\xi), \qquad \varphi = x + \gamma, \\ &e = \left(1 + 2EN^2\right)^{1/2}, \qquad p = N^2, \qquad \omega(E) = (-2E)^{3/2}. \end{aligned} \tag{6.2.2}$$

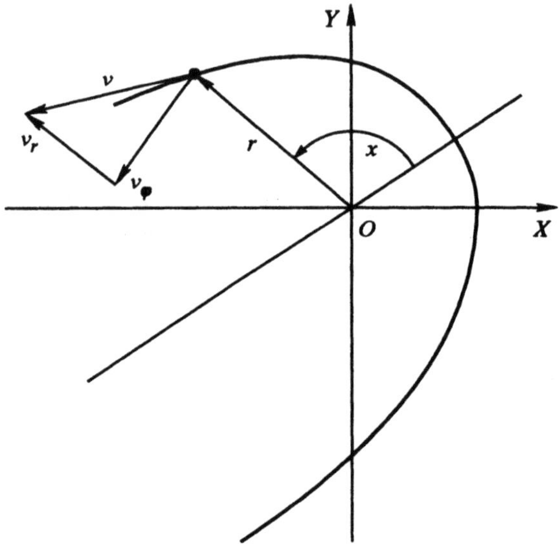

Fig. 6.3.

Here, E is the total energy, N is the angular momentum, γ is an angular constant characterising the twist of the apsis [72, 108] (see Fig. 6.3); and δ is a time constant.

The eccentricity e and the focal parameter p are expressed through E and N, and the parameter ξ connects the variables r and t; the connection between ξ and the phase $\psi = \omega(t+\delta)$ is one-to-one:

$$\begin{aligned} &\psi = \xi - e\sin\psi, \qquad \xi = \xi(\psi, e), \\ &\xi(\psi + 2\pi, e) \equiv 2\pi + \xi(\psi, e). \end{aligned} \tag{6.2.3}$$

Differentiating the integrals (2) according to system (1) we obtain the set of equations of controlled motion in the standard form:

$$\begin{aligned} &\dot{E} = \varepsilon(u_r v_r + u_\varphi v_\varphi), \qquad \dot{N} = \varepsilon u_\varphi r, \\ &\dot{\gamma} = \varepsilon\frac{N}{e}\left(-u_r\cos x + u_\varphi\frac{2 + e\cos x}{1 + e\cos x}\sin x\right), \\ &\dot{\psi} = \omega(E) + \varepsilon u_r f_{\psi r}(x, e, N) + \varepsilon u_\varphi f_{\psi\varphi}(x, e, N), \\ &E(0) = E^0 < 0, \quad N(0) = N^0, \quad \gamma(0) = \gamma^0, \quad \psi(0) = \psi^0. \end{aligned} \tag{6.2.4}$$

The connection between the true anomaly x and the phase ψ is given by the differential relationship

$$d\psi = \left(1 - e^2\right)^{3/2}(1 + e\cos x)^{-2}\,dx. \tag{6.2.5}$$

In Equations (4), the variables r, v_r, and v_φ are known 2π-periodic functions of x depending also on E and N (or e and p). The functions $f_{\psi r}$ and $f_{\psi\varphi}$ are similar; their explicit form is not essential for what follows. Often, it is more convenient to analyze, instead of two first equations (4), the equivalent equations for e and p:

$$\begin{gathered}\dot e = \varepsilon p^{1/2}\left[u_r \sin x + u_\varphi \frac{e(1+\cos^2 x) + 2\cos x}{1 + e\cos x}\right],\\ \dot p = 2\varepsilon u_\varphi p^{3/2}(1 + e\cos x)^{-1}; \qquad e(0) = e^0, \quad p(0) = p^0.\end{gathered} \tag{6.2.6}$$

For the system (4)–(6) we state the problem of the optimal control of the slow variables E, N (or e and p), and γ on a fixed time interval $[0, T]$ with a performance index of the form

$$J[u] = \frac{\varepsilon}{2}\int_0^T u^2\,dt \longrightarrow \min, \quad u = (u_r, u_\varphi), \quad |u| \leqslant \infty, \quad T = \frac{\Theta}{\varepsilon}. \tag{6.2.7}$$

Expression (7) has a physical sense for a low jet thrust [1, 46, 57–59, 71, 72, 76, 86, 88, 99, 107, 108, 111, 116, 153]. According to (3.2.2) it can be reduced to the form of a terminal functional:

$$J[u] = z(T) \longrightarrow \min, \quad \dot z = \frac{\varepsilon}{2}u^2, \quad z(0) = 0. \tag{6.2.8}$$

The boundary conditions in the general case have the form (3.2.1):

$$L_i(E, N, \gamma)\big|_T = 0; \qquad M_i(e, p, \gamma)\big|_T = 0 \qquad (i = 1, 2, 3). \tag{6.2.9}$$

Here, one or two relationships may be absent. In particular, the following statements of the problem are of applied interest:

– changing all elements of an orbit (shape, dimension and orientation)

$$\begin{gathered}E(T) = E^* \leqslant 0, \qquad N(T) = N^*, \qquad \gamma(T) = \gamma^*;\\ e(T) = e^* \in (0, 1), \qquad p(T) = p^* \neq 0, \qquad \gamma(T) = \gamma^*;\end{gathered} \tag{6.2.10}$$

– partially changing the shape and dimension, and twisting the apsis

$$L(E,N)\big|_T = 0; \qquad M(e,p)\big|_T = 0, \qquad \gamma(T) = \gamma^*. \tag{6.2.11}$$

More special statements are possible as well when the terminal values for some elements of the orbit are specified.

6.2.2. CONSTRUCTION OF THE AVERAGED HAMILTONIAN

Using the technique of Section 2.2, we construct the Hamiltonian function for the problem (4) and (7) and determine the control u^* as a function of the state variables E, N, γ, z, and ψ and the variables adjoint with them. Taking the cyclicity of γ and z into account we immediately find, according to (3.2.4), the expressions p_γ = const and $p_z = -1$. Then the following expressions are obtained for the control u_r^*, u_φ^* and the Hamiltonian H^*:

$$\begin{aligned} u_r^* &= p_E v_r + p_\gamma f_{\gamma r} + q f_{\psi r}, \qquad p_\gamma = \text{const}, \\ u_\varphi^* &= p_E v_\varphi + p_N r + p_\nu f_{\gamma\varphi} + q f_{\psi\gamma}; \end{aligned} \tag{6.2.12}$$

$$H^* = \frac{\varepsilon}{2}(u_r^{*2} + u_\varphi^{*2}) + \omega q. \tag{6.2.13}$$

Using the averaging scheme from Sections 3.2 and 3.4 and the relationship (5) one can obtain the following expression for the averaged Hamiltonian function K, retaining former notations for the averaged variables.

$$\varepsilon^{-1}K = \omega\beta + k_0, \quad \omega(E) = (-2E)^{3/2}, \quad E = \frac{1}{2}p^{-1}(1-e^2). \tag{6.2.14}$$

The function k_0 in the variables E, N, and γ has the form

$$\begin{aligned} k_0 &= -Ep_E^2 + \frac{2+3e^2}{4(1-e^2)}p^2p_N^2 + \frac{5-4e^2}{4e^2(1-e^2)}pp_\gamma^2 + Np_Ep_N, \\ e^2 &= 1 + 2EN^2, \qquad p = N^2. \end{aligned} \tag{6.2.15}$$

In the variables e, p, and γ, the function k_0 takes the form

$$k_0 = (5/4)pp_e^2 + \frac{2+3e^2}{(1-e^2)^2}p^3p_p^2 + \frac{5-4e^2}{4e^2(1-e^2)}pp_\gamma^2 - \frac{5ep^2}{1-e^2}p_ep_p. \tag{6.2.16}$$

In accordance with Section 2.1.3, we can use formulae (2.1.18) when proceeding from one set of variables (E, N, γ, ψ) to another (e, p, γ, ψ).

The quantity k, due to (13), (8), is

$$k_0 = \dot{z}, \quad z = z(\tau), \quad \tau = \varepsilon t \in [0, \Theta], \quad (\dot{\ }) \equiv (d/d\tau). \tag{6.2.17}$$

The averaged system has the integral (see (3.2.18))

$$\omega\beta + k_0 = \text{const}. \tag{6.2.18}$$

If $\beta = 0$ is the point of minimum for J_0, then $k_0 = \text{const}$ and, moreover,

$$\Theta k_0 = z(\Theta) = J_0[0], \qquad J_0 = J_0[\beta]. \tag{6.2.19}$$

The solution of the boundary-value problem to a first approximation with conditions of the form (9)–(11) for any admissible β and the subsequent minimization of $J_0[\beta]$ can be carried out using numerical methods. We have to integrate, over the slow time τ, the set of four equations (β and p_γ are parameters) admitting the first integral (18), instead of eight original equations. To construct the solution in a small neighbourhood of the value $\beta = 0$ we can use perturbation methods, i.e. expansions in terms of powers of β or successive approximations.

6.2.3. OPTIMAL GATHERING THE SPECIFIED ENERGY

Consider the problem of control of the total energy E:

$$E(T) = E^* \qquad (E^* \leqslant 0,\ T = \Theta\varepsilon^{-1}) \tag{6.2.20}$$

which admits a complete analytical solution. Terminal values for the remainder variables are not fixed. The equations and boundary conditions immediately imply for γ, p_γ, and p_N that

$$\begin{gathered} \dot{\gamma} = \partial k_0/\partial p_\gamma|_{p_\gamma=0} = 0, \qquad \gamma = \gamma^0, \quad p_\gamma = \text{const} = 0, \\ \dot{p}_N = -p_N\left(p_E + \frac{2+3e^2}{1-e^2}N^3 p_N\right), \quad p_N(\Theta) = 0, \quad p_N(\tau) \equiv 0. \end{gathered} \tag{6.2.21}$$

As a result, the averaged boundary-value problem is reduced to the form

$$\begin{aligned} &\dot{E} = -2Ep_E, \quad \dot{N} = Np_E, \quad \dot{p}_E = 3(-2E)^{1/2}\beta + p_E^2, \\ &E(0) = E^0, \quad N(0) = N^0; \quad E(\Theta) = E^*. \end{aligned} \tag{6.2.22}$$

Equations (18), (21), and (22) imply the integrals

$$(-2E)^{3/2}\beta - Ep_E^2 = h = \text{const}, \qquad 1 + EN^2 = e^2 = \text{const}. \tag{6.2.23}$$

Having solved the first relationship (23) with respect to p_E and substituting p_E into Equation (22) for E we obtain a separable equation. Integration of this equation leads to the complete set of integrals of the system. However, without integration of system (22) one can prove that the value $\beta = 0$ is optimal and the assumptions of Section 3.2 are fulfilled. At first, the solution of boundary-value problem (22) is constructed for $\beta = 0$:

$$\begin{aligned} &E(\tau) = E^0\sigma^2(\tau), \qquad N(\tau) = N^0\sigma^{-1}(\tau), \\ &p_E(\tau) = \frac{1}{\Theta\sigma(\tau)}\left[1 - \left(\frac{E^*}{E^0}\right)^{1/2}\right], \qquad \sigma(\tau) = 1 - \frac{\tau}{\Theta}\left[1 - \left(\frac{E^*}{E^0}\right)^{1/2}\right], \\ &J_0[0] = \Theta^{-1}[(-E^*)^{1/2} - (-E^0)^{1/2}]^2. \end{aligned} \tag{6.2.24}$$

Then, the conditions of rapid oscillations (3.2.22) and the local minimum (3.2.30) are checked. Expanding $J_0[\beta]$ and the solution of problem (22) in terms of powers of β we obtain the expression

$$J_0''[0] = -\frac{3}{20}E^0\Theta[3 + 4\sigma(\Theta) + 3\sigma^2(\Theta)] > 0. \tag{6.2.25}$$

Indeed, the quantity $E^0 < 0$ and the expression within the square brackets is positive for any real σ, since the discriminant is negative. Thus, the local optimal solution is constructed in the form (12), (19), (21), and (24). We also succeed in establishing its absolute (global) optimality for all β for which there exists a solution of the problem (22). From (15) and (17) follows the expression for $J[\beta]$:

$$J_0[\beta] = z(\Theta) = \int_0^\Theta k_0\, d\tau = -\int_0^\Theta Ep_E^2\, d\tau. \tag{6.2.26}$$

Substituting $p_E = -\dot{E}/2E$ from (22) into (26) we obtain

$$J_0[\beta] = -\int_0^\Theta \dot{E}^2 \frac{d\tau}{4E} = -\int_0^\Theta \left[\frac{d(-E)^{1/2}}{d\tau}\right] d\tau. \quad (6.2.27)$$

Using the Cauchy–Bunyakowski inequality for the pair of functions $d(-E)^{1/2}/d\tau$ and 1 we obtain, according to (24), the inequality [46]

$$J_0[\beta] \geqslant \frac{1}{\Theta}\left[\int_0^\Theta \frac{d(-E)^{1/2}}{d\tau} d\tau\right]^2 = J_0[0]. \quad (6.2.28).$$

Substitution of $q = p_\gamma = p_N = 0$ into (12) results in the feedback optimal control

$$\begin{aligned} u_r^* &= \frac{v_r}{\Theta - \tau}\left[1 - \left(\frac{E^*}{E}\right)^{1/2}\right] = \\ &= \frac{1}{\Theta - \tau}\left[1 - \left(\frac{E^*}{E}\right)^{1/2}\right]\frac{e}{N}\sin(\varphi - \gamma), \\ u_\varphi^* &= \frac{v_\varphi}{\Theta - \tau}\left[1 - \left(\frac{E^*}{E}\right)^{1/2}\right] = \\ &= \frac{1}{\Theta - \tau}\left[1 - \left(\frac{E^*}{E}\right)^{1/2}\right]\frac{1 + e\cos(\varphi - \gamma)}{N}. \end{aligned} \quad (6.2.29)$$

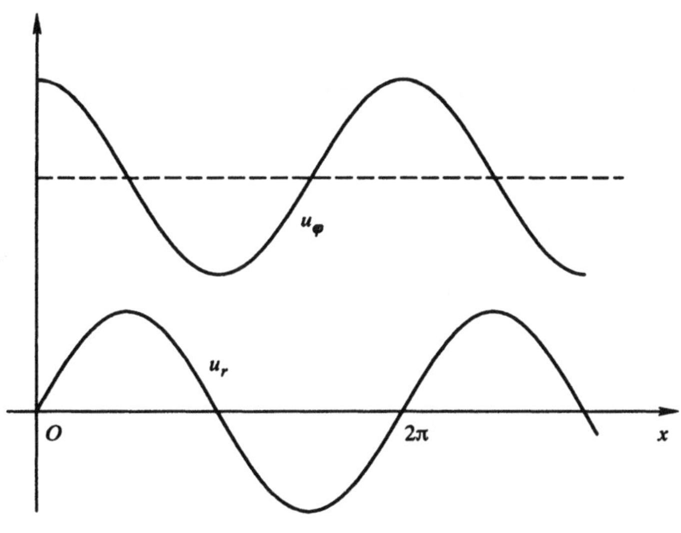

Fig. 6.4.

Thus, the optimal control is a tangential thrust whose magnitude is periodically changing during the orbital motion of the particle. As a function of the anomaly x it has the shape shown in Fig. 6.4. The eccentricity of the orbit remains constant to a first approximation and the curve of the energy *vs* time is represented as an increasing or decreasing branch of the parabola (24). One should be aware of a singularity of the feedback optimal control (29) as $\tau \to \Theta$.

6.2.4. QUASIOPTIMAL TWIST OF THE APSIS

The form of the averaged Hamiltonian function (14)–(16) implies that the boundary-value problem to a first approximation admits the solution

$$\begin{gathered} E(\tau) \equiv E^0, \qquad N(\tau) \equiv N^0; \qquad p_E(\tau) = p_N(\tau) \equiv 0 \\ \left(e(\tau) \equiv e^0,\ p(\tau) \equiv p^0;\ p_e(\tau) = p_p(\tau) \equiv 0\right). \end{gathered} \tag{6.2.30}$$

The solution of the problem of the apsis twist under the requirement that the equalities (30) are satisfied is found easily, since $p_\gamma = \text{const}$. As a result, the apsidal angle γ changes linearly in $\tau \in [0, \Theta]$:

$$\begin{gathered} \gamma(\tau) = \gamma^0 + (\gamma^* - \gamma^0)\tau\Theta^{-1}, \qquad p_\gamma = (\gamma^* - \gamma^0)/\chi^0\Theta, \\ \chi^0 = \frac{1}{2}(p^0/e^{02})(5 - 4e^{02})/(1 - e^{02}). \end{gathered} \tag{6.2.31}$$

Substituting p_γ from (31) into (12) when $p_E = p_\gamma = q \equiv 0$ we obtain the feedback optimal control $u^*_{r,\varphi}$:

$$\begin{gathered} u_r^* = (\gamma^* - \gamma)\chi^{-1}(\Theta - \tau)^{-1} f_{\gamma r}(e, p, x), \\ u_\varphi^* = (\gamma^* - \gamma)\chi^{-1}(\Theta - \tau)^{-1} f_{\gamma\varphi}(e, p, x), \\ \chi = \chi(e, p) = \frac{1}{2}(p/e^2)(5 - 4e^2)/(1 - e^2), \\ f_{\gamma r} = -\frac{N}{e}\cos x, \quad f_{\gamma\varphi} = \frac{N}{e}\,\frac{2 + e\cos x}{1 + e\cos x}\sin x, \quad x = \varphi - \gamma. \end{gathered} \tag{6.2.32}$$

The constructed feedback control (32) also has a singularity as $\tau \to \Theta$. This is caused by the statement of the control problem.

The value of the functional (8) is obtained on the basis of (16) and (19):

$$J_0 = \frac{1}{2}\frac{\gamma^* - \gamma^0}{\chi^0\Theta}. \tag{6.2.33}$$

Expressions (32) imply that the optimal control (u_r^*, u_φ^*) approaches zero as does the quantity e as $e \to 0$. Expenditures described by the quadratic functional (7) are equal to the value (33) and decrease as Θ increases. They are smaller, the smaller is the eccentricity e of the orbit. It evidently follows from the equation for γ (4) and corresponds to physical intuition. The given solution for the problem of twist of the apsis is not optimal in the case of free variables E and N (e and p). This solution is of interest, however, if during the twist of the apsis the dimension and the shape of the ellipse are to be kept unchanged according to conditions (30). It is worth noting that the formulae fail to be valid for small e comparable with ε. In this case, we can apply the results of the analysis of the optimal control problem for the motion on the orbit close to the circular one, analogously to Section 6.1.

The methods developed in Sections 6.1 and 6.2 can be extended to the spatial variant of the control problem for the elements of an elliptic orbit.

Next Sections 6.3 and 6.4 are devoted to the analysis of some problems of control of spacecraft rotations with respect to the centre of mass by "low thrust" jet engines. These problems are considered within the framework of the model of a rigid body with an ideal control system.

6.3. Control of Rotations of a Rigid Body Close to That of Dynamic Symmetry by Means of Small Torques

6.3.1. STATEMENT OF THE MINIMUM-TIME CONTROL PROBLEM

Consider the controlled system (1.1.24) under the conditions

$$\begin{gathered} M_i = \varepsilon b_i u_i, \qquad I_2 = I_1(1+\varepsilon\varkappa), \qquad I_3 I_1^{-1} = d \neq 1, \\ \varepsilon \ll 1, \qquad b_i, d, \varkappa \sim 1, \qquad i = 1,2,3. \end{gathered} \tag{6.3.1}$$

The general generating solution

$$\omega_1 = a\cos\psi, \quad \omega_2 = a\sin\psi, \quad \omega_3 = c \quad (a > 0,\ c \neq 0) \tag{6.3.2}$$

is used for reducing to the standard system of the form (3.1.3)

$$\begin{aligned}
&\dot{a} = \varepsilon(l_1 u_1 \cos\psi + l_2 u_2 \sin\psi) + \frac{1}{2}\varepsilon\varkappa(2-d)ac\sin 2\psi,\\
&\dot{c} = \varepsilon l_3 u_3 - \frac{1}{2}\varepsilon\varkappa d^{-1}a^2 \sin 2\psi; \qquad l_i = b_i I_i^{-1}, \quad i = 1,2,3,\\
&\dot{\psi} = (d-1)c + \varepsilon a^{-1}\big[l_2 u_2 \cos\psi - l_1 u_1 \sin\psi +\\
&\qquad + \varkappa ac\big(\cos 2\psi - d\cos^2\psi\big)\big],\\
&a(0) = a^0 > 0, \quad c(0) = c^0 \neq 0, \quad \psi(0) = \psi^0.
\end{aligned} \tag{6.3.3}$$

For (3) we pose the minimum-time problem

$$a(T) = a^*, \quad c(T) = c^*, \quad T \longrightarrow \min, \quad u \in U \quad \big(a^* \geqslant 0,\ c^0 c^* > 0\big). \tag{6.3.4}$$

This problem is studied below under different constraints on the control vector u, $u \in U$.

(1) *The system of control with restricted total power.* Let the domain U be specified by the relationship

$$U = \big\{u\colon\ u^2 = u_1^2 + u_2^2 + u_3^2 \leqslant 1\big\}. \tag{6.3.5}$$

Then the maximality condition for the Hamiltonian function to a first approximation in terms of ε implies the expressions for the optimal control (see Section 2.3)

$$\begin{aligned}
&u_1 = \frac{pl_1}{R}\cos\psi, \qquad u_2 = \frac{pl_2}{R}\sin\psi, \qquad u_3 = \frac{pl_3}{R},\\
&R = \big[p^2(l_1^2\cos^2\psi + l_2^2 \sin\psi) + r^2 l_3^2\big]^{1/2}.
\end{aligned} \tag{6.3.6}$$

Here, p and r are the slow variables adjoint with a and c. The averaging over ψ, taking into account the fact that the average of gyroscopic terms is equal to zero, allows one to obtain $h_0 = \langle R\rangle$. For $\beta = 0$ (see Section 3.3), we obtain the boundary-value problem to a first approximation:

$$\begin{aligned}
&\dot{a} = \dot{p}^{-1}\langle R\rangle - l_3^2 p^{-1} r^2 \langle R^{-1}\rangle, \qquad a(0) = a^0, \quad a(\Theta) = a^*,\\
&\dot{c} = l_3^2\langle R^{-1}\rangle, \qquad c(0) = c^0, \quad c(\Theta) = c^*, \quad (\dot{\ }) = (d/d\tau),\\
&\langle R\rangle = 1; \qquad p, r = \text{const}, \quad \tau = \varepsilon t, \quad \tau \in [0,\Theta], \quad \Theta = \varepsilon T.
\end{aligned} \tag{6.3.7}$$

Average values $\langle R\rangle$ and $\langle R^{-1}\rangle$ in (7) are

$$\begin{aligned}
&\langle R^{\pm 1}\rangle = \frac{2}{\pi}(l_1^2 p^2 + l_3^2 r^2)G^{\pm}(k), \qquad -\infty < k \leqslant 1;\\
&G^+ = \mathbf{E}(\sqrt{k}), \qquad G^- = \mathbf{K}(\sqrt{k}), \quad 0 \leqslant k < 1,\\
&G^+ = k'\mathbf{E}(\sqrt{k/k'}), \qquad G^- = \mathbf{K}(\sqrt{k/k'})/k', \quad -\infty < k \leqslant 0;\\
&k = (\beta_1^2 - \beta_2^2)(\beta_1^2 + \lambda^2)^{-1}, \quad k' = k - 1 \leqslant -1;\\
&\beta_1 = l_1/l_3, \quad \beta_2 = l_2/l_3, \quad \lambda = r/p.
\end{aligned} \tag{6.3.8}$$

Here, $\mathbf{K}$ and $\mathbf{E}$ are complete elliptic integrals of the first and second kinds. System (7) is integrated easily, since its right-hand side is constant. The boundary conditions lead to the set of three transcendental equations with respect to the unknowns p, r, and Θ. Eliminating Θ, we obtain one transcendental equation with respect to λ. This equation always has the root defining the solution of the control problem to be found. Determination of the parameter λ as a function of the known quantities l_i and the ratio $(a^* - a^0)/(c^* - c^0)$ is simplified when having introduced the new parameters

$$\begin{aligned}
&\Lambda = \lambda/\beta_2, \quad |\Lambda| < \infty; \qquad \delta = \beta_1/\beta_2 = l_1/l_2, \quad 0 < \delta < \infty;\\
&\gamma = (a^* - a^0)/(c^* - c^0)\beta_2, \qquad |\gamma| < \infty;\\
&k = (\delta^2 - 1)/(\delta^2 + \Lambda^2), \qquad -\infty < k \leqslant 1.
\end{aligned} \tag{6.3.9}$$

According to (8), the sign of the parameter k, i.e. the choice of expressions $G^{\pm}(k)$, is determined by the quantity δ^2 ($\delta^2 \geqslant 1$ or < 1). In particular, for $\delta^2 = 1$, both expressions coincide and $\Lambda = 1/\gamma$. In the general case, it is convenient to represent the dependence $\Lambda(\gamma, \delta^2)$ as a family of curves with the parameter δ^2. As a result, we obtain the following expressions for this family:

$$\begin{aligned}
&\gamma = \mathbf{E}(\sqrt{k})(\Lambda^2 + \delta^2)/\mathbf{K}(\sqrt{k})\Lambda - \Lambda, \qquad 0 \leqslant k \leqslant 1;\\
&\gamma = \mathbf{E}((\sqrt{k/k'})(\Lambda^2 + 1)/\mathbf{K}(\sqrt{k/k'})\Lambda - \Lambda, \qquad -\infty < k \leqslant 0.
\end{aligned} \tag{6.3.10}$$

The family of curves $\Lambda(\gamma, \delta^2)$ corresponding to different values of δ^2 is shown in Fig. 6.5. It follows from (10) that the function $\Lambda(\gamma, \delta^2)$ is odd with respect to γ. Therefore, it is sufficient to construct the curves corresponding to positive γ and Λ.

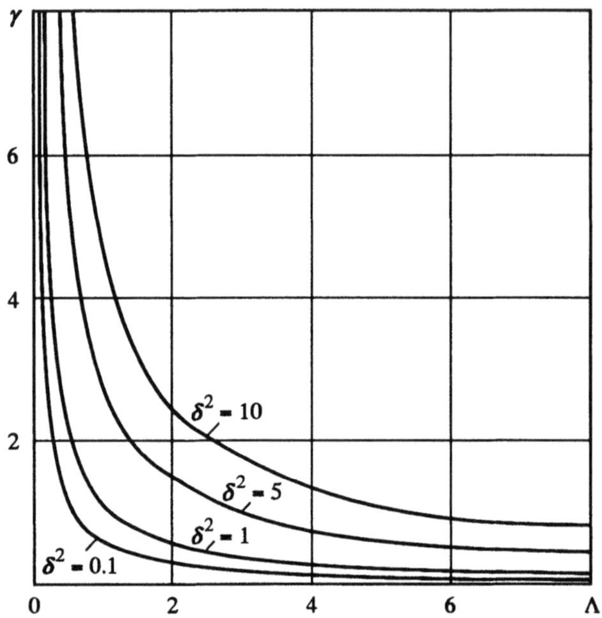

Fig. 6.5.

Having determined $|\lambda|$, according to (6), we obtain the feedback control, the trajectory, and the minimal time:

$$\begin{aligned}
&u_{1,2} = \beta_{1,2}\omega_{1,2}R_*^{-1}\operatorname{sign}(a^* - a), \qquad u_3 = |\lambda| a R_*^{-1}\operatorname{sign}(c^* - c),\\
&R_* = \left[(\beta_1^2 + \lambda^2)\omega_1^2 + (\beta_2^2 + \lambda^2)\omega_3^2\right]^{1/2},\\
&a(\tau) = (a^* = a^0)\tau\Theta^{-1} + a^0, \qquad c(\tau) = (c^* - c^0)\tau\Theta^{-1} + c^0,\\
&\Theta = (a^* - a^0)p + (c^* - c^0)r, \qquad p = \varphi^{-1}\operatorname{sign}(a^* - a^0),\\
&r = \lambda/\varphi, \qquad \varphi = \varphi(\lambda) = (2/\pi)(l_1^2 + l_2^2\lambda^2)G^+(k).
\end{aligned} \tag{6.3.11}$$

For the specific case when $a(T)$ is not fixed in (4), the solution is obtained from (11) when $|\lambda| \to \infty$:

$$\begin{aligned}
&u_1 = u_2 \equiv 0, \qquad u_3 = \operatorname{sign}(\omega_3^* - \omega_3),\\
&a(\tau) = a^0, \qquad \omega_3 = c(\tau), \qquad \Theta = \Theta_c = |\omega_3^* - \omega_3^0|/l_3.
\end{aligned} \tag{6.3.12}$$

If the quantity $\omega_3(\tau)$ is not fixed, then $r = \lambda = 0$ and $u_3 \equiv 0$ ($\omega_3 \equiv \omega_3^0$); $u_{1,2}(\omega)$ are determined by (11) where $\lambda = 0$;

$$\Theta = \Theta_a = (\pi/2)|a^* - a|/(l_3 G^+(k)), \qquad k = 1 - l_2^2/l_1^2. \tag{6.3.13}$$

When encountering troubles in implementing the control (11) which requires a knowledge of λ, one can use the laws for $\lambda = 0$ and $|\lambda| = \infty$ in succession. The necessary total time will be $\Theta_c + \Theta_a \geqslant \Theta$.

(2) *Control using restricted torques.* Consider problem (3) and (4) with the constraints

$$U = \{u\colon\ |u_i| \leqslant 1,\ i = 1, 2, 3\}. \tag{6.3.14}$$

This corresponds to three pairs of limited thrust engines rigidly attached to the body. The posed time-optimal control problem leads to singular controls [28, 31, 66, 155]. The conditions of the maximum principle imply that optimal control is not determined uniquely, one of the adjoint variables identically vanishes and the corresponding control is not determined (see Section 2.6). However, an approximate control can be constructed comparatively easily by using the technique of Section 3.3. Analogously to Section 6.3.1, item (1), we obtain the system to a first approximation

$$\begin{aligned} &\dot a = (2/\pi)(l_1 + l_2)v(\tau), \qquad \dot c = l_3 w(\tau); \\ &v = \operatorname{sign} p, \quad p \neq 0; \qquad w = \operatorname{sign} r, \quad r \neq 0. \end{aligned} \tag{6.3.15}$$

Taking into account the boundary conditions (7) we find the minimal time Θ:

$$\Theta = \max\{\Theta_a, \Theta_c\}, \quad \Theta_a = \frac{\pi}{2}\frac{|a^* - a^0|}{l_1 + l_2}, \quad \Theta_c = \frac{|c^* - c^0|}{l_3}. \tag{6.3.16}$$

The functions $v(\tau)$, $|v| \leqslant 1$, and $w(\tau)$, $|w| \leqslant 1$, in (15) are determined, nonuniquely in the general case, as follows:

$$\int_0^\Theta v(\tau)\,d\tau = \frac{\pi}{2}\frac{a^* - a^0}{l_1 + l_2}, \qquad \int_0^\Theta w(\tau)\,d\tau = \frac{c^* - c^0}{l_3}. \tag{6.3.17}$$

Based on the relationships (17), the feedback optimal control u has the form

$$u_{1,2} = v_s(a, c)\operatorname{sign}\omega_{1,2}, \qquad u_2 = w_s(a, c), \tag{6.3.18}$$

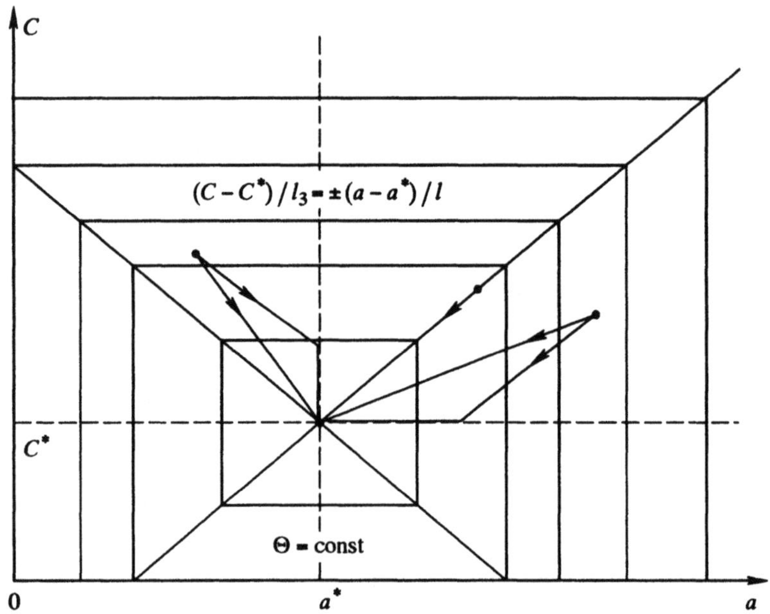

Fig. 6.6.

where $v_s(a,c)$ and $w_s(a,c)$ are defined for the current time instant, i.e. from (17) where $a^0 = a$ and $c^0 = c$. Figure 6.6 gives cross-sections of Bellman's function which has the shape of a part of a pyramid with the vertex at the point $a = a^* > 0$, $c = c^* \neq 0$ cut by the planes $a = 0$, $c = 0$; $l = (2/\pi)(l_1 + l_2)$.

(3) *Control by means of a couple of turning engines and a couple of fixed ones.* Consider the case when the domain for the control torque is bounded by a circular cylinder, i.e.

$$U = \left\{u\colon\ u_1^2 + u_2^2 \leqslant 1,\ |u_3| \leqslant 1\right\} \qquad (l_{1,2} = l). \tag{6.3.19}$$

The posed problem, as in Section 6.3.1, item (2), gives rise to singular controls:

$$\begin{aligned} &u_{1,2} = \omega_{1,2}a^{-1}v, \quad |v| \leqslant 1; \qquad u_3 = w, \quad |w| \leqslant 1; \\ &v = v_p(\tau), \quad v = v_s(a,c); \quad w = w_p(\tau), \qquad w = w_s(a,c). \end{aligned} \tag{6.3.20}$$

The minimal time Θ and the functions v, $|v| \leqslant 1$, and w, $|w| \leqslant 1$, from (20) are determined by the relationships (16) and (17) where $(2/\pi)(l_1 + l_2) = l$.

The case of the elliptic cylinder when $l_1 \neq l_2$ is studied analogously.

6.3.2. CONTROL OPTIMAL WITH RESPECT TO ENERGY CONSUMPTION

Consider the control problem (3) and (4) on a fixed time interval T with the performance index

$$J[\mathbf{u}] = \frac{\varepsilon}{2}\int_0^T u^2\,dt \longrightarrow \min, \qquad u = |\mathbf{u}| < \infty, \qquad T = \frac{\Theta}{\varepsilon}. \tag{6.3.21}$$

Additional constraints are not imposed on u. Using the methods of Section 3.2 we find, to a first approximation, the optimal control, the functional, and the slow variables

$$\begin{aligned} &u_{1,2} = \frac{2l_{1,2}}{l_1^2 + l_2^2}\,\frac{a^* - a}{\Theta - \tau}\,\frac{\omega_{1,2}}{a}, \qquad u_3 = \frac{\omega_3^* - \omega_3}{\Theta - \tau}, \\ &J_0[0] = \Theta^{-1}\left[\left(a^* - a^0\right)\left(l_1^2 + l_2^2\right)^{-1} + \frac{1}{2}\left(c^* - c^0\right)^2 l_3^{-2}\right], \\ &a(\tau) = a^0 + \left(a^* - a^0\right)\tau\Theta^{-1}, \qquad c(\tau) = c^0 + \left(c^* - c^0\right)\tau\Theta^{-1}. \end{aligned} \tag{6.3.22}$$

It stems from (22) that $u \sim \Theta^{-1}$, i.e. for sufficiently large values of Θ, the controls do not reach the imposed bounds if these take place. The expression for J_0 allows one to derive Bellman's function in the original variables using the well-known mode $(\Theta \to \Theta - \tau,\ a^0 \to a,\ c^0 \to c)$.

One should again pay attention to the singularity of the feedback control and Bellman's function as $\tau \to \Theta$. The singularity is caused by the statement of the problem.

6.4. Control of Rotations of a Nonsymmetrical Rigid Body

6.4.1. TIME-OPTIMAL BRAKING

Consider the more general case of different moments of inertia. To be definite, let $I_1 < I_2 < I_3$. System (1.1.24) is reduced to the standard form (3.1.3) by using known integrals, viz. the energy E and the magnitude of the angular momentum L, for uncontrolled motion

$$\begin{aligned} &2E = I_1\omega_1^2 + I_2\omega_2^2 + I_3\omega_3^2, \\ &L^2 = I_1^2\omega_1^2 + I_2^2\omega_2^2 + I_3\omega_3^2, \qquad 2EI_1 \leqslant L^2 \leqslant 2EI_3. \end{aligned} \tag{6.4.1}$$

Projections $\omega_i(t)$ are expressed (not uniquely) in terms of the elliptic functions of Jacobi [87, 91, 154]. As one of the integrals it is convenient to take the quantity k^2 ($k^2 = k^2(E, L^2)$, $k^2 \in [0,1]$); k is a modulus of elliptic functions characterizing the motion of the end point of the vector of angular momentum $\mathbf{L}$ on the sphere $L^2 = \text{const}$. The parameter k^2, as well as the aforenamed projections ω_i, is defined nonuniquely; for $L^2 \geqslant 2EI_2$, we have

$$\begin{aligned}
&k^2 = (I_2 - I_1)(I_3 - I_2)^{-1}(2EI_3 - L^2)(L^2 - 2EI_1)^{-1},\\
&\omega_1 = a_1(E, L^2)\,\mathrm{cn}(\theta, k), \qquad \omega_2 = a_2(E, L^2)\,\mathrm{sn}(\theta, k),\\
&\omega_3 = \pm a_3(E, L^2)\,\mathrm{dn}(\theta, k); \qquad \theta = 4(\mathbf{K}(k)/2\pi)\psi,\\
&a_{1,2}(E, L^2) = \left[(2EI_3 - L^2)/I_{1,2}(I_3 - I_{1,2})\right]^{1/2},\\
&a_3(E, L^2) = \left[(L^2 - 2EI_1)/I_3(I_3 - I_1)\right]^{1/2},\\
&\psi = (2\pi/T_0)(t - t_0) + \psi^0,\\
&T_0 = 4\mathbf{K}(k)(I_1I_2I_3)^{1/2}(I_3 - I_2)^{-1/2}(L^2 - 2EI_1)^{1/2}.
\end{aligned} \tag{6.4.2}$$

Here, T_0 is a period of the functions ω_i in t, and $\mathbf{K}$ is the complete elliptic integral of the first kind. Expressions for k^2 and ω_i when $L^2 \leqslant 2EI_2$ are obtained from (2) by the change of indices $1 \to 3$, $2 \to 2$, $3 \to 1$.

Differentiation of $L(E)$ and k^2 from (1) and (2) according to the controlled system (1.1.24) leads to the equations

$$\begin{aligned}
&\dot{L} = \frac{3}{L}\sum_{i=1}^{3} b_i I_i \omega_i u_i \qquad \left(\dot{E} = \varepsilon \sum_{i=1}^{3} b_i \omega_i u_i\right),\\
&\dot{k}^2 = (\partial k^2/\partial L)\dot{L} + (\partial k^2/\partial E)\dot{E}.
\end{aligned} \tag{6.4.3}$$

The equation for rotating phase ψ is not used to a first approximation and is not given here. We consider a control system with restricted power, i.e. restrictions have the form (3.5), and pose the time-optimal control problem for the slow variables L and k^2:

$$L(T) = L^* \geqslant 0, \quad k^2 = k^{*2} \in [0,1], \quad T \longrightarrow \min, \quad u^2 \leqslant 1. \tag{6.4.4}$$

We compose the Hamiltonian function for problem (3) and (4), and from the maximality conditions find the controls to a first approximation

$$u_i = \rho_i/\rho, \quad \rho_i = \omega_i \mu_i, \quad \mu_i = b_i I_i p/L + b_i \chi_i(k^2) r/L^2, \quad i = 1,2,3. \tag{6.4.5}$$

Here, p and r are the variables adjoint with L and k^2; χ_i are known functions; e.g. for $L^2 > 2EI_2$, we have

$$\chi_i = 2\big[(I_2 - I_1)I_3 + (I_3 - I_2)I_1k^2\big](1 - I_i\chi)/\big[(I_3 - I_2)(1 - \chi I_i)\big],$$
$$\chi = \big[I_2 - I_1 + (I_3 - I_2)k^2\big]/\big[(I_2 - I_1)I_3 + (I_3 - I_2)I_1k^2\big].$$

Substitution of the expressions (5) into the Hamiltonian function H reduces this function to the form $H^* = [A + B\,\mathrm{sn}^2(\theta, k)]^{1/2}$; $A, A + B > 0$, where A and B are known functions of L^2, k^2, p, r, and the parameters I_i, b_i. The averaging of H^*, an implicit function of ψ, is replaced by averaging over the elliptic amplitude φ $(\beta = 0)$:

$$H_0(L^2, k^2, p, r) = \langle H^* \rangle = \frac{1}{4\mathrm{K}(k)} \int_0^{2\pi} \frac{(A + B\sin^2\varphi)^{1/2}}{(1 - k^2\sin^2\varphi)^{1/2}}\,\mathrm{d}\varphi,$$
$$\mathrm{sn}(\theta, k) = \sin\varphi, \qquad \mathrm{d}\psi = (2\pi/(4\mathrm{K}))(1 - k^2\sin^2\varphi)^{-1/2}\,\mathrm{d}\varphi. \tag{6.4.6}$$

Using the methods of Sections 3.3 and 3.4, we obtain the averaged boundary-value problem to a first approximation

$$\dot{L} = \frac{\partial H_0}{\partial p} \equiv f_L, \qquad \dot{k}^2 = \frac{\partial H_0}{\partial r} \equiv f_{k^2},$$
$$\dot{p} = -\frac{\partial H_0}{\partial L}, \qquad \dot{r} = -\frac{\partial H_0}{\partial p^2}, \qquad (\dot{\ }) \equiv \left(\frac{\mathrm{d}}{\mathrm{d}\tau}\right);$$
$$L(0) = L^0, \quad k^2(0) = k^{02}; \quad L(\Theta) = L^* \geqslant 0, \quad k^2(\Theta) = k^{*2},$$
$$H_0\big|_\Theta = H_0 = \mathrm{const} > 0, \quad \tau = \varepsilon t \in [0, \Theta]. \tag{6.4.7}$$

Its solution, in the general case, can be constructed numerically. From the applied point of view, the brake problem $(L^*/L^0 \ll 1)$ is interesting when the value of $k^2(\Theta)$ is not essential (and is not defined). Then the solution of the maximum principle boundary-value problem of the form (7) is reduced to the construction of a family of solutions for the initial-value problem (k^{*2} is a parameter of the family)

$$\frac{\mathrm{d}k^2}{\mathrm{d}L} = \frac{f_k^2}{f_L}, \qquad \frac{\mathrm{d}p}{\mathrm{d}L} = -\frac{\partial H_0/\partial L}{f_L}, \qquad \frac{\mathrm{d}r}{\mathrm{d}L} = -\frac{\partial H_0/\partial k^2}{f_L},$$
$$k^2(L^*) = k^{*2} \in [0, 1], \qquad p(L^*) = -1, \qquad r(L^*) = 0. \tag{6.4.8}$$

The boundary condition for p is reduced to the form (8) by normalization. A more detailed description of the algorithm for constructing the optimal feedback control and an example of calculations for concrete values of the parameters I_i and b_i of the problem can be found in [46].

The implementation of this algorithm is rather tedious. The calculation of the optimal control is interesting for comparing with the results of using locally optimal controls. The determination and implementation of locally optimal controls are substantially easier (see below).

6.4.2. LOCALLY OPTIMAL AND QUASIOPTIMAL LAWS OF BRAKING

Consider the problem of braking rotations of a rigid body having arbitrary moments of inertia $I_1 \leqslant I_2 \leqslant I_3$. The available optimal laws are rather complex in the general case. Therefore, simple control schemes close to optimal ones are of interest, for example locally optimal controls giving the maximal rate of decreasing integrals (4.1).

(1) *Special cases of control domains.* In the case of constraints (3.5), the minimality condition $\min_u \dot{L}$, $|u| \leqslant 1$, (see (4.3)) implies that

$$u_i^L = -b_i L_i \left(\sum_{i=1}^{3} b_i^2 L_i^2 \right)^{-1/2}, \quad \dot{L} = -\varepsilon \left(\sum_{i=1}^{3} b_i^2 \frac{L_i^2}{L^2} \right)^{1/2} \leqslant -\varepsilon b;$$
$$b = \min\{b_1, b_2, b_3\}, \quad b > 0; \qquad 0 \leqslant L(\tau) \leqslant L^0 - b\tau. \tag{6.4.9}$$

The law of braking (9) brings rotations to a full stop in the time $\Theta \leqslant \Theta_L = L^0/b$.

Analogously, one can obtain from the condition $\dot{E} \to \min_u$, $|u| \leqslant 1$, (see (4.3)) that

$$u_i^E = -b_i \omega_i \left(\sum_{i=1}^{3} b_i^2 \omega_i^2 \right)^{-1/2}, \quad \dot{E} = -\varepsilon \left(\sum_{i=1}^{3} b_i^2 \omega_i^2 \right)^{1/2} \leqslant -\varepsilon \beta \sqrt{E};$$
$$\beta = \min_{i=1,2,3} \left(\sqrt{2/I_i b_i}\right) > 0, \quad 0 \leqslant E(\tau) \leqslant \left(\sqrt{E^0} - \frac{1}{2}\beta\tau\right)^2, \tag{6.4.10}$$
$$\Theta \leqslant \Theta_E = 2\sqrt{E^0}/\beta.$$

In the case of the constraint (3.14), locally optimal laws with respect to L and E

coincide:

$$
\begin{aligned}
&u_i^{L,E} = -\operatorname{sign}\omega_i, \qquad \dot{L} = -\varepsilon\sum_{i=1}^{3} b_i I_i|\omega_i|L^{-1} \leqslant \varepsilon b,\\
&\dot{E} = -\varepsilon\sum_{i=1}^{3} b_i|\omega_i| \leqslant \varepsilon\beta\sqrt{E}; \qquad \Theta \leqslant \min\{\Theta_L, \Theta_E\}.
\end{aligned}
\tag{6.4.11}
$$

The obtained control laws (9)–(11) for any $\varepsilon > 0$ guarantee bringing rotations to a full stop in time $T \sim \varepsilon^{-1}$. These laws are optimal for some particular cases, for instance u_i^L is time-optimal for equal $b_i = b$: $u_i = -L_i L^{-1}$.

For the investigation of the evolution of rotations under some specified braking laws for a nonsymmetrical body, see [46]. It has been established that the magnitude of the angular momentum L strictly decreases and the averaged equation for k^2 admits stationary points determined by the parameters b_i and I_i, $i = 1, 2, 3$; one or two of these points being stable in the sense of averaged equations. The quantity k^2 has been shown to achieve the stable stationary point k^2: $k^2 - k_*^2 \sim (\Theta - \tau)^\gamma$, $\gamma > 0$ at the stop time. The question of bringing rotations to a full stop in a time close to the optimal or quasioptimal value obtained by means of asymptotic methods has also been investigated.

(2) *Control domain of a general form.* The estimates for the time of braking rotations of a rigid body can also be constructed efficiently for some more general cases. For example, let the admissible domain U for the control vector $\mathbf{u}$ be rather arbitrary: $\mathbf{u} \in U = U(\mathbf{L})$. We only suppose that there exist two concentric balls $U_*(L)$ and $U^*(L)$ having radii $R_*(L)$ and $R^*(L)$, respectively, such that

$$
\begin{aligned}
&U_*(L) \subseteq U(\mathbf{L}) \subseteq U^*(L), \qquad \mathbf{L} \in D, \qquad \mathbf{u} \in U(\mathbf{L}),\\
&0 < R_0 \leqslant R_*(L) \leqslant R(L) \leqslant R^0 < \infty, \qquad L \in [0, L^0].
\end{aligned}
\tag{6.4.12}
$$

The functions $R_*(L)$ and $R^*(L)$ are assumed to be sufficiently smooth. Then, using the Schwartz inequality [28, 49, 75] and taking inclusions and inequalities (12) into account, we can obtain for the time-optimal control $\mathbf{u} = \mathbf{u}(\mathbf{L}, \varepsilon)$ the following inequalities (see also Chapter 7)

$$
\begin{aligned}
&-\varepsilon R_*(L) \geqslant \dot{L} = \varepsilon\left(\frac{\mathbf{L}}{L} \cdot \mathbf{u}\right) \geqslant -\varepsilon R^*(L),\\
&\mathbf{u} \in U(\mathbf{L}), \qquad \mathbf{L} \in D, \qquad L \in [0, L^0].
\end{aligned}
\tag{6.4.13}
$$

Here, we assume $b_i = 1$, $i = 1, 2, 3$, without loss of generality.

Let the functions $R_*(L)$ and $R^*(L)$ satisfy the Lipschitz conditions in L. Then there exist unique solutions $L_*(\tau, L^0)$ and $L^*(\tau, L^0)$ for the following initial-value problems:

$$\begin{aligned} \dot{L} &= -\varepsilon R_*(L), \quad L = L_*(\tau, L^0); \qquad \tau = -\int_{L^0}^{L_*} \frac{dl}{R_*(l)}, \\ \dot{L} &= -\varepsilon R^*(L), \quad L = L^*(\tau, L^0); \qquad \tau = -\int_{L^0}^{L^*} \frac{dl}{R^*(l)}. \end{aligned} \tag{6.4.14}$$

According to (14) we determine the quantities Θ_* and Θ^*:

$$\Theta_*(L^0) = \int_0^{L^0} \frac{dL}{R_*(L)}, \qquad \Theta^*(L) = \int_0^{L^0} \frac{dL}{R^*(L)}, \qquad \Theta_* \geqslant \Theta^*.$$

Due to the monotonicity of the functions $L_*(\tau, L^0)$ and $L^*(\tau, L^0)$ in τ, the following bilateral estimate is valid for the minimal time $T = T(L^0, \varepsilon)$:

$$T_* = \Theta_*(L^0)\varepsilon^{-1} \geqslant T(L^0\varepsilon) \geqslant \Theta^*(L^0)\varepsilon^{-1}. \tag{6.4.15}$$

The expounded approach based on the separation of slow and fast variables allows one to simplify the analysis and solution of control problems for rotations of a spacecraft acted upon by small torques of various physical natures: gravitational, gyroscopic, air dynamical, electromagnetic, and others. An approximate solution of the problem of optimal stabilization of a spacecraft on an elliptic orbit using small control torques comparable with gravitational ones is given in [1, 46].

CHAPTER 7

Approximate Synthesis of Optimal Control for Perturbed Systems with Invariant Norm

We develop an effective method of constructing Bellman's function and the optimal control for some classes of perturbed systems on a restricted time interval. The approach is based on the perturbation technique for solving the initial-value problem for the of Hamilton–Jacobi–Bellman equation. The efficiency of the method is essentially determined by the possibility of a complete solution of the corresponding unperturbed optimal control problem by using Schwartz's functional inequality.

7.1. Time-Optimal Stabilization of a Perturbed System with Invariant Norm

7.1.1. STATEMENT OF THE OPTIMAL CONTROL PROBLEM

Consider a controlled system of the following form:

$$\dot{y} = f_0(y) + \varepsilon f(y) + \big[I + \varepsilon F(y)\big]u, \qquad y(0) = y^0. \tag{7.1.1}$$

Here, y, u, f_0, and f are n-vectors; I is the unit $(n \times n)$-matrix, F is an arbitrary $(n \times n)$-matrix; and $\varepsilon > 0$ is a scalar numerical parameter, $\varepsilon \in [0, \varepsilon_0]$. The functions f_0, f, and F are supposed to be defined in some convex domain D_y including the points $y = y^0$ and $y = 0$ and have derivatives of sufficient orders. They can also depend continuously on ε. The domain D_y can be considered as an n-dimensional ball of radius ρ ($\rho = \text{const} > |y^0|$), having its centre at the origin.

The control vector function u is supposed to be subject to the constraint $u \in U$:

$$U = \big\{u\colon\ \big|u^2\big| \leqslant u^{02},\ u^0 = \text{const} > 0\big\}. \tag{7.1.2}$$

Without loss of generality, we can consider that $u^0 = 1$ in (2) (after the change of time $t = t_*/u^0$). We suppose that the uncontrolled unperturbed system (1), i.e. when

$u \equiv 0$ and $\varepsilon = 0$, is a system with invariant norm [28]:

$$\begin{aligned} &(\eta \cdot f_0(y)) \equiv 0, \qquad y \in D_y, \\ &\eta = y/h, \qquad h = |y| \in \{0, h^0\}, \qquad h^0 = |y^0|. \end{aligned} \tag{7.1.3}$$

Here, and in what follows, η is the unit vector of the vector y and $(\cdot)$ means scalar (dot) product. It follows from (3) that $h(t) \equiv h^0$ when $\varepsilon = 0$ and $u \equiv 0$; therefore $|y_i(t)| \leqslant h^0$, $i = 1, \ldots, n$. In mechanics, forces possessing the property of type (3) are called gyroscopic [1, 7, 9, 11, 21, 91, 97, 111, 127]; their "power" at each time instant is equal to zero.

Feedback time-optimal control bringing the unperturbed system (1) to the origin is determined by using the functional inequality of Schwartz [28, 49]

$$\begin{aligned} &-1 \leqslant \dot{h} = (\eta \cdot u) \leqslant 1, \qquad u_0(y) = -\eta; \\ &h_0 = h_0(t, h^0) = h^0(1 - t/T_0), \qquad T_0 = T_0(h^0) = h^0. \end{aligned} \tag{7.1.4}$$

Here, $u_0(y)$ is the optimal stabilizing control in the feedback form, T_0 is the minimal time. Integrating the initial-value problem after substitution of the function $u_0(y) = -\eta$ into system (1), where $\varepsilon = 0$, we obtain the unperturbed optimal state trajectory and open-loop control

$$y = y_0(t, y^0), \qquad u_0(t, y^0) = -\eta(t, y^0) = -y_0/h_0. \tag{7.1.5}$$

Below, the function $y_0(t, y^0)$ is assumed known.

The change $y = zh_0$ results in the initial-value problem for the unknown z:

$$\dot{z} = h_0^{-1} f_0(h_0 z), \qquad z(0) = \eta^0 \quad (\dot{h}_0 = -1). \tag{7.1.6}$$

Due to the assumptions of smoothness adopted above, $f_0(0) = 0$ and, therefore, the initial-value problem (6) has a unique solution in D_y. If f_0 is a homogeneous function of degree $m \geqslant 1$, then (6) is reduced to the form

$$\frac{dz}{ds} = f_0(z), \qquad z(0) = \eta^0, \qquad s = \int_0^t h_0^{m-1}(\tau, h^0)\, d\tau, \tag{7.1.7}$$

i.e. to the system with invariant norm. Thus, in this case it is sufficient to know the general solution of the system with $u \equiv 0$ and $\varepsilon = 0$ for which the norm is the first integral.

Bellman's function for the unperturbed control problem, i.e. the non-negative smooth solution of the corresponding initial-value problem for the Hamilton–Jacobi equation [127] (or Bellman's equation [40]) derived on the basis of sufficient conditions of the dynamic programming method [32, 40, 43, 55, 65, 105, 106, 111, 131, 144], is $T_0(y) = h$. The optimal control $u_0(y)$ is obtained on the basis of $T_0(y)$: $u_0(y) = -(\partial T_0/\partial y)|\partial T_0/\partial y|^{-1} = -\eta$ (see below).

The control $u_0(y)$ (4) provides full stabilization for the perturbed system (1) when ε is sufficiently small. This assertion follows from the equation for $h = |y|$:

$$\dot{h} = -1 + \varepsilon(\eta \cdot (f - F\eta)), \qquad h(0) = h^0, \tag{7.1.8}$$

and the boundedness of the factor of ε when $y \in D_y$. The error with respect to the minimal time and the state trajectory in the general case is a quantity $O(\varepsilon)$. In some applied problems, it may be necessary to take perturbation into account. In this case, the optimal control problem is to be solved with a higher accuracy in terms of ε.

We pose the following problem of the optimal control of the perturbed system (1). It is required to find, with a given accuracy in terms of ε, the feedback time-optimal control law $u = u(y, \varepsilon)$, the minimal time $T = T_0(y^0, \varepsilon)$ necessary to stabilize the system, and to construct the state trajectory $y(t, y^0, \varepsilon)$ such that

$$\begin{aligned} &y(0, y^0, \varepsilon) = y^0, \qquad y(T(y^0, \varepsilon), y^0, \varepsilon) = 0, \\ &y(t, y^0, 0) = y_0(t, y^0), \end{aligned} \tag{7.1.9}$$

and the corresponding open-loop control $u(t, y^0, \varepsilon)$.

7.1.2. CONTROLLED ROTATIONS OF A RIGID BODY

As an important example of the system of type (1) consider the dynamic Euler equations (1.1.24) in the component representation:

$$I_1\dot{\omega}_1 + (I_3 - I_2)\omega_2\omega_3 = M_1, \qquad \omega_1(0) = \omega_1^0 \quad (1, 2, 3). \tag{7.1.10}$$

Here, numbers in parentheses mean the cyclic permutation of indices, $M_i = b_i u_i$ ($i = 1, 2, 3$) are control torques with respect to axes fixed within the body, $b_i = \text{const} > 0$, and u_i are control variables constrained by the inequality (2). We state the problem of transferring the state point ω_i of system (10) from the initial state ω_i^0 to the origin ($\omega_i(T) = 0$, $i = 1, 2, 3$) in the minimal possible time T, i.e. the time-optimal stabilization problem. In the general case, we do not succeed in constructing an analytical solution. To find the relationships between the parameters I_i and b_i for which equality (3) takes place we introduce the variables $z_i = L_i b_i^{-1}$ where $L_i = I_i\omega_i$ are components of the angular momentum vector. Equations (10) are reduced to the form

$$\dot{z}_1 + (I_3 - I_2)I_2^{-1}I_3^{-1}b_2b_3b_1^{-1}z_2z_3 = u_1, \quad z_1^0 = I_1\omega_1^0 b_1^{-1} \quad (1,2,3). \qquad (7.1.11)$$

The invariance condition (3) for system (11) is fulfilled if parameters I_i and b_i are connected by the relationships

$$\sum_{i=1}^{3} \frac{I_{i+2} - I_{i+1}}{b_i} \frac{b_{i+1}}{I_{i+1}} \frac{b_{i+2}}{I_{i+2}} = 0, \qquad I_{i+3} \equiv I_i, \qquad b_{i+3} = b_i, \qquad (7.1.12)$$

or

$$I_1(I_3 - I_2)b_2^2b_3^2 + I_2(I_1 - I_3)b_1^2b_3^2 + I_3(I_2 - I_1)b_1^2b_2^2 = 0.$$

Using the "triangle inequality" for I_i we ascertain that the set of such values of parameters I_i and b_i is nonvoid. Equality (12) can be represented in a more convenient form ($I_i \neq I_j$, $i \neq j$):

$$b_i = I_i^{1/2}(\mu I_i + \nu)^{-1/2}, \qquad i = 1, 2, 3. \qquad (7.1.13)$$

Here, μ and ν are arbitrary constants such that the expression under the square root sign is positive. In applications, the following special cases of conditions (12) and (13) are encountered.

1. A rigid body with arbitrary moments of inertia $I_3 \geqslant I_2 \geqslant I_1 > 0$: a) $b_i = b$ ($i = 1, 2, 3$), vector $\mathbf{z}$ is collinear to $\mathbf{L}$; $\mathbf{L} = b\mathbf{z}$; b) $b_i = bI_i^{1/2}$, $z_i = I_i^{1/2}b^{-1}\omega_i$ ($b_3 \geqslant b_2 \geqslant b_1 > 0$), the square of the norm of vector $\mathbf{z}$ is proportional to the kinetic energy: $E = (1/2)bz^2$.

2. For a rigid body with dynamic symmetry ($I_1 = I_2 = I$), equality (12) is fulfilled when $b_1 = b_2 = b$; I_3 and b_3 being arbitrary. If $Ib^{-1} = I_3b_3^{-1}$, then vector $\mathbf{z}$ is collinear to the angular velocity: $\mathbf{z} = Iv^{-1}\omega$.

3. For a rigid body with spherical symmetry ($I_i = I$), relationship (12) is fulfilled for any $b_i > 0$, $i = 1, 2, 3$.

Thus, if equalities (12) and (13) are valid, then, according to (4), the feedback control which brakes rotations and the minimal time are

$$\begin{aligned} &u_i = -z_i z^{-1}, \quad T_0(z) = z, \quad z = |\mathbf{z}| = \left(z_1^2 + z_2^2 + z_3^2\right)^{1/2}, \\ &z = z_0\left(t, z^0\right) = z^0\left(1 - t/T_0\left(z^0\right)\right), \quad z_i = b_i^{-1}I_i\omega_i, \qquad i = 1, 2, 3. \end{aligned} \qquad (7.1.14)$$

The optimal controlled motion $z_{i,0}(t)$ is constructed on the basis of the motion of a free rigid body (Euler and Poinsot); this motion being described in terms of elliptic functions of Jacobi [1, 87, 127, 154] in Case 1 of a nonsymmetrical rigid body. Omitting details, we can represent the angular velocity components as follows (for details, see Section 6.4)

$$\omega_i = \omega_{i0}\big(t+\tau, L^0, E^0\big); \qquad \tau, L^0, E^0 = \text{const}, \quad i = 1,2,3. \tag{7.1.15}$$

Here, ω_{i0} are expressed in terms of Jacobi elliptic functions sn, cn, and dn [87, 154]. The solution for the controlled system (10) has the form $L_i = zl_i$ where the unknowns l_i satisfy the set of Euler's equations governing free rotations:

$$\begin{gathered}\frac{dl_1}{ds} + \frac{I_3 - I_2}{I_2 I_3} l_2 l_3 = 0, \qquad l_1(0) = \frac{L_1^0}{z^0} \quad (1,2,3),\\ s = \int_0^t z_0\big(\theta, z^0\big)\,d\theta = z^0 t\left(1 - \frac{t}{2T_0}\right), \qquad T_0 = z^0.\end{gathered} \tag{7.1.16}$$

As a result, we obtain the following representation for the optimal trajectory ω_i

$$L_i = z_0\omega_{i0}\big(s + s^0, L^0/z^0, E^0/z^{02}\big), \qquad s^0 = \text{const}. \tag{7.1.17}$$

It is evident that $L_i|_{T_0} = 0$, since $z_0(T_0, z^0) = 0$ according to (14).

7.1.3. CONSTRUCTION OF APPROXIMATE TIME-OPTIMAL FEEDBACK CONTROL ON THE BASIS OF THE DYNAMIC PROGRAMMING METHOD

We construct a non-negative differentiable Bellman function $T(y,\varepsilon)$ which becomes $T_0(y)$ when $\varepsilon = 0$. This satisfies the equation and the boundary condition

$$\begin{gathered}\left(\frac{\partial T}{\partial y}\cdot\big[f_0(y) + \varepsilon f(y)\big]\right) + \min_{u^2 \leqslant 1}\left(\frac{\partial T}{\partial y}\cdot\big[I + \varepsilon F(y)\big]u\right) = -1,\\ T(0,\varepsilon) \equiv 0.\end{gathered} \tag{7.1.18}$$

The minimization of the linear (in u) function in (18) leads to the closed initial-value problem for the nonlinear partial differential equation

$$\begin{gathered}\left(\frac{\partial T}{\partial y}\cdot\big[f_0(y) + \varepsilon f(y)\big]\right) - \left|\frac{\partial T}{\partial y}\big[I + \varepsilon F(y)\big]\right| = -1,\\ T(0,\varepsilon) \equiv 0,\end{gathered} \tag{7.1.19}$$

and to the expression for optimal control

$$u^* = -\frac{\partial T}{\partial y}(I+\varepsilon F)\left|\frac{\partial T}{\partial y}(I+\varepsilon F)\right|^{-1}, \qquad u^{*2} = 1. \tag{7.1.20}$$

The solution of problem (19) has a "cone-type" singular point [1, 6, 7] at the origin. It is constructed in the form of a series expansion in terms of powers of small parameter ε:

$$T(y,\varepsilon) = T_0(y) + \varepsilon T_1(y) + \cdots + \varepsilon^j T_j(y) + \cdots, \qquad T_j(0) = 0. \tag{7.1.21}$$

Here, unknown coefficients T_j are determined as a result of the consecutive solution of the linked initial-value problems

$$\begin{aligned} &\left(\frac{\partial T_0}{\partial y}\cdot f_0(y)\right) - \left|\frac{\partial T_0}{\partial y}\right| = -1, \qquad T_0(0) = 0;\\ &\left(\frac{\partial T_i}{\partial y}\cdot f_0(y)\right) - \left(\frac{\partial T_i}{\partial y}\cdot\frac{\partial T_0}{\partial y}\right)\left|\frac{\partial T_0}{\partial y}\right|^{-1} = V_j(y),\\ &T_j(0) = 0, \qquad j \geqslant 1. \end{aligned} \tag{7.1.22}$$

The first equation (22) is nonlinear but all subsequent equations are linear and have known right-hand sides $V_j(y)$ expressed through $T_0(y), T_1(y), \ldots, T_{j-i}(y)$ calculated in previous steps, for example

$$V_1(y) = \left(\frac{\partial T_0}{\partial y}\cdot\frac{\partial T_0}{\partial y}F\right)\left|\frac{\partial T_0}{\partial y}\right|^{-1} - \left(\frac{\partial T_0}{\partial y}\cdot f\right). \tag{7.1.23}$$

For arbitrary values of the index $j = 1, 2, \ldots,$ the functions $V_j(y)$ are determined by

$$\begin{aligned} &V_j = W_j - \left(\frac{\partial T_{j-1}}{\partial y}\cdot f\right), \qquad W_j = W_j\left(F, \frac{\partial T_0}{\partial y}, \ldots, \frac{\partial T_{j-1}}{\partial y}\right),\\ &\left|\frac{\partial T}{\partial y}(I+\varepsilon F)\right| = \left|\frac{\partial T_0}{\partial y}\right| + \varepsilon\left[\left(\frac{\partial T_1}{\partial y}\cdot\frac{\partial T_0}{\partial y}\right)\left|\frac{\partial T_0}{\partial y}\right|^{-1} + W_1\right] + \cdots +\\ &\quad + \varepsilon^j\left[\left(\frac{\partial T_j}{\partial y}\cdot\frac{\partial T_0}{\partial y}\right)\left|\frac{\partial T_0}{\partial y}\right|^{-1} + W_j\right] + \varepsilon^{j+1} + \cdots. \end{aligned} \tag{7.1.24}$$

It is immediately ascertained that $T_0(y) = h = |y|$ is the solution of the initial-value problem (22) for $j = 0$. Sufficient conditions for the optimal feedback control [40, 65] imply that $T_0(y)$ is a Bellman function for the unperturbed optimal control problem. It should be noted that the solution of the initial-value problem (22) for T_0 is not unique. For example, the function $T_0(y) = (c \cdot y)$, $|c| = 1$ (when $f_0 \equiv 0$) is also a solution but it does not meet sufficient conditions of optimality.

The scalar product $(\partial T_0/\partial y \cdot f_0) \equiv 0$ in (22) on the basis of (3), and the quantity $|\partial T_0/\partial y| \equiv 1$. Therefore, formal expansions (21)–(24) are valid for sufficiently small $\varepsilon > 0$. The function $T_0(y)$ determines the feedback control according to (20): $u_0(y) = -(\partial T_0/\partial y)|\partial T_0/\partial y|^{-1} = -\eta$.

The solution of the initial-value problem for $j \geqslant 1$ is constructed by using the method of characteristics [75]. This is possible, since the sum of the squares of the coefficients of the components of the vector $\partial T_j/\partial y$ is positive:

$$\left(f_0(y) - \eta\right)^2 = 1 + f_0^2(y) \geqslant 1 > 0.$$

Based on the above considerations, we can reduce equations of characteristics to the form

$$\frac{\mathrm{d}h}{-1} = \frac{\mathrm{d}y_1}{f_{01}(y) - \eta_1} = \cdots = \frac{\mathrm{d}y_n}{f_{0n}(y) - \eta_n} = \frac{\mathrm{d}T_j}{V_{j(y)}}. \tag{7.1.25}$$

Here, the first ratio is the derivative proportion with respect to n subsequent ones. The general integral of (25) can be represented, according to (5), in the form

$$y^* = y_0\left(t, y^0\right) = y_0\left(h^0 - h, y^0\right) = y^*\left(h, y^0\right). \tag{7.1.26}$$

According to (25) and (26), the expression for $T_j(y)$, vanishing at $y = 0$, is given by

$$T_j(y) = -\int_0^h V_j\left(y^*(l, y)\right)\mathrm{d}l, \qquad j = 1, 2, \ldots. \tag{7.1.27}$$

Thus, Bellman's function is determined uniquely in the form (21) and (27) with a given accuracy in terms of ε. The accuracy is limited by the smoothness properties of system (1). To give a foundation for the developed mode of constructing the approximate solution of the initial-value problem (19) one has to consider the problem of bringing the state point to a μ-vicinity of the origin in the minimal time T_μ and then to proceed to the limit as $\mu \downarrow 0$. It turns out (see [1, 75]) that the approach to the limit is uniform in ε and y ($|y| \geqslant 0$) and, as a result, we obtain expression (27) for $T_{0j}(y)$.

The optimal control $y^*(y,\varepsilon)$ is found as a result of substitution of $T(y,\varepsilon)$, according to (21) and (17), into (20). The function $u^*(y,\varepsilon)$ satisfies the constraint (2) exactly. But if one represents it in the form of expansion

$$\begin{aligned} u^*(y,\varepsilon) &= -\eta + \varepsilon u_1(y) + \cdots + \varepsilon^j u_j(y) + \varepsilon^{j+1} + \cdots \\ &= -\eta + \varepsilon \bar{u}_{(j)}(y,\varepsilon) + \varepsilon^{j+1} + \cdots, \end{aligned} \tag{7.1.28}$$

then the constraint (2) is satisfied with the error $O(\varepsilon^{j+1})$. The coefficients $u_1, \ldots, u_j$ are determined by differentiation of $T_j(y)$, $j \geqslant 1$, in particular

$$u_1(y) = \eta\left(\frac{\partial T_1}{\partial y} \cdot \eta\right) - \eta \frac{\partial T_1}{\partial y}\eta + U_1(y), \qquad U_1 \equiv -\eta F + \eta W_1. \tag{7.1.29}$$

The form similar to (29) also takes place for the coefficients u_j when $j \geqslant 1$; $U_j(y)$ being determined analogously to the scalar functions W_j in (24).

7.1.4. CONSTRUCTION OF APPROXIMATE OPTIMAL TRAJECTORY

Having substituted the control $u^*(y,\varepsilon)$ (28) into system (1), we obtain the initial-value problem for determining the optimal state trajectory $y = y(t, y^0, \varepsilon)$. The function y is constructed with a given accuracy in terms of ε, by using the method of successive approximations, in the form $y = y_0 + \varepsilon x$ where the unknown x satisfies the quasilinear system

$$\begin{aligned} \dot{x} &= \left(\frac{\partial f_0}{\partial y_0} h_0 - I\right)\frac{x}{h_0} + \eta_0 \frac{(\eta_0 \cdot x)}{h_0} + p(t) + \varepsilon P(t,x,\varepsilon), \\ p(t) &= f(y_0) + f(y_0)\eta_0 + u_1(y_0), \qquad x(0,\varepsilon) \equiv 0, \\ &\quad t \in [0, T_{(j)}(y^0,\varepsilon)], \quad T_{(j)} = T_0 + \varepsilon T_1 + \cdots + \varepsilon^j T_j. \end{aligned} \tag{7.1.30}$$

Here, $P(t,x,\varepsilon)$ is a known function, smooth in x in the domain considered. The solution of the initial-value problem (30) is constructed by successive approximations according to the following scheme:

$$\begin{aligned} x_k(t,\varepsilon) &= x_0(t) + \varepsilon X(t) \int_0^t X^{-1}(\tau) P(\tau, x_{k-1}, \varepsilon)\, d\tau, \\ &\quad k = 1, \ldots, j-1; \\ x_0(t) &= X(t) \int_0^t X^{-1}(\tau) p(\tau)\, d\tau, \qquad X(t) = \frac{\partial y_0}{\partial y^0}, \quad X(0) = I. \end{aligned} \tag{7.1.31}$$

Here, $X(t)$ is the fundamental matrix for the unperturbed system for variations. Successive approximations (31) determine the unique solution $x_{(j)}(t,\varepsilon)$ with the error $O(\varepsilon^j)$, and $y_{(j)} = x_0 + \varepsilon x_{(j)}$ give the optimal state trajectory with the error $O(\varepsilon^{j+1})$, $y_{(j)}(T_{(j)}(y^0,\varepsilon), y^0, \varepsilon) = O(\varepsilon^{j+1})$. It is worth noting that a singularity does not appear in (31) as $t \to T_{(j)}$.

One can also use another technique for the integration of the initial-value problem (1). For a system with $\varepsilon = 0$, suppose that the "complete set" of integrals of the form

$$c = C(y_0), \qquad \psi = \sigma(y_0) = t + \tau \quad (c, \tau = \text{const};\ y_0 = \hat{y}(\psi, c))$$

is known. Differentiating c and ψ according to the perturbed system (1) we obtain the equations in osculating variables for $t \in [0, T]$:

$$\begin{aligned}
&\dot{c} = \varepsilon \frac{\partial C}{\partial y_0} \varphi(y_0, \varepsilon), \qquad c^0 = C(y^0),\\
&\dot{\psi} = 1 + \varepsilon\left(\frac{\partial \sigma}{\partial y_0} \cdot \varphi(y_0, \varepsilon)\right), \qquad \psi^0 = \sigma(y^0), \quad y_0 = \hat{y}(\psi, c),\\
&\varphi(y_0, \varepsilon) \equiv f(y_0) + F(y_0) u^*_{(j)}(y_0, \varepsilon) + \bar{u}_{(j)}(y_0, \varepsilon).
\end{aligned} \tag{7.1.32}$$

The equation for the variable h has a form similar to Equation (32) for ψ:

$$\begin{aligned}
\dot{h} = &-1 + \varepsilon\left(\eta_0 \cdot \left[f(y_0) + \bar{u}_{(j)}(y_0, \varepsilon)\right]\right) +\\
&+ \varepsilon\left(\eta_0 \cdot F(y_0) u^*_{(j)}(y_0, \varepsilon)\right), \qquad h^0 = \left|y^0\right|, \quad \eta_0 = y_0 |y_0|^{-1}.
\end{aligned} \tag{7.1.33}$$

The unknowns c and ψ are constructed following the scheme of type (31):

$$\begin{aligned}
&c_k(t, \varepsilon) = c_0 + \varepsilon \int_0^t \frac{\partial C_{k-1}}{\partial y_0} \varphi_{k-1}\, d\tau,\\
&c_0 = C(y^0), \qquad k = 1, \ldots, j-1,\\
&\psi_k(t, \varepsilon) = \psi_0 + \varepsilon \int_0^t \left(\frac{\partial \sigma_{k-1}}{\partial y_0} \cdot \varphi_{k-1}\right) d\tau, \quad \psi_0 = t + \sigma(y^0),\\
&C_k \equiv C\big(\hat{y}(\psi_k, c_k)\big), \quad \sigma_k \equiv \sigma\big(\hat{y}(\psi_k, c_k)\big), \quad \psi_k \equiv \varphi\big(\hat{y}(\psi_k, c_k), \varepsilon\big).
\end{aligned} \tag{7.1.34}$$

Equation (33) for h is integrated analogously.

Substitution of the obtained approximate optimal trajectory $y_{(j)}(t, y^0, \varepsilon)$ into (20) or (28) leads to the open-loop control $u_{(j)}(t, y^0, \varepsilon)$ with the error $O(\varepsilon^{j+1})$ for $t \in [0, T_{(j)}(y^0, \varepsilon)]$.

Thus, we have solved the problem of bringing the state point of system (1) to the origin in minimum-time, i.e. the minimum-time stabilization problem, with the accuracy $O(\varepsilon^j)$. The issues concerning the foundation of the developed approach are considered in [1].

7.1.5. BRAKING ROTATIONS OF A RIGID BODY CLOSE TO THAT WITH SPHERICAL SYMMETRY

To illustrate this, consider the time-optimal stabilization problem for Case 3 in Section 7.1.2.

Let $I_i = I(1 + \varepsilon\varkappa_i)$, $i = 1, 2, 3$, where $\varepsilon \ll 1$, $\varkappa_i \sim 1$ and $b_i > 0$ are arbitrary ($b_i \sim 1$). Then the system (11) has the form

$$\begin{aligned} &\dot{z}_1 = u_1 - \varepsilon a_1 z_2 z_3, \qquad z_1(0) = z_1^0 \quad (z_1 \equiv b_1^{-1} I_1 \omega_1), \\ &a_1 = (b_2 b_3 / I b_1)(\varkappa_2 - \varkappa_3)(1 + \varepsilon\varkappa_3)^{-1}(1 + \varepsilon\tau_3)^{-1} \quad (1, 2, 3). \end{aligned} \tag{7.1.35}$$

Denote by r_i the derivatives $r_i = \partial T/\partial z_i$, where $T = T(\mathbf{z}, \varepsilon)$ is a Bellman function determined according to (19) and (20):

$$\begin{aligned} &R + \varepsilon(a_1 r_1 z_2 z_3 + a_2 z_1 r_2 z_3 + a_3 z_1 z_2 r_3) = -1, \qquad T(0, \varepsilon) \equiv 0, \\ &R = |\mathbf{r}|, \qquad \mathbf{r} = (r_1, r_2, r_3), \qquad \mathbf{u} = -\mathbf{r}R^{-1}. \end{aligned} \tag{7.1.36}$$

Using the technique of Section 7.1.3, at the first step ($j = 0$) we obtain the initial-value problem corresponding to the unperturbed system; as a result, we have

$$R_0 = 1, \quad T_0(0) = 0, \quad T_0(z) = |\mathbf{z}| = z, \quad \mathbf{r}_0 = \eta = \mathbf{z}/z. \tag{7.1.37}$$

The equation for T_1 is linear with a known nonhomogeneous term

$$\begin{aligned} &(\eta \cdot \mathbf{r}_1) = -a_1 \eta_1 z_2 z_3 - a_2 z_1 \eta_2 z_3 - a_3 z_1 z_2 \eta_3, \qquad T_1(\mathbf{0}) = 0, \\ &\eta = (\eta_1, \eta_2, \eta_3), \qquad \eta_i = z_i/z. \end{aligned} \tag{7.1.38}$$

The solution of the initial-value problem (38) is constructed uniquely by the method of characteristics [1, 75], since $\eta^2 = 1$ ($\neq 0$). As a result, we have

$$T_1(\mathbf{z}) = A z_1 z_2 z_3, \qquad A = -(1/3)(a_1 + a_2 + a_3). \tag{7.1.39}$$

Here and below, all values of the indices are equivalent. Next, the coefficients $T_j(\mathbf{z})$ ($j \geqslant 2$) are determined analogously; for instance,

$$\begin{aligned} T_2(\mathbf{z}) = -(1/30)Az^{-1}\{a_1\big[12(z_1z_2z_3)^2 + \big(5z_3^2 - z_1^2\big)z_2^4 + \\ + \big(5z_2^2 - z_1^2\big)z_3^4 - \big(z_2^2 + z_3^2\big)z_1^4\big] + a_2\big[12(z_1z_2z_3)^2 + \\ + \big(5z_3^2 - z_2^2\big)z_1^4 + \big(5z_1^2 - z_2^2\big)z_3^4 - \big(z_1^2 + z_3^2\big)z_2^4\big] + \\ + a_3\big[12(z_1z_2z_3)^2 + \big(5z_1^2 - z_3^2\big)z_2^4 + \big(5z_2^2 - z_3^2\big)z_1^4 - \\ - \big(z_1^2 + z_2^2\big)z_3^4\big]\}. \end{aligned} \tag{7.1.40}$$

The length of the expressions for $T_j(\mathbf{z})$ rapidly increases as j increases. The estimates $|T_j| \sim (|a_1| + |a_2| + |a_3|)z^{2j+1}$ imply that the expansion of the type (21) will uniformly converge to the unique function $T = T(\mathbf{z}, \varepsilon)$, smooth everywhere apart from the point $\mathbf{z} = 0$, for sufficiently small ε:

$$0 \leqslant \varepsilon \leqslant \varepsilon_0 < \big(|a_1| + |a_2| + |a_3|\big)^{-1}\big(z^0\big)^{-2}. \tag{7.1.41}$$

It is worth noting that $T_j(\mathbf{z}) \equiv 0$ for $A = 0$, since in this case, the invariance condition (13) is satisfied exactly and $T(\mathbf{z}) = z$.

The optimal control, with the error $O(\varepsilon^2)$, is given by

$$u^*_{(1)1}(\mathbf{z}, \varepsilon) = -\eta_1 + \varepsilon\big[\eta_1(\eta_2 r_{2,1} + \eta_3 r_{3,1}) - \big(\eta_2^2 + \eta_3^2\big)r_{1,1}\big] \quad (1, 2, 3). \tag{7.1.42}$$

According to (8), the equation for z after substitution of the control of the first approximation $\mathbf{u}^*_{(1)} = -\boldsymbol{\eta} + \varepsilon\mathbf{u}_1$ becomes

$$\dot{z} = -1 + 3\varepsilon T_1 z^{-1} + \varepsilon(\boldsymbol{\eta} \cdot \mathbf{u}_1) + \varepsilon^2 + \cdots, \qquad z(0) = z^0. \tag{7.1.43}$$

Since $(\boldsymbol{\eta} \cdot \mathbf{u}_1) \equiv 0$, (43) implies that

$$z = z^0 - t + \varepsilon T_1\big(z^0\big)\big(z^0\big)^{-3}\big[(z^0)^3 - \big(z^0 - t^3\big)\big] + \varepsilon^2 + \cdots.$$

For the value $t = T_{(1)} = z^0 + \varepsilon T_1(\mathbf{z}^0)$, the quantity $z(T_{(1)}, \varepsilon) = O(\varepsilon^2)$, i.e. the vector $\mathbf{z}$, and, hence, $\boldsymbol{\omega}$, comes to an ε^2-vicinity of the origin. If we use the equation of the first approximation (42) to provide bringing system (35) to a state of rest, the error will be $O(\varepsilon^2)$ in terms of the functional: the necessary time will be greater by the amount mentioned. A similar reasoning is valid for higher approximations

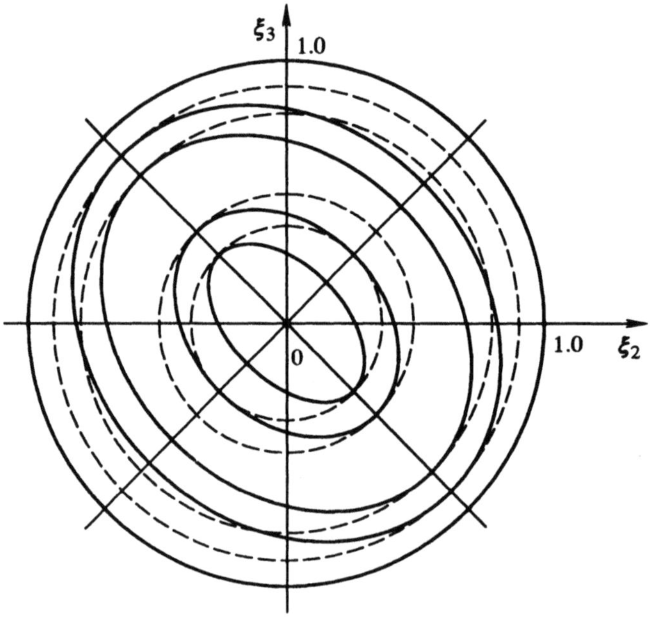

Fig. 7.1.

as well. It should be noted that $T_j(\mathbf{z}) \equiv 0$, $j = 1, 2, \ldots$, when $b_i = b$, i.e. $A = 0$, $T = z = b^{-1}L$ is Bellman's function of the "unperturbed" optimal control problem ($\mathbf{L} = (L_1, L_2, L_3)$, $L_i = I_i\omega_i$).

Figure 7.1 shows cross-sections of the level surfaces for the function T to the first approximation

$$T_{(1)} = z + \varepsilon T_1(\mathbf{z}) = c > 0$$

by planes $z_1 = \text{const}$. By introducing new variables $\xi_i = z_i/c$, the equation of the level surface is reduced to the form

$$\xi_1^2 + \xi_2^2 + \xi_3^2 - 2\alpha\xi_1\xi_2\xi_3 = 1, \qquad \text{where} \quad \alpha = \varepsilon A c^2.$$

Cross-sections $\xi_1 = \text{const}$ $(1 \geqslant \xi_1 > 0)$ are ellipses with small eccentricities $O(\varepsilon)$. The axes of these ellipses are rotated through the angle $\pi/4$ with respect to the coordinate axes. The values of the parameter α for Fig. 7.1 were taken as $\alpha = 0$ and $\alpha = 0.3$. The cross-sections by planes $z_{2,3} = \text{const}$ coincide with those given in Fig. 7.1.

7.2. Approximate Synthesis of Optimal Control of Motion with Respect to a Part of the Variables

7.2.1. STATEMENT OF THE CONTROL PROBLEM

Consider a perturbed controlled system of a form more general than in Section 7.1:

$$\begin{aligned}
&\dot{x} = F(x) + u + \varepsilon f(x, y, u), \qquad x(0) = x^0, \\
&\dot{y} = G(x, y) + \varepsilon g(x, y, u), \qquad y(0) = y^0, \\
&\quad u \in U_\varepsilon(x, y), \quad (x, y, \varepsilon) \in D_x \times D_y \times [0, \varepsilon_0].
\end{aligned} \tag{7.2.1}$$

Here, x is a controlled n-vector, the control $u \in U_\varepsilon$ has the same dimension, y is an m-vector ($m \geqslant 0$), $\varepsilon \in [0, \varepsilon_0]$ is a numerical parameter, $t \in [0, T]$ is time; and x^0 and y^0 are initial data. The functions F, f, G, and g are supposed to be smooth in the domain of the arguments x, y, and u, and may continuously depend on ε. Moreover, the function F satisfies the condition

$$\big(\eta \cdot F(x)\big) = \varphi(h), \qquad \eta = x/h, \quad h = |x|. \tag{7.2.2}$$

As to the closed bounded domain U_ε, we suppose that it may smoothly depend on x, y, and ε, i.e. $U_\varepsilon = U_\varepsilon(x, y)$; for $\varepsilon = 0$, the unperturbed domain $U_\varepsilon = U_0$ having the form considered above, in Section 7.1:

$$U_0 = \{u\colon\ |u| \leqslant 1\}, \qquad |u| = \left(u_1^2 + \cdots + u_n^2\right)^{1/2}.$$

For system (1) we pose the time-optimal control problem with respect to part of the variables (the vector x):

$$x(T) = 0, \qquad T \longrightarrow \min_u, \qquad u \in U_\varepsilon(x, y). \tag{7.2.3}$$

The value $y(T)$ is not fixed.

If x is a generalized velocity vector and y is a vector, comprising coordinates, variable parameters and time, then we have the time-optimal stabilization problem for system (1) considering perturbations from a wider class than in Section 7.1. This generalization allows us to study a number of applied problems of controlling the velocity of the centre of mass of a spacecraft (or the motion with respect to the centre of mass) taking into account perturbations of a different physical nature.

7.2.2. ANALYSIS OF AN UNPERTURBED CONTROL PROBLEM

The investigation of the original minimum-time problem is carried out on the basis of the sufficient condition of the method of dynamic programming and, from a methodical point of view, relates to Section 7.1 This approach essentially uses the solution of the generating problem, i.e. (1) and (3) for $\varepsilon = 0$:

$$\begin{aligned} &\dot{x} = F(x) + u, \qquad x(0) = x^0; \\ &x(T) = 0, \qquad T \longrightarrow \min, \qquad |u| \leqslant 1. \end{aligned} \tag{7.2.4}$$

Based on the Schwartz inequality [28, 49] (see Section 7.1), the following relationships are established for the construction of the unperturbed solution

$$\begin{aligned} &\dot{x} = F(x) + u_0(x), \quad x(0) = x^0; \quad u_0(x) = -\eta, \\ &(\dot{x} \cdot \eta) = \dot{h} = -1 + \varphi(h), \quad h = |x|, \quad h(0) = h^0 = |x^0|, \\ &t = \theta(h, h^0) = \int_h^{h^0} \frac{\mathrm{d}l}{1 - \varphi(l)}, \quad T_0(h) = \theta(0, h). \end{aligned} \tag{7.2.5}$$

Here, $T_0(h^0)$ is the minimal time; it being suggested that $\varphi(h) < 1$ for $h \in [0, h^0]$.

By the substitution $x = vh_0(t, h^0)$, where v is an unknown variable, Equation (5) for x is reduced to the form of systems with an invariant norm:

$$\dot{v} = h_0^{-1} F(vH_0) - h_0^{-1} \varphi(h_0) v, \qquad v(0) = \eta^0. \tag{7.2.6}$$

Indeed, multiplying (in the sense of the dot product) Equation (6) by v and taking (2) and (6) into account one can ascertain that $|v(t)| \equiv 1$. If the difference $F - \varphi v$ is a homogeneous function of h of degree $\alpha > 0$, then Equation (6) is reduced to the time-invariant form:

$$\frac{\mathrm{d}v}{\mathrm{d}s} = F(v) - \varphi(1) v, \qquad s = \int_0^t h_0^{\alpha-1}(\tau, h^0)\, \mathrm{d}\tau, \quad s \in [0, \sigma(h^0)]. \tag{7.2.7}$$

Here, s is a new argument, $s = s(t, h^0)$, $\sigma(h^0) = s(\theta(0, h^0), h^0)$. In what follows, the general solution $x_0^*(t, x^0)$ of the initial-value problem (5) is supposed to be known for any $x^0 \in D_x$. The equation for y (1) for $\varepsilon = 0$ and known x_0 has the form

$$\dot{y} = G(x_0^*(t, x^0), y), \qquad y(0) = y^0. \tag{7.2.8}$$

The general solution of the initial-value problem (8) for $y^0 \in D_y$ is also supposed to be known: $y = y^*(t, x^0, y^0)$.

Thus, the solution of the unperturbed system (1) for $u = -\eta$ can be represented in two forms:

$$\begin{aligned} &x = x_0^*(t, x^0), \quad y = y_0^*(t, x^0, y^0), \quad t \in [0, T_0]; \\ &x = x_0(h, h^0, \eta^0), \quad y = y_0(h, h^0, \eta^0, y^0), \quad h = h_0(t, h^0). \end{aligned} \tag{7.2.9}$$

The functions x_0 and y_0 are obtained from x_0^* and y_0^* when substituting $t = \theta(h, h^0)$ according to (5); they satisfy the initial conditions when $h = h^0$ and, moreover, x_0 satisfies the terminal condition when $h = 0$. In the important special case of Equation (8), $\dot{y} = x$ ($G \equiv x$), i.e. when y is a coordinate and x is a velocity, the function y_0 is expressed in terms of x_0 by the quadrature

$$y_0(h, h^0, \eta^0, y^0) = y^0 - \int_{h^0}^{h} \frac{x_0(l, h^0, \eta^0)}{1 - \varphi(l)} \, dl. \tag{7.2.10}$$

Thus, the solution of the unperturbed minimum-time problem (4) is completely determined in the form (5) and (9) or (10).

7.2.3. CONSTRUCTION OF THE PERTURBED BELLMAN'S FUNCTION

Consider the control problem (1)–(3) for $\varepsilon > 0$. It is required to find, with a given accuracy in terms of ε, the optimal control in the feedback form $u(x, y, \varepsilon)$, the corresponding value of the minimal time $T(x^0, y^0, \varepsilon)$ (Bellman's function $T(x, y, \varepsilon)$) as well as the optimal state trajectory, $x = x(t, x^0, y^0, \varepsilon)$, $y = y(t, x^0, y^0, \varepsilon)$, and the open-loop control $u(t, x^0, y^0, \varepsilon)$.

The smoothness of the system implies an ε-proximity of the perturbed and unperturbed trajectories. Therefore, for an approximate construction of the open-loop control it is required that the smoothness conditions are fulfilled only in a small vicinity of the unperturbed optimal motion. When constructing the feed-back control, we suppose that the smoothness condition is fulfilled in all the domain $(x, y) \in D_x \times D_y$ where the state point of the system may find itself and for all $\varepsilon \in [0, \varepsilon_0]$, $u \in U_\varepsilon(x, y)$.

The Hamilton–Jacobi–Bellman equation and the condition on the hyperplane $x = 0$ have the form

$$\begin{aligned} &\min_{u \in U_\varepsilon} \left[(P \cdot (F + u + \varepsilon f)) + (Q \cdot (G + \varepsilon g)) \right] = -1, \\ &P = \partial V / \partial x, \quad Q = \partial V / \partial y, \quad V = V(x, y, \varepsilon), \quad V(0, y, \varepsilon) = 0. \end{aligned} \tag{7.2.11}$$

The construction of the closed initial-value problem for Bellman's function V leads, according to (11), to the minimization of a nonlinear function of u over the closed bounded domain $U_\varepsilon(x, y)$. We admit that U_ε has a sufficiently smooth boundary but may be specified in rather a sophisticated way. To determine this, we suppose that it is described by a single inequality of the form

$$U_\varepsilon = \{u\colon\ u^2 + \varepsilon R(u, x, y) \leqslant 1\}, \qquad U_0 = \{u\colon\ u^2 \leqslant 1\}, \tag{7.2.12}$$

where R is a sufficiently smooth function of u, x, and y. To solve the minimization problem in (11) we reduce the domain U_ε (12) to the form U_0 by using a smooth nonsingular transformation of the vector u. Indeed, assuming $u = \rho\nu$ where $\rho = |u|$ is the modulus of the vector u, and ν is the vector of "direction cosines", we obtain the equation of the boundary of the domain U_ε (12):

$$\rho = \left[1 - \varepsilon r(\rho)\right]^{1/2}, \qquad r(\rho) \equiv R(\rho\nu, x, y). \tag{7.2.13}$$

Equation (13) is then solved with respect to ρ; ν, x, y, and ε being parameters. To determine the root ρ_* as a function of ν, x, y, and ε one may apply the usual method of expanding in terms of powers of ε:

$$\begin{aligned}\rho_{(i+1)} &= \rho_0 + (\rho_{(1)} - \rho_0) + (\rho_{(2)} - \rho_{(1)}) + \cdots + \\ &\qquad + (\rho_{(i)} - \rho_{(i-1)}) + (\rho_{(i+1)} - \rho_{(i)}), \qquad \rho_{(0)} = \rho_0, \\ \rho_{(i+1)} - \rho_{(i)} &= \varepsilon^{i+1}\rho_{i+1}, \quad \rho_0 = 1, \qquad i = 0, 1, \ldots .\end{aligned} \tag{7.2.14}$$

It is also possible to use the recursive scheme of the method of successive approximations which does not require a high smoothness of the function $R = r(\rho)$ and is suitable for computer implementation. This approach gives a power-type convergence in terms of ε for sufficiently small $\varepsilon > 0$:

$$\begin{aligned}&\rho_{(i+1)} = \left[1 - \varepsilon r(\rho_{(i)})\right]^{1/2}, \quad \rho_{(0)} = \rho_0 = 1, \qquad i = 0, 1, \ldots, \\ &\rho_{(i+1)} - \rho_{(i)} = O(\varepsilon^{i+1}), \qquad \varepsilon|dr/d\rho| < 2.\end{aligned} \tag{7.2.15}$$

The method of Newton (or the method of tangents [49]) leads to quadratic convergence in terms of ε:

$$\begin{aligned}&\rho_{(i+1)} = \left[\mu(\rho_{(i)}) - \rho_{(i)} d\mu(\rho_{(i)})/d\rho\right]\left[1 - d\mu(\rho_{(i)})/d\rho\right]^{-1}, \\ &\mu(\rho) \equiv \left[1 - \varepsilon r(\rho)\right]^{1/2}, \qquad \rho_{(i+1)} - \rho_{(i)} = O\left(\varepsilon^{\zeta(i)}\right), \quad \zeta(i) = 2^i.\end{aligned} \tag{7.2.16}$$

Thus, let a unique smooth function $\rho_*(\nu, x, y, \varepsilon)$ be constructed. This can be used for the transformation $u \to \tilde{u}$ in the control space

$$u = \rho_* \tilde{u}, \qquad \rho_* = 1 + \varepsilon\delta(\nu, x, y, \varepsilon), \quad \tilde{u} = l\nu, \quad |l| \leqslant 1. \tag{7.2.17}$$

Here, $\tilde{u}$ is a control taking its values from U_0. Substitution of u into (11), according to (17), gives the relationship where minimization is carried out over $\tilde{u} \in U_0$; the left-hand side of the equation is "deformed" by an amount of $O(\varepsilon)$. The function $\tilde{u}$, $\tilde{u} \in U_0$, is determined from the minimality condition for the expression in (11) (the tilde is dropped below). Dividing by $|P| \neq 0$ we obtain the minimization problem of the form

$$\begin{gathered} J(p, \varepsilon) = \min_{u \in U_0} \left[(p \cdot u) + \varepsilon\Phi(u)\right], \qquad U_0 = \{u\colon\ |u| \leqslant 1\}, \\ p \equiv P|P|^{-1}, \quad |p| = 1; \qquad \Phi(u) \equiv (p \cdot f) + (Q \cdot g)|P|^{-1}. \end{gathered} \tag{7.2.18}$$

Here, Φ is a smooth function of u, $u \in U_0$; the functions J and u are to be defined. The dependence on other arguments is not indicated yet for compactness. The division by $|P|$ is justified for sufficiently small ε, since for $\varepsilon = 0$ this quantity is bounded and does not vanish. For the unperturbed problem (18) it follows that

$$u_0 = -p, \qquad J(p, 0) = -1. \tag{7.2.19}$$

Determination of the unknown functions $u(p, \varepsilon)$ and $J(p, \varepsilon)$ is carried out by using expansions in terms of powers of the parameter ε:

$$\begin{gathered} u = -p + \varepsilon u_1 + \cdots + \varepsilon^j u_j + \cdots, \\ J = -1 + \varepsilon J_1 + \cdots + \varepsilon^j J_j + \cdots. \end{gathered} \tag{7.2.20}$$

The coefficients u_j and J_j ($j \geqslant 1$) as functions of the unit vector p and other parameters are obtained as a result of substituting (20) into (18) and comparing the coefficients of like powers ε^j. In particular, for $j = 1$

$$\begin{gathered} J_1 = \min_{u_1}(p \cdot u_1) + \Phi_0, \qquad (u_0 \cdot u_1) \leqslant 0; \qquad \Phi_0 \equiv \Phi(u_0); \\ J_1 = \Phi(-p) = \Phi_0, \qquad (p \cdot u_1) = 0, \end{gathered} \tag{7.2.21}$$

since $u_0 = -p$. The vector u_1 is orthogonal to p, i.e. $(u_1 \cdot p) = 0$ $((u_0 \cdot u_1) = 0)$ and is defined from (21) nonuniquely. For $j = 2$,

$$J_2 = \min_{u_1,u_2}\left[(p \cdot u_2) + \left(\frac{\partial \Phi_0}{\partial u} \cdot u_1\right)\right]; \qquad \frac{\partial \Phi_0}{\partial u} \equiv \frac{\partial \Phi_0(-p)}{\partial u};$$
$$(p \cdot u_1) = 0, \qquad 2(u_0 \cdot u_2) + u_1^2 \leqslant 0. \tag{7.2.22}$$

The last constraint in (22) is equivalent to the inequality $(p \cdot u_2) \geqslant u_1^2/2$; therefore, we obtain the constrained minimization problem for the quadratic form of u_1:

$$J_2 = \min_{u_1}\left[\frac{1}{2}u_1^2 + \left(\frac{\partial \Phi_0}{\partial u} \cdot u_1\right)\right], \qquad (p \cdot u_1) = 0. \tag{7.2.23}$$

Using the method of Lagrange's multipliers [43, 63] we find the solution of problem (23) in explicit form:

$$u_1 = \left(\frac{\partial \Phi_0}{\partial u} \cdot p\right)p - \frac{\partial \Phi_0}{\partial u},$$
$$J_2 = -\frac{1}{2}\left[\left(\frac{\partial \Phi_0}{\partial u}\right)^2 - \left(\frac{\partial \Phi_0}{\partial u} \cdot p\right)^2\right] \leqslant 0. \tag{7.2.24}$$

As a result, we have found the functions $J_1(p)$, $J_2(p)$, $u_1(p)$, and u_2 is still to be determined. For $j = 3$, the following minimization problem takes place:

$$J_3 = \min_{u_2,u_3}\left[(p \cdot u_3) + \left(\frac{\partial \Phi_0}{\partial u} \cdot u_2\right)\right] + \frac{1}{2}\left(\frac{\partial^2 \Phi_0}{\partial u^2} u_1 \cdot u_1\right),$$
$$(p \cdot u_3) \geqslant (u_1 \cdot u_2), \qquad (p \cdot u_2) = \frac{1}{2}u_1^2. \tag{7.2.25}$$

Minimization over u_3, taking into account the equality (25) for u_2 and the expression (24) for u_1, allows one to determine J_3 and the constraints for u_2 and u_3:

$$J_3 = \frac{1}{2}\left(\frac{\partial \Phi_0}{\partial u} \cdot p\right)u_1^2 + \frac{1}{2}\left(\frac{\partial^2 \Phi_0}{\partial u^2} u_1 \cdot u_1\right), \qquad (p \cdot u_3) = (u_1 \cdot u_2). \tag{7.2.26}$$

The unknowns u_2 and u_3 are found from the minimality conditions for coefficients J_j ($j \geqslant 4$). In particular, from the condition for J_4 we have

$$J_4 = \min_{u_2,u_3,u_4} \left[(p \cdot u_4) + \left(\frac{\partial \Phi_0}{\partial u} \cdot u_3 \right) + \left(\frac{\partial^2 \Phi_0}{\partial u^2} u_1 \cdot u_2 \right) + \frac{1}{6} \frac{\partial^3 \Phi_0}{\partial u^3} u_1^3 \right],$$
$$(p \cdot u_4) \geqslant (u_1 \cdot u_3) + \frac{1}{2} u_2^2. \tag{7.2.27}$$

Minimizing over u_4 and eliminating the dependence on u_3 using the relationships (26) we obtain for J_4 a quadratic form of u_2. The quantities u_2 and J_4 are then determined by the method of Lagrange's multipliers:

$$u_2 = \frac{1}{2} u_1^2 p + (z \cdot p)p - z,$$
$$z \equiv \left(\frac{\partial \Phi_0}{\partial} \cdot p \right) u_1 + \frac{\partial^2 \Phi_0}{\partial u^2} u_1, \tag{7.2.28}$$
$$J_4 = \frac{1}{2} \left[\frac{1}{2} u_1^2 + (z \cdot p) \right]^2 - \frac{1}{2} z^2 + \frac{1}{6} \frac{\partial^3 \Phi_0}{\partial u^3} u_1^3.$$

In such a way we can determine any number of the coefficients u_j and J_j and construct, with required accuracy in terms of ε, the closed initial-value problem for (11). In the particular case when the form (18) is linear in u, the solution of the minimization problem is found exactly in the form

$$u = u(p, \varepsilon) = -(p + \varepsilon \Phi^1) \left| p + \varepsilon \Phi^1 \right|^{-1},$$
$$J = -\left| p + \varepsilon \Phi^1 \right| + \varepsilon \Phi^0; \tag{7.2.29}$$
$$\Phi(u) \equiv \Phi^0 + (\Phi^1 \cdot u).$$

In the general nonlinear case, the control u has the form

$$u = -p + \varepsilon \left[\left(\frac{\partial \Phi_0}{\partial u} \cdot p \right) p - \frac{\partial \Phi_0}{\partial u} \right] + \cdots + \varepsilon^j u_j(p) + \cdots,$$
$$p \equiv P|P|^{-1}, \quad \Phi_0 \equiv (p \cdot f(x, y, -p)) + (Q \cdot g(x, y, -p))|P|^{-1}. \tag{7.2.30}$$

The closed initial-value problem for Bellman's function V is obtained by substituting the known, according to (21)–(28), coefficients $J_j(p)$ into (18) and then into (11). As a result, we obtain

$$\begin{aligned}&(P\cdot F)-|P|+(Q\cdot G)+\\&\varepsilon\big[(P\cdot f(x,y,-p))+(Q\cdot(x,y,-p))\big]+\ldots+\varepsilon^j|P|J_j(p)+\cdots=-1,\\&V=V(x,y,\varepsilon),\qquad V(0,y,\varepsilon)\equiv 0.\end{aligned}\tag{7.2.31}$$

It should be noted that the reduction of the functional Equation (11) to the form (31) can be conducted on the basis of the expounded mode without the preliminary transformation (17) in the control space.

7.2.4. THE METHOD OF LAGRANGE'S MULTIPLIERS

The scheme of successive minimization of a quasilinear function developed in Section 7.2.3 is, essentially, equivalent to the application of the method of Lagrange's multipliers [43, 63]. Immediate application of this method to problem (18) leads to similar simple manipulations. For sufficiently small $\varepsilon > 0$, the minimum is achieved on the boundary of the domain U_0, since the gradient (with respect to u) of the function to be minimized does not vanish if $\varepsilon = 0$. As a result, we obtain the constrained minimization problem

$$J(p,\varepsilon)=\min_u\big[(p\cdot u)+\varepsilon\Phi(u)\big],\qquad u^2=1.\tag{7.2.32}$$

Considering Φ as a function continuously differentiable on the sphere $u^2 = 1$, one can write down the following equations:

$$p+\varepsilon\partial\Phi/\partial u+\lambda u=0,\qquad u^2=1,\tag{7.2.33}$$

where $\lambda/2$ is a scalar Lagrange's multiplier. Eliminating the unknown λ from (33) we obtain the equation relative to the vector u

$$u=-(p+\varepsilon\partial\Phi/\partial u)|p+\varepsilon\partial\Phi/\partial u|^{-1}.\tag{7.2.34}$$

Using one of the approximate methods given above we can determine from (34) the unknown function $u = u(p,\varepsilon)$ with required accuracy and then, according to (32), find $J = J(p,\varepsilon)$. If, in particular, Φ is linear in u, the expression (34) immediately gives $u(p,\varepsilon)$.

Minimization in (11) and (12) can be carried out by using the method of Lagrange's multipliers without preliminary transformation of the vector u according to (13)–(17). As a result, the constrained minimization problem of the type (32) is obtained:

$$J(p,\varepsilon) = \min_u \left[(p \cdot u) + \varepsilon\Phi(u)\right], \qquad u^2 + \varepsilon r(u) = 1, \tag{7.2.35}$$

which is reduced to the set of $(n+1)$ equations relative to u and λ:

$$p + \varepsilon\frac{\partial\Phi}{\partial u} + \lambda u + \frac{1}{2}\varepsilon\lambda\frac{\partial r}{\partial u} = 0, \qquad u^2 + \varepsilon r = 1. \tag{7.2.36}$$

Elimination of the parameter $\lambda > 0$ from (36) leads to the expression

$$\lambda = \left\{\frac{\varepsilon}{2}\left(\frac{\partial r}{\partial u}\cdot\left(p+\varepsilon\frac{\partial\Phi}{\partial u}\right)\right) + \left[\left(\frac{\varepsilon}{2}\right)^2\left(\frac{\partial r}{\partial u}\left(p+\varepsilon\frac{\partial\Phi}{\partial u}\right)\right)^2 + \right.\right.$$
$$\left.\left. + \left(1 - \varepsilon r - \left(\frac{\varepsilon}{2}\right)^2\left(\frac{\partial r}{\partial u}\right)\right)^2\left(p+\varepsilon\frac{\partial\Phi}{\partial u}\right)^2\right]^{1/2}\right\} \times$$
$$\times \left[1 - \varepsilon r - \left(\frac{\varepsilon}{2}\right)^2\left(\frac{\partial r}{\partial u}\right)^2\right]^{-1}.$$

Substitution of $\lambda = \lambda(p,u,\varepsilon) = 1 + O(\varepsilon)$ into the vector equation (36) for u leads to the quasilinear equation of the form

$$u = -\left(p + \varepsilon\frac{\partial\Phi}{\partial u}\right)\lambda^{-1} - \frac{\varepsilon}{2}\frac{\partial r}{\partial u} \equiv p - \varepsilon\Pi(p,u,\varepsilon). \tag{7.2.37}$$

The solution $u(p,\varepsilon)$ of Equation (37) exists, is unique, and is a smooth function of p ($|p| = 1$), since the function Π satisfies the condition of existence of the implicit function u. To calculate $u(p,\varepsilon)$ one may use one of the schemes of the small parameter method given above. This does not cause any difficulties. The exact solution of system (36) is found by elementary operations if Φ is linear and r is a quadratic function of u. In this case, U_ε is an ellipsoid close to the unit ball. Thus, the initial-value problem (31) is explicitly constructed with a given accuracy in terms of ε.

7.2.5. CONSTRUCTION OF BELLMAN'S FUNCTION AND THE OPTIMAL CONTROL

The unknown Bellman function V of the minimum-time problem is determined as a smooth nonnegative solution of the initial-value problem (31) in the form

$$V(x, y, \varepsilon) = V_0(x, y) + \varepsilon V_1(x, y) + \cdots + \varepsilon^j V_j(x, y) + \cdots. \tag{7.2.38}$$

As usual, substituting (38) into (31) and comparing the coefficients of like powers ε^j, one can obtain the linking sequence of the initial-value problems. At the first step, for $j = 0$, we have

$$(P_0 \cdot F) - |P_0| + (Q_0 \cdot G) = -1, \qquad V_0(0, y) \equiv 0. \tag{7.2.39}$$

Since the control u_0 does not depend on y when $\varepsilon = 0$, see (5), it implies that $V_0 = V_0(x) = T_0(h)$ and $u_0(x) = -p = -\eta$. Moreover, the norm $|P_0(x)|$ is positive and bounded for $h \in [0, h^0]$. Therefore, the conducted constructions are justified for sufficiently small $\varepsilon > 0$.

The initial-value problems for subsequent coefficients V_j ($j \geqslant 1$) are linear and have the form

$$(P_j \cdot [F(x) - \eta]) + (Q_j \cdot G(x, y)) = W_j(x, y), \qquad V_j(0, y) \equiv 0. \tag{7.2.40}$$

Here, the right-hand sides W_j are known; they are defined by f, g, and the coefficients $V_0(x), V_1(x, y), \ldots, V_{j-1}(x, y)$ calculated on previous steps. In particular,

$$\begin{aligned} W_1(x, y) &= \big(P_0(x) \cdot f(x, y, -\eta)\big), \\ P_0 &= \partial V_0/\partial x = \eta/(1 - \varphi(h)). \end{aligned} \tag{7.2.41}$$

Solutions of the initial-value problems (40) are constructed, taking expressions (9) into account, by the method of characteristics generalizing the technique of Section 7.1

$$V_j(x, y) = -\int_0^h W_j\big(x_0(l, h, \eta), y_0(l, h, y)\big) \frac{dl}{1 - \varphi(l)}. \tag{7.2.42}$$

Thus, the construction of the unknowns $V_j(x, y)$ is reduced to integration of some functions $W_j(x, y)$ along families of trajectories (9). The accuracy of the calculations is determined by the degree of a system's smoothness, i.e. by the differential properties of (31) with respect to x, y, and p. Then, based on the known expression (38) for Bellman's function $V(x, y, \varepsilon)$, the feedback optimal control

$$\begin{aligned} &u(x, y, \varepsilon) = -\eta + \varepsilon u_1(x, y) + \cdots + \varepsilon^j u_j(x, y) + \cdots, \\ &u_1(x, y) = -\left[P_1 - (\eta \cdot P_1)\eta\right]\left[1 - \varphi(h)\right] + \left(\frac{\partial \delta}{\partial u} \cdot \eta\right)\eta - \frac{\partial \delta}{\partial u}, \\ &\eta = x/h, \quad h = |X|, \quad P_1 = \partial V_1/\partial x, \quad \delta \equiv \left(\eta \cdot f(x, y, -\eta)\right). \end{aligned} \tag{7.2.43}$$

is determined with the same accuracy. Subsequent coefficients $u_j(x, y)$ ($j \geqslant 2$) are rather bulky. They are not given here because one usually confines oneself to the first approximation when estimating the influence of perturbations.

It should be noted that the approximate control (43) satisfies the constraint $u \in U_\varepsilon$ (12) not exactly but with the error $O(\varepsilon^{j+1})$. Compulsory constraint of the function leads to the same error in terms of the functional and state trajectory. This error does not exceed the limits of accuracy of the calculations. The substantiation of the method is contained in [1, 11].

7.2.6. CONSTRUCTION OF THE STATE TRAJECTORY

After substitution of the control $u = u(x, y, \varepsilon)$ (43) into system (1) we obtain the closed initial-value problem for determining the optimal trajectory, $x = x(t, x^0, y^0, \varepsilon)$ and $y = y(t, x^0, y^0, \varepsilon)$. The unknowns x, y are to be calculated with the accuracy ε^j on the time interval $t \in [0, T(x^0, y^0, \varepsilon)]$. The solution to be found is constructed by numerical or analytical small parameter methods. For example, one may apply the technique of Section 7.1 using the reduction to osculating variables on the basis of known integrals of the unperturbed system:

$$\begin{aligned} &c_x = C_x(x), \qquad t + \tau = \int \frac{dh}{1 - \varphi(h)}, \qquad c_y = C_y(x, y); \\ &x = \hat{x}(t + \tau, c_x, \qquad y = \hat{y}(t + \tau, c_x, c_y). \end{aligned} \tag{7.2.44}$$

As a result, system (1) is reduced to the form [1, 11]

$$\begin{aligned} &\dot{c}_x = \varepsilon \frac{\partial C_x}{\partial x} X(\hat{x}, \hat{y}), \qquad c_x(0) = C_x(x^0), \\ &\dot{c}_y = \varepsilon \frac{\partial C_y}{\partial x} X(\hat{x}, \hat{y}) + \varepsilon \frac{\partial C_y}{\partial x} Y(\hat{x}, \hat{y}), \qquad c_y(0) = C_y(x^0, y^0), \\ &\dot{h} = -1 + \varphi(h) + \varepsilon\left(\eta \cdot X(\hat{x}, \hat{y})\right), \qquad h(0) = \left|x^0\right|. \end{aligned} \tag{7.2.45}$$

Approximate integration of system (45) is carried out by using successive approximations or expansions in terms of powers of ε (see Section 7.1). As a result, we obtain continuously differentiable functions $x_{(j)}(t,\varepsilon)$, $h_{(j)}(t,\varepsilon)$ and $y_{(j)}(t,\varepsilon)$ for $t \in [0,T]$ at each step.

Substitution of the constructed functions $x = x_{(j)}(t, x^0, y^0, \varepsilon)$, $y = y_{(j)}(t, x^0, y^0, \varepsilon)$ into expression (43) for the control $u(x, y, \varepsilon)$ allows one to find the approximate optimal control in the open-loop form $u_{(j)} = (t, x^0, y^0, \varepsilon)$ which satisfies the constraint $u \in U_\varepsilon$ (12) with the error $O(\varepsilon^{j+1})$ for $t \in [0, T_{(j)}(x^0, y^0, \varepsilon)]$. It is important to note that the application of the approximate feedback control (43) brings the state point of system (1) to the hyperplane (3). But the time necessary for this will be slightly greater than the minimal time, in the general case by an amount $O(\varepsilon^{j+1})$, i.e. by an amount of the error of calculating Bellman's function (38) and the optimal control $u(x, y, \varepsilon)$ (43). The open-loop control, evidently, does not provide exact stabilization. The issues of numerical implementation of algorithms of constructing the feedback control on the basis of the methods developed in Sections 7.1 and 7.2 are discussed below in Section 7.3.

The method of the approximate construction of the feedback control developed above in Section 7.2 is equally applicable to more general optimal control problems. For example, it allows considering the following generalization of equations of motion, terminal conditions and the performance index:

$$\begin{aligned}
&\dot{x} = F(x,y,z) + S(x,y,z)u + \varepsilon f(x,y,z,u), \qquad x(t_0) = x^0,\\
&\dot{y} = G(x,y,z) + \varepsilon g(x,y,z,u), \qquad y(t_0) = y^0,\\
&\dot{z} = \varepsilon r(x,y,z,u), \qquad z(t_0) = z^0;\\
&c^2(z)\left[h - h_*(z)\right]\big|_T + \varepsilon m(x,y,z)\big|_T = 0,\\
&J[u] = \varepsilon \varkappa(x,y,z)\big|_T + \int_{t_0}^{T} \left[K(h,z,|u|) + \varepsilon k(x,y,z,u)\right]dt \longrightarrow \min.
\end{aligned} \tag{7.2.46}$$

Here, x, y, and z are vectors of any dimensions, the function F satisfies the condition of the type (2): $(\eta \cdot F(x,y,z)) \equiv \varphi(h,z)$, where $\eta = x/h$, $h = |x|$; S is an orthogonal matrix. The introduced functions are supposed to be sufficiently smooth, and the control domain $U_\varepsilon(x,y,z)$ is close to sphere of radius $u_0(h,z)$ and has a smooth boundary. The instant T of the process termination may be either specified (fixed) beforehand or determined as a result of the solution of the control problem.

7.2.7. MODEL EXAMPLE

As an illustration, we consider the problem of the control of the motion of the centre of mass of a flying vehicle in the atmosphere without taking the air dynamical lifting force into account. The equations of motion are taken in the form [1, 72, 111]

$$\dot{\mathbf{r}} = \mathbf{v}, \quad \mathbf{r}(0) = \mathbf{r}^0; \quad m\dot{\mathbf{v}} = \mathbf{P} - \frac{1}{2}c(v)\rho(\mathbf{r})v\mathbf{v} - m\mathbf{g}(\mathbf{r}), \quad \mathbf{v}(0) = \mathbf{v}^0. \tag{7.2.47}$$

Here, m is the vehicle mass, $\mathbf{P}$ is a controllable thrust force, $|\mathbf{P}| \leqslant P^0$; c is a drag coefficient, ρ is the density of the atmosphere, and $\mathbf{g}$ is the acceleration due to gravity. Suppose, to a first approximation, that the mass of the vehicle is constant and that the angle is small, i.e. ρ and $\mathbf{g}$ are approximately constant. Then, using the notations adopted in (46), $\mathbf{v}$ is the vector x; $(\mathbf{r}, t)$ is the vector y, m is the "quasiconstant" parameter z, and the unperturbed drag force is the vector F. For an approximate solution of the problem of controlling the quantity $v(T)$ one may use the perturbation technique developed above.

To avoid bulky calculations we assume that $m = \text{const}$ and the gravity and drag forces are regarded as perturbations of the order ε. At the same time, any dependence of these forces on $\mathbf{v}$ and $\mathbf{r}$ is admitted. As a result, the system of type (1) is obtained:

$$\dot{\mathbf{r}} = \mathbf{v}, \quad \mathbf{r}(0) = \mathbf{r}^0; \quad \dot{\mathbf{v}} = \mathbf{u} + \varepsilon \mathbf{f}(\mathbf{v}, \mathbf{r}), \quad \mathbf{v}(0) = \mathbf{v}^0. \tag{7.2.48}$$

Here, $\mathbf{u} = \mathbf{P}/m$ is the control, $|\mathbf{u}| \leqslant u^0 = P^0/m$; $\varepsilon \mathbf{f}$ is the perturbation.

We pose the minimum-time problem of the type (46):

$$v(T) = v^* \geqslant 0, \qquad T \longrightarrow \min, \qquad |\mathbf{u}| \leqslant u^0. \tag{7.2.49}$$

According to Section 7.2.2, the control u_0 and the minimal time for $\varepsilon = 0$ are given by

$$\begin{aligned} &\mathbf{u}_0(\mathbf{v}) = -\eta u^0 \operatorname{sign}(v - v^*), \\ &\eta = \frac{\mathbf{v}}{v}, \quad T_0(v^0) = \frac{|v^0 - v^*|}{u^0}. \end{aligned} \tag{7.2.50}$$

The family of unperturbed trajectories of the optimal motion has the form

$$\begin{aligned} &\mathbf{v}(t, v^0) = v(t, v^0)\eta^0, \\ &v(t, v^0) = v^0 + (v^* - v^0)t/T_0, \\ &\mathbf{r}(t, \mathbf{v}^0, \mathbf{r}^0) = \mathbf{r}^0 + \frac{1}{2}\eta^0(v^{02} - v^2)(u^0)^{-1} \operatorname{sign}(v^0 - v^*). \end{aligned} \tag{7.2.51}$$

Taking into account formula (42) for the coefficient V_1 of the addition to $V_0 = T_0(v)$ (Bellman's function for the unperturbed problem), we obtain the expression

$$\begin{aligned} &V_1(\mathbf{v}, \mathbf{r}) = \frac{1}{u^{02}} \int_{v^*}^{v} (\eta \cdot \mathbf{f}_0(\mathbf{v}, \mathbf{r}, h))\, dh, \\ &\mathbf{f}_0(\mathbf{v}, \mathbf{r}, h) \equiv \mathbf{f}\Big(\eta h, \mathbf{r} + \frac{1}{2}\eta(v^2 - h^2)(u^0)^{-1} \operatorname{sign}(v - v^*)\Big). \end{aligned} \tag{7.2.52}$$

It follows from (52) and (47) that the drag decreases the minimal time if $v^* < v$ and increases it if $v^* > v$, which is evident. Formula (52) provides the possibility of a simple investigation of the influence of the gravity force on the system's motion as well as of the dependence of the minimal time on other parameters of the problem. An approximate calculation of the control and the trajectory is carried out by means of standard modes.

Note that the approach developed in Sections 7.1 and 7.2 is transferred to time-varying systems with a fixed time of the process termination [1, 9, 34, 70].

7.3. Remarks on Implementation of Approximate Feedback Control

7.3.1. SUCCESSIVE APPROXIMATIONS FOR THE DOMAIN OF THE CONTROL VECTOR

It is of interest to investigate in more detail the possibilities of using small parameter techniques in the case when the domain U_ε has a sophisticated boundary. We assume that there exist two concentric balls U^- and U^+ such that

$$U^- \subseteq U_\varepsilon \subseteq U^+, \quad U^\mp = \{u\colon\ |u| \leqslant u^\mp\}, \quad u^+ - u^- \ll (u^- + u^+)/2. \quad (7.3.1)$$

Therefore, the domain U_ε is close to the ball U_0 with centre at the origin and radius $u^0 = (1/2)(u^- + u^+)$, see Fig. 7.2.

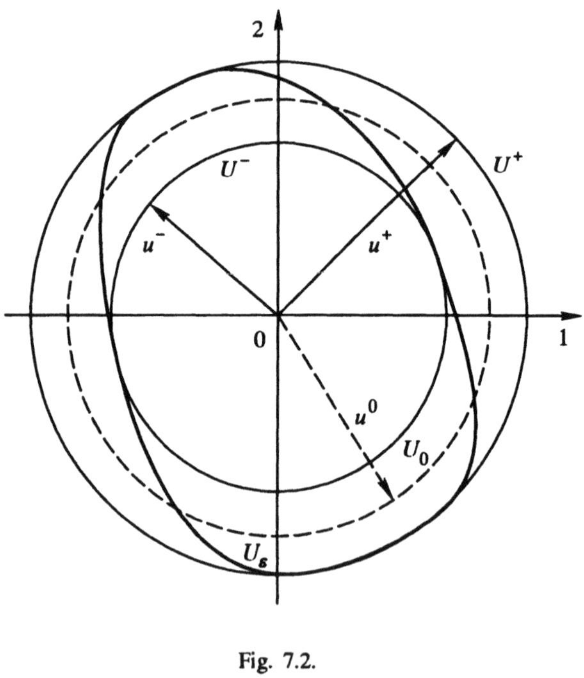

Fig. 7.2.

The dimensionless quantity

$$\varepsilon = (1/2)(u^+ - u^-)/u^0 \ll 1$$

is taken as a measure of the closeness. When considering the problem of minimizing the dominant term $(\mathbf{P} \cdot \mathbf{u})$ in (2.11) over the domain U_ε subject to conditions (1), we make the identity transformation

$$\min_{u\in U_\varepsilon}(P \cdot u) \equiv -u^0|P| + \left[\min_{u\in U_\varepsilon}(P \cdot u) + u^0|P|\right]. \tag{7.3.2}$$

Since the following estimates, based on (1), are valid

$$-u^+|P| \leqslant \min_{u\in U_\varepsilon}(P \cdot u) \leqslant -u^-|P|,$$

the expression in square brackets in (2) has the order of ε. By introducing the notation

$$\varepsilon\Phi(P, x, y, \varepsilon) \equiv u^0|P| + \min_{u\in U_\varepsilon}(P \cdot u), \tag{7.3.3}$$

Equation (2.11) is reduced to the form

$$-u^0|P| + (P \cdot F) + (Q \cdot G) = -1 + \varepsilon\Psi(P, Q, x, y, \varepsilon). \tag{7.3.4}$$

To a first approximation in terms of ε, one can minimize the terms $(p \cdot \varepsilon f)$, $(Q \cdot \varepsilon g)$ included in Ψ over the ball U_0. Further investigation of the initial-value problem (4) is carried out analogously to Section 7.2.

Below, some examples of the domain U_ε are considered.

1) Let U_ε be a cube with its centre at the origin, the dimension of the space $n = 3$; then $u^- = 1$, $u^+ = \sqrt{3}$ and $u^0 = (1 + \sqrt{3})/2 \approx 1.36$, i.e. $\varepsilon \approx 0.27$. If $n = 2$, then $u^0 = (1 - \sqrt{2})/2 \approx 1.21$ and $\varepsilon \approx 0.17$.

2) Let U_ε be an ellipsoid close to the ball of radius u^0:

$$\sum_{i=1}^{n} \frac{(u_i - \varepsilon l_i)^2}{(1 + \varepsilon\beta_i)^2} \leqslant u^{02}, \qquad l_i = \alpha_{i0} + \sum_{i=1}^{n} \alpha_{ij} u_j, \tag{7.3.5}$$

$$\varepsilon \ll 1; \qquad \alpha_{ij}, \beta_i = \text{const}, \quad i = 1, \ldots, n, \quad j = 0, 1, \ldots, n.$$

It seems more convenient to represent the solution of the minimization problem (2) over domain (5) in vector form. We introduce the vector $v = (v_1, \dots, v_n)$ restricted by the sphere of radius u^0 and connected with $u = (u_1, \dots, u_n)$ by the relationship $\varepsilon\alpha_0 + (I + \varepsilon B)v = (I + \varepsilon A)u$. Here, B is a diagonal matrix whose elements are β_i. The vector u is determined by inverting the matrix $(I + \varepsilon A)$:

$$\begin{aligned} &u = \varepsilon(I + \varepsilon D)\alpha_0 + (I + \varepsilon C)v, \qquad v \in U_0, \\ &D = -A(1 + \varepsilon A)^{-1}, \qquad C = B + D + \varepsilon DB. \end{aligned} \tag{7.3.6}$$

Then, expression (6) for u is substituted into (2) and the minimization is performed with respect to v over the ball $|v| \leqslant u^0$. This results in

$$\begin{aligned} &v = -u^0 P(I + \varepsilon C)\left|P(I + \varepsilon C)\right|^{-1}, \\ &\min_{u \in U_\varepsilon} (P \cdot u) = -u^0\left|P(I + \varepsilon C)\right| + \varepsilon\big(P \cdot (I + \varepsilon D)\alpha_0\big). \end{aligned} \tag{7.3.7}$$

It follows from (7) that the value of the minimum to be found differs from $-u^0|P|$ by an amount $O(\varepsilon)$ (see Section 7.2). After having determined Bellman's function V, the control v is found according to (7) and the original control u is obtained by using (6).

To illustrate this mode we consider the problem of quenching the equatorial component of the angular velocity vector for a rigid body with dynamic symmetry by using torques with restricted moduli (see Sections 1.1 and 1.4). The equations of motion are reduced to the form

$$\begin{aligned} &\dot{x}_1 + \alpha x_2 = u_1, \qquad x_1(0) = x_1^0 \qquad (x_i = I\omega_i/u^0), \\ &\dot{x}_2 - \alpha x_1 = u_2, \qquad x_2(0) = x_2^0, \qquad |u_i| \leqslant 1, \quad i = 1, 2. \end{aligned} \tag{7.3.8}$$

For $\alpha \neq 0$, transformation of variables x_i and t reduces the equations (8) to the form with $\alpha = 1$. We write down the initial-value problem for Bellman's function V according to (2)–(4):

$$\begin{aligned} &(-r_1 x_2 + r_2 x_1) - u^0 r = -1 + \varepsilon\big(|r_1| + |r_2| - u^0 r\big), \\ &r_i = \partial V/\partial x_i, \quad i = 1, 2; \qquad V(0, \varepsilon) \equiv 0 \quad (\varepsilon = 1). \end{aligned} \tag{7.3.9}$$

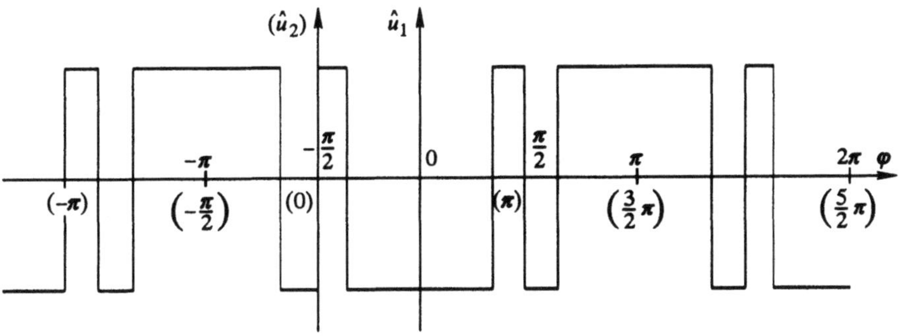

Fig. 7.3.

Based on the methods of Section 7.1, we obtain expressions corresponding to the unperturbed problem:

$$\begin{aligned} &V_0 = h/u^0, \qquad r_{0i} = \eta_i/u^0, \qquad h = h^0 - u^0 t, \\ &\eta_1 = \eta_1^0 \cos\frac{h^0 - h}{u^0} - \eta_2^0 \sin\frac{h^0 - h}{u^0}, \qquad \cos\delta^0 = \eta_1^0, \\ &\eta_2 = \eta_1^0 \sin\frac{h^0 - h}{u^0} + \eta_2^0 \cos\frac{h^0 - h}{u^0}, \qquad \sin\delta^0 = \eta_2^0. \end{aligned} \tag{7.3.10}$$

By using formula (1.27) the function V_1 is found in the form

$$\begin{aligned} &V_1(x_1, x_2) \\ &= \frac{h}{u^0} - \frac{1}{u^{02}} \int_0^h \left[\left| \cos\left(\frac{h-l}{u^0} + \delta\right) \right| + \left| \sin\left(\frac{h-l}{u^0} + \delta\right) \right| \right] dl. \end{aligned} \tag{7.3.11}$$

Then we estimate the quantity V_1 for the value $u^0 = (1+\sqrt{2})/2 \approx 1.21$ taken according to (1). Since $1 \leqslant |\cos\varphi| + |\sin\varphi| \leqslant \sqrt{2}$, $|V_1/V_0| \leqslant 0.17$ for all h. For large h ($h \gg u^0$), the relative error is even less ($\lesssim 0.05$). The contribution of subsequent approximations is rather less.

Let $\alpha = 0$ in (8) which corresponds to $x_3(I - I_3) = 0$. Then, based on the approach given above, we obtain the approximate representation for the unknown Bellman's function V:

$$\widehat{V}(x_1, x_2) = \frac{h}{u^0} + \left(h - \frac{|x_1| + |x_2|}{u^0}\right) + \cdots. \tag{7.3.12}$$

The exact expression for V is $V^* = \max(|x_1|, |x_2|)$. The maximal relative error $\max_{x_1,x_2} |V^* - \widehat{V}|/\widehat{V}$ is equal to $|2\sqrt{2}/(1+\sqrt{2}) + \sqrt{2} - 4/(1+\sqrt{2}) - 1| \approx 0.07$. This is achieved at $|x_1| = |x_2|$. The approximate optimal control corresponding to the approximate Bellman's function $\widehat{V}$ from (12) is given by

$$\hat{u}_i = -\operatorname{sign}\frac{\partial \widehat{V}}{\partial x_i} = -\operatorname{sign}\eta_i \operatorname{sign}\left[|\eta_i|(1+u^0) - 1\right]. \tag{7.3.13}$$

A graphical description of the functions u_i (13) is given in Fig. 7.3; for convenience, the denotations $\eta_1 = \cos\varphi$ and $\eta_2 = \sin\varphi$, $\varphi \in [-\pi, \pi]$ are introduced.

7.3.2. NUMERICAL IMPLEMENTATION OF APPROXIMATE FEEDBACK CONTROLS

Consider the problem of constructing Bellman's function according to the method of Section 7.2. The problem of calculating the functions $W_j(x, y)$ in the initial-value problem (2.40) is assumed to have been solved. The calculation of the coefficients $V_j(x, y)$ according to (2.42) requires determining a multiparametric family of functions $x_0(l, h, \eta)$ and $y_0(l, h, \eta, y)$ (see (2.9)) which can be represented in a more convenient form as $\tilde{x}_0(l, x)$ and $\tilde{y}_0(l, x, y)$. This family of functions must be rich enough to define the unknown coefficients V_j on a sufficiently dense set of points (x, y). The domain to which the parameter x belongs is restricted by the inequality $|x| \leqslant h^0$, where h^0 is the radius of the ball of possible initial values for the variable x. The domain for the parameter y is determined as a set of all $y = y(t, x^0, y^0)$ such that $|x^0| \leqslant h^0$ and y^0 belongs to the set of possible initial values, $t \in [0, \theta(0, h^0)]$.

Within these domains we specify sets of points (x_α, y_γ) where α and γ are vector indices. The components α_i $(i = 1, 2, \ldots, n)$ and γ_k $(k = 1, 2, \ldots, m)$ of these vectors take values from 0 to N. The variable h is defined on the set of $n(N+1)$ points: $h_{\{\alpha\}} = |x_\alpha|$. Construction of the set of trajectories consists in integrating the system of equations with the values of variables specified at the points mentioned:

$$\begin{gathered}\frac{dx}{dh} = \frac{F(x) - \eta}{1 - \varphi(h)}, \qquad \frac{dy}{dh} = \frac{G(x, y)}{1 - \varphi(h)}, \\ x\big|_{h=h_{\{\alpha\}}} = x_\alpha, \qquad y\big|_{h=h_{\{\alpha\}}} = y_\gamma.\end{gathered} \tag{7.3.14}$$

Let the family of trajectories be constructed according to (14):

$$\begin{gathered}x = x_0(h, h_{\{\alpha\}}, x_\alpha), \qquad y = y_0(h, h_{\{\alpha\}}, x_\alpha, y_\gamma), \\ h \in \{h_{\{\beta\}}\}, \qquad \{\beta\} \subseteq \{\alpha\}.\end{gathered}$$

Then the integral in (2.42) is replaced by the sum over the discrete set of points $h_{\{\beta\}}$ where the function W_j under the sign of the integral is specified. In this case, $h_{\{\beta\}} \leqslant h_{\{\alpha\}}$ and the quantities $h_{\{\alpha\}}$, x_α, and y_γ are regarded as parameters. They define the points where the function V_j is specified: $V_j(h_{\{\alpha\}}, x_\alpha, y_\gamma)) = V_{j(\alpha,\gamma)}$.

The given discretization procedure can also be implemented on the basis of integration of the original system over the time t:

$$\begin{aligned} &\dot{x} = F(x) - \eta, \qquad x(0) = x_\alpha, \qquad t_{\alpha,\beta} = \int_{|x_\alpha|}^{|x_\beta|} \frac{dl}{\varphi(l) - 1}, \\ &\dot{y} = G(x, y), \qquad y(0) = y_\gamma, \qquad |x_\beta| \leqslant |x_\alpha|. \end{aligned} \tag{7.3.15}$$

The constructed discrete families of solutions of problem (15) $x_{\alpha,\beta} = x_0(t_{\alpha,\beta}, x_\alpha)$ and $y_{\alpha,\beta,\gamma} = y_0(t_{\alpha,\beta}, x_\alpha, y_\gamma)$, are then substituted into formula (2.42)

$$\begin{aligned} &V_j(x_\alpha, y_\gamma) \\ &= -\int_0^{|x_\alpha|} W_j\left(x_0\left(\int_{|x_\alpha|}^h \frac{dl}{\varphi(l)-1}, x_\alpha\right), y_0\left(\int_{|x_\alpha|}^h \frac{dl}{\varphi(l)-1}, x_\alpha, y_\gamma\right)\right) \times \\ &\times \frac{dl}{1-\varphi(l)}. \end{aligned}$$

Here, integration can be carried out by using any numerical method [43, 48]. The optimal feedback control is determined by differentiation of Bellman's function V. This differentiation can be implemented either analytically or numerically.

7.4. Time-Optimal Stabilization of a Spacecraft

7.4.1. BRAKING PERTURBED ROTATIONS OF A SPACECRAFT WITH DYNAMICS SYMMETRY

Consider system (1.11) under the conditions

$$\begin{aligned} &I_1 = I_2 = I, \qquad b_1 = b_2 = b \neq 0; \\ &M_i = b_i u_i + \varepsilon N_i(\omega), \quad i = 1, 2, 3, \qquad \mathbf{u}^2 \leqslant 1. \end{aligned} \tag{7.4.1}$$

The perturbing torque $\varepsilon \mathbf{N}$ takes account of the influence of external and internal (gyroscopic) forces. Transformation to new variables z_i results in the control problem for the system of the form (1.1)–(1.3):

$$\begin{aligned}
&\dot{z}_{1,2} \pm dz_3 z_{2,1} = u_{1,2}, \qquad \dot{z}_3 = u_3 + \varepsilon f_3, \qquad z_i(0) = z_i^0;\\
&z_i = I_i \omega_i b_i^{-1}, \qquad f_i = N_i b_i^{-1}, \qquad i = 1,2,3;\\
&d = b_3(I_3 - I)(II_3)^{-1};\\
&\mathbf{z}(T) = 0, \quad T \longrightarrow \min, \quad |\mathbf{u}| \leqslant 1.
\end{aligned} \tag{7.4.2}$$

For $\varepsilon = 0$, the solution of the optimal control problem is known, according to Equations (1.4)–(1.7):

$$\begin{aligned}
&\mathbf{u}_0 = -\eta, \quad h_0 = h^0\left(1 - tT_0^{-1}\right), \quad T_0 = h^0, \quad \mathbf{z}_0 = h_0\eta_0\left(t, \mathbf{z}^0\right);\\
&\eta_{10} = \frac{h_\perp^0}{h^0}\cos\left(\sigma + \sigma^0\right), \quad \eta_{20} = \frac{h_\perp^0}{h^0}\sin\left(\sigma + \sigma^0\right), \quad \eta_{30} = \frac{z_3^0}{h^0};\\
&h_\perp = \left(z_1^2 + z_2^2\right)^{1/2}, \quad \sigma = dz_3^0 t\left(1 - \frac{1}{2}\frac{t}{T_0}\right),\\
&\cos\sigma^0 = \frac{z_1^0}{h_\perp^0}, \quad \sin\sigma^0 = \frac{z_2^0}{h_\perp^0}.
\end{aligned} \tag{7.4.3}$$

The quantities z_1 (or z_2), z_3, h_0, $h_\perp$, and σ are qualitatively shown in Fig. 7.4. As has been noted, $\dot{\sigma} = 0$ when $t = T_0$ [1, 46].

Bellman's function $T(z, \varepsilon)$, which becomes $T_0 = h$ for $\varepsilon = 0$, is constructed as a solution of the initial-value problem for the nonlinear partial differential equation (see (1.20)–(1.22)):

$$\begin{aligned}
&R + dr_1 z_2 z_3 - dz_1 r_2 z_3 - \varepsilon(\mathbf{r} \cdot \mathbf{f}) = 1, \qquad T(0, \varepsilon) \equiv 0;\\
&R = |\mathbf{r}|, \qquad \mathbf{r} = \partial T/\partial z, \qquad \mathbf{u} = -\mathbf{r}R^{-1};\\
&T = h + \varepsilon T_1(\mathbf{z}) + \varepsilon^2 + \cdots.
\end{aligned} \tag{7.4.4}$$

Based on the generating solution (3), and using the method of characteristics, one obtains an expression of type (1.27) for the unknown coefficients $T_j(\mathbf{z})$, in particular

$$T_1(\mathbf{z}) = -\int_0^h (\eta_0 \cdot \mathbf{f}(l\eta_0))\, dl, \qquad \eta_0 = \eta_0(h - l, \mathbf{z}). \tag{7.4.5}$$

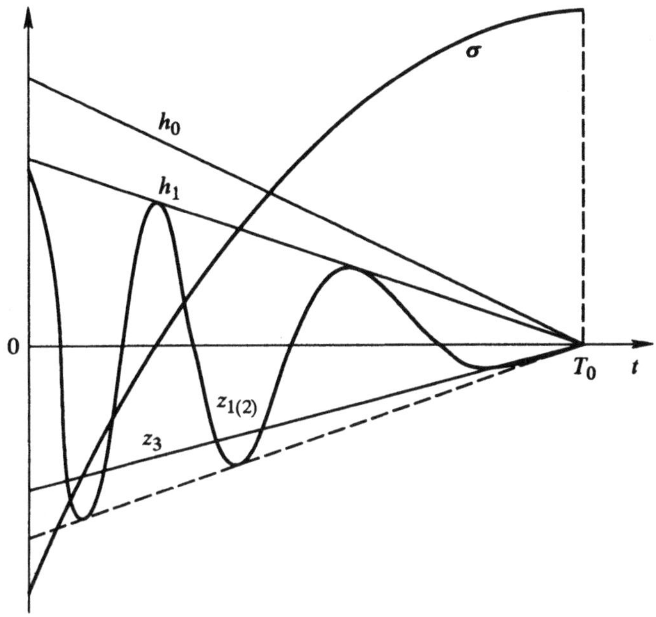

Fig. 7.4.

Subsequent coefficients are calculated analogously. The optimal control is found by using expressions (4):

$$\mathbf{u}_{(1)}(\mathbf{z}, \varepsilon) = -(\boldsymbol{\eta} + \varepsilon \mathbf{r}_1)|\boldsymbol{\eta} + \varepsilon \mathbf{r}_1|^{-1}. \tag{7.4.6}$$

This satisfies the constraint (2) and brings the vector $\mathbf{z}$ to an ε^2-vicinity of the origin: $\mathbf{z}(h^0 + \varepsilon T_1(z^0), \mathbf{z}^0, \varepsilon) = O(\varepsilon^2)$. The approximate optimal trajectory $\mathbf{z}_{(1)}(t, \mathbf{z}^0, \varepsilon)$ is obtained by integrating system (2) when $\mathbf{u} = \mathbf{u}_{(1)}(\mathbf{z}, \varepsilon)$.

Below, we consider some concrete examples of perturbations. We suppose the relationship of the axial symmetry (1) to be fulfilled approximately with the error $O(\varepsilon)$: $I_{1,2}(\varepsilon) = I(1 + \varepsilon\delta_{1,2})$ and $b_{1,2}(\varepsilon) = b(1 + \varepsilon\beta_{1,2})$; the system being acted upon by the dissipative torque, linear in velocity, $-\varepsilon\Lambda\boldsymbol{\omega}$ (Λ is a non-negative-definite matrix). Transformations (2) reduce the function $\mathbf{f}$ to the form

$$f_i(\mathbf{z}) = \gamma_i z_{i+1} z_{i+2} - \sum_{j=1}^{3} \lambda_{ij} z_j, \qquad \lambda_{ij} = \Lambda_{ij} b_j / b_i I_j, \quad i = 1, 2, 3;$$
$$\gamma_1 = -(\partial/\partial\varepsilon)\left[(I_3 - I_2(\varepsilon)) I_3^{-1} I_2^{-1}(\varepsilon) b_2(\varepsilon) b_3 b_1^{-1}(\varepsilon)\right]\Big|_{\varepsilon=0} \qquad (1, 2, 3). \tag{7.4.7}$$

The expression for the coefficient $T_1(\mathbf{z})$ consists, according to (5), of two terms corresponding to different perturbations (7) (gyroscopic and dissipative):

$$
\begin{aligned}
&T_1(\mathbf{z}) = T_1^{\gamma}(\mathbf{z}) + T_1^{\lambda}(\mathbf{z}); \qquad T_1^{\lambda}(\mathbf{z}) = -\sum_{i,j=1}^{3} \lambda_{ij}\alpha_{ij}(\mathbf{z}),\\
&T_1^{\gamma}(z) = -\frac{1}{2}(\gamma_1+\gamma_2+\gamma_3)z_3h^{-3}\Big\{\left(z_2^2-z_1^2\right)\cos(dz_3h)+\\
&\quad + 2z_1z_2\sin(dz_3h)\int_0^h l^2\sin\left(d\eta_3 l^2\right)\mathrm{d}l+\\
&\quad + \left[2z_1z_2\cos(dz_3h)-\left(z_2^2-z_1^2\right)\sin(dz_3h)\right]\times\\
&\quad \times\int_0^h l^2\cos\left(d\eta_3 l^2\right)\mathrm{d}l\Big\}.
\end{aligned}
\tag{7.4.8}
$$

Here, the function $T_1^{\gamma}(\mathbf{z})$ is expressed in terms of Fresnel's integrals when $d\eta_3 \neq 0$ [87]. The coefficients $\alpha_{ij}(\mathbf{z})$ are found in terms of elementary functions. Their explicit expression and the analysis of controlled motion can be found in [1, 7, 46], see also (15).

7.4.2. STABILIZATION OF A NONSYMMETRIC SPACECRAFT

The problem of minimum-time braking the perturbed rotations of system (1.11) is considered:

$$
\begin{aligned}
&\dot{\mathbf{L}} + \boldsymbol{\omega}\times\mathbf{L} = \mathbf{M}, \qquad \mathbf{M} = \mathbf{u} + \varepsilon\mathbf{f}(\boldsymbol{\omega},\mathbf{u}), \qquad \mathbf{u}\in U_\varepsilon(\boldsymbol{\omega}),\\
&\mathbf{L} = I\boldsymbol{\omega}, \qquad I = \operatorname{diag}(I_1, I_2, I_3).
\end{aligned}
\tag{7.4.9}
$$

The domain U_ε is close to the sphere of the radius u^0, which can be assumed to be equal to unity, with its centre at the origin, and has a sufficiently smooth boundary Γ_ε. To be more definite we assume $\Gamma_\varepsilon = \{u\colon\ u^2 + \varepsilon g(\mathbf{u},\boldsymbol{\omega}) = 1\}$ where g is a sufficiently smooth function of $\mathbf{u}$ and $\boldsymbol{\omega}$ (see Section 7.2); $\boldsymbol{\omega}\sqrt{u_0}\to\boldsymbol{\omega}$, $\sqrt{u_0}\,t\to t$.

When $\varepsilon = 0$, the system has the form (1.11)–(1.14) where $\mu = 1$, $\nu = 0$, and $b_i = 1$, $i = 1,2,3$. The optimal control (see (1.15)) in this case is $\mathbf{u}_0 = -\mathbf{L}L^{-1}$, $T_0 = L^0$ and the optimal trajectory $\omega_0(t,\omega^0)$ is determined on the basis of the known solution for $\mathbf{u}_0 \equiv 0$, i.e. on the base of the motion of Euler–Poinsot. For controlled motion, one can rather easily find the quantities L_0 and E_0 which are integrals of free rotations. Taking dot products of (9), when $\varepsilon = 0$, by $\mathbf{L}_0$ and ω_0 one can obtain the expressions $L_0 = L^0 - t$ and $E_0 = E^0L_0^2/L^{02}$. It follows that for all $0 \leqslant t \leqslant T_0$, the inequalities $2E_0I_1 \leqslant L_0^2 \leqslant 2E_0I_3$ are valid and, moreover, $L_0^2 = 2E_0I_*$, where

$I_1 \leqslant I_* \leqslant I_3$. Taking account of the integrals, system (9) is reduced, by a change of variables as usual, to one separable equation which can be integrated in terms of elliptic functions. Another way to construct $\omega_0(t, \omega^0)$ consists in the substitution $\mathbf{L}_0 = I\omega_0 = L_0 \mathbf{v}$, where $\mathbf{v}$ is the unknown unit vector of $\mathbf{L}_0$; for $\mathbf{v}$ one obtains the equation

$$\begin{aligned} &d\mathbf{v}/ds + I^{-1}\mathbf{v} \times \mathbf{v} = 0, \qquad \mathbf{v}(0) = \mathbf{L}^0/L^0 \equiv \mathbf{v}^0, \\ &s = \frac{1}{2}(L^{02} - L_0^2), \qquad s \in \left[0, \frac{1}{2}L^{02}\right]. \end{aligned} \tag{7.4.10}$$

Its solution $\mathbf{v}(L_0, L^0, \mathbf{v}^0)$ has the form (6.4.2) within the region $h_3 \leqslant h \leqslant h_2$:

$$\begin{aligned} &v_1 = a_1(\mathbf{v})\,\mathrm{cn}(\sigma, \varkappa), \quad v_2 = a_2(\mathbf{v})\,\mathrm{sn}(\sigma, \varkappa), \quad v_3 = a_3(\mathbf{v})\,\mathrm{dn}(\sigma, \varkappa), \\ &a_{1,2} = I_{1,2}\left[(2hI_3 - 1)/I_{1,2}(I_3 - I_{1,2})\right]^{1/2}, \\ &a_3 = \pm I_3\left[(1 - 2hI_1)/I_3(I_3 - I_1)\right]^{1/2}, \qquad h_i = \frac{1}{2}I_i^{-1}, \quad i = 1, 2, 3, \\ &h = h(\mathbf{v}) = h_1 v_1^2 + h_2 v_2^2 + h_3 v_3^2 = E_0/L_0^2 = E^0/L^{02} = \text{const}, \\ &\varkappa^2 = \frac{(I_2 - I_1)(2hI_3 - 1)}{(I_3 - I_2)(1 - 2hI_1)} = \frac{(I_2 - I_1)(2E_0 I_3 - L_0^2)}{(I_3 - I_2)(L_o^2 - 2E_o I_1)} = k^2 = \text{const}, \\ &\varkappa^2, k^2 \in [0, 1]; \qquad \sigma = \chi s + \sigma^0, \\ &\chi = (I_1 I_2 I_3)^{-1/2}(I_3 - I_2)^{1/2}(1 - 2hI_1)^{-1/2}. \end{aligned} \tag{7.4.11}$$

Here, $\varkappa$ is the modulus of elliptic functions, the phase constant σ^0 is to be chosen from the initial condition (10) with accuracy up to the period $4\mathbf{K}(\varkappa)$ (see (6.4.2)). For the complementary region $h_2 \leqslant h \leqslant h_1$, the function $\mathbf{v}$ is obtained from (11) by changing the indices $1 \rightleftarrows 3$. Thus, the unperturbed controlled motion is completely determined. It is the motion of the Euler–Poinsot type ($\varkappa = \text{const}$) with the quantity L_0 linearly decreasing in time. The optimal open-loop control $\mathbf{u}_0(t)$ is obtained by substituting (11) into the expression $\mathbf{u}_0 = -\mathbf{L}_0/L_0$; this results in $\mathbf{u}_0(t) = -\mathbf{v}_0(t, \omega^0)$.

To construct the approximate control when $\varepsilon = 0$, we use the technique of Section 7.2. For an unknown Bellman's function V, an initial-value problem of the (2.31) occurs

$$\begin{aligned} &\left|\frac{\partial V}{\partial \mathbf{L}}\right| + \left(\frac{\partial V}{\partial \mathbf{L}} \cdot [\omega \times \mathbf{L}]\right) = 1 + \varepsilon F\left(\mathbf{L}, \frac{\partial V}{\partial \mathbf{L}}, \varepsilon\right), \\ &V = V(\mathbf{L}, \varepsilon) = L + \varepsilon V_1(\mathbf{L}) + \varepsilon^2 + \cdots, \qquad V(\mathbf{0}, \varepsilon) \equiv 0. \end{aligned} \tag{7.4.12}$$

Here, F is a known, sufficiently smooth, function. For the coefficients V_j we obtain the expression in the form of quadratures (2.42), in particular

$$V_1(\mathbf{L}) = \int_0^L F\big(l\mathbf{v}_0(l, L, \mathbf{v}), \mathbf{v}_0(l, L, \mathbf{v}), 0\big)\, \mathrm{d}l. \tag{7.4.13}$$

The function $\mathbf{v}_0$ in (13) is determined according to (11). Some calculations for concrete types of perturbations are given below.

(1) Let $U = U_0$ and the perturbation εf be linear in $\boldsymbol{\omega}$: $\mathbf{f} = N\boldsymbol{\omega}$, $N = (n_{ij}) = \text{const}$ $(i, j = 1, 2, 3)$. Then from a condition similar to (1.18) one finds the control $\mathbf{u} = -(\partial V/\partial \mathbf{L})|\partial V/\partial \mathbf{L}|^{-1}$. The function F in (12) in this case is $F = (\partial V/\partial \mathbf{L} \cdot N\boldsymbol{\omega})$. The coefficient V_1 according to (13) is

$$V_1(\mathbf{L}) = \sum_{i,j=1}^{3} \frac{n_{ij}}{I_j} \int_0^L v_{0i}(l, L, \mathbf{v}) v_{0j}(l, L, \mathbf{v}) l\, \mathrm{d}l. \tag{7.4.14}$$

The integrals in (14) are calculated in terms of elliptic functions (for comparison see (8)):

$$\begin{aligned}
&\alpha_{ij}(\mathbf{L}) = \int_0^L v_{0i} v_{0j} l\, \mathrm{d}l = \frac{1}{\chi(\mathbf{v})} \int_{\sigma^0}^{\sigma_0} v_{0i} v_{0j}\, \mathrm{d}\sigma, \qquad \sigma_0 = \frac{1}{2}\chi L^2 + \sigma^0;\\
&\alpha_{11} = \frac{1}{2} a_1^2 (\varkappa^2 - 1)\varkappa^{-2} L^2 + \varkappa^{-2}\big[E(\operatorname{am}\sigma_0, \varkappa) - E(\operatorname{am}\sigma^0, \varkappa)\big],\\
&\alpha_{22} = \frac{1}{2} a_2^2 \varkappa^{-2} L^2 + \varkappa^{-2}\big[E(\operatorname{am}\sigma^0, \varkappa) - E(\operatorname{am}\sigma_0, \varkappa)\big],\\
&\alpha_{33} = a_3^2 \varkappa^{-2}\big[E(\operatorname{am}\sigma_0, \varkappa) - E(\operatorname{am}\sigma^0, \varkappa)\big],\\
&\alpha_{12} = \alpha_{21} = a_1 a_2 \chi^{-1} \varkappa^{-2}\big(\operatorname{dn}\sigma^0 - \operatorname{dn}\sigma_0\big),\\
&\alpha_{13} = \alpha_{31} = a_1 a_3 \chi^{-1}\big(\operatorname{sn}\sigma_0 + \operatorname{sn}\sigma^0\big),\\
&\alpha_{23} = \alpha_{32} = a_2 a_3 \chi^{-1}\big(\operatorname{cn}\sigma^0 - \operatorname{cn}\sigma_0\big).
\end{aligned} \tag{7.4.15}$$

Here, E is an elliptic integral of the second kind, am is the elliptic amplitude, and the coefficients a_i, $\varkappa$, and χ are known functions of $\mathbf{L}$. Though parameter σ^0 is determined from (10) not uniquely (but with an accuracy up to the period $4\mathbf{K}(\varkappa)$), however, the first three expressions (as well as subsequent ones) are determined uniquely. One can note that if N is a dissipative matrix, then $V_1 < 0$ for all $\mathbf{L}$ $(L > 0)$; if N is a positive-definite matrix, then $V_1 > 0$, i.e. the minimal time increases. For a rigid body with dynamic symmetry, the coefficients α_{ij} in (8) are obtained from formulae (15) by proceeding to the limit as $\varkappa \to 0$ and are expressed in terms of elementary functions.

(2) Let small constant torques m_i, $|\mathbf{m}| \ll 1$, act with respect to each axis fixed within the body. Then (5) implies that

$$V_1(\mathbf{L}) = \sum_{i=1}^{3} m_i \int_0^L v_{0i}(l, L, \mathbf{v})\,\mathrm{d}l. \tag{7.4.16}$$

The expressions for the coefficients in (16) for $i = 1, 2$ are generalizations of Fresnel's integrals (see (8)) for the case of Jacobi elliptic functions. The properties of these integrals imply that for large L, the main contribution is due to the torque m_3 (for $h_3 \leqslant h \leqslant h_2$) or m_1 (in the complementary region $h \geqslant h_2$); contributions of $m_{1,2}$ or $m_{2,3}$ are limited.

(3) The function $V_1(\mathbf{L})$ is calculated under the assumption that the disturbance $\mathbf{m}$, $|m| \leqslant m_0 \ll 1$, tends to increase the quantity T, the time of braking (the game approach), i.e. $\mathbf{m} = m_0 \partial V/\partial \mathbf{L}$. In this case, Bellman's function increases by the amount $V_1 = m_0 L$. But if each torque m_i is restricted: $m_i^* \leqslant m_i \leqslant m_i^{**}$, $i = 1, 2, 3$, then using the equation of Isaacs and Bellman one finds the expression [32, 40, 43, 55, 65, 83, 105]

$$V_1(\mathbf{L}) = \frac{1}{2} \sum_{i=1}^{3} \int_0^L \left[(m_i^* + m_i^{**}) v_{0i} + (m_i^{**} - m_i^*) |v_{0i}| \right] \mathrm{d}l. \tag{7.4.17}$$

It follows from (17) that $V \to \infty$ as $L \to \infty$.

(4) Consider perturbations of the form $\mathbf{f} = A(\omega)\mathbf{u}$, $|\mathbf{u}| \leqslant 1$, where $A = (a_{ij})$. For the coefficient V_1 we obtain the expression

$$V_1(\mathbf{L}) = \sum_{i,j=1}^{3} \int_0^L a_{ij} \left(l I^{-1} \mathbf{v}_0(l, L, \mathbf{v}) v_{0i} v_{0j} \right) \mathrm{d}l. \tag{7.4.18}$$

Therefore, if perturbations widen the domain of the control

$$\mathbf{M} = (I + \varepsilon A)\mathbf{u},$$

i.e. A is a positive-definite matrix for all $|\omega| \leqslant \omega^0$, then (18) implies $V_1 < 0$ (the minimal time decreases); if $A < 0$, then $V_1 > 0$.

Possible generalizations of this approach are given in [1, 34, 70]. It follows from the above that the perturbation technique, developed in this chapter (Sections 7.1 and 7.2), for solving the problem of the optimal control synthesis on the basis of dynamic programming is an efficient tool for the approximate construction of stabilization laws for nonsymmetric rigid bodies.

7.4.3. STABILIZATION OF BODIES WITH INTERNAL DEGREES OF FREEDOM

(1) *Braking rotations of a rigid body with a moving particle inside.* The Schwartz inequality also turns out to be useful when constructing feedback braking control laws for "quasirigid" bodies whose deformation (displacements of particles from their equilibrium positions) due to inertia forces are small, in a definite sense. In this case, we suppose that possible large initial displacements and relative oscillations rapidly decay. Then, assuming control torques are smooth functions of time

$$\mathbf{M} = b(t)\mathbf{u}, \qquad |\mathbf{u}| \leqslant 1, \qquad 0 < b^* \leqslant b(t) \leqslant b^{**}, \tag{7.4.19}$$

the application of the "boundary layer" asymptotic technique [45, 122, 125, 126, 129, 148, 152], developed in [1, 21] for the model of a particle with a strong damper, leads to approximate equations of motion of a rigid body.

For a body with axial symmetry (when unperturbed), these equations are reduced to the form [21]

$$\begin{gathered}\omega_{1,2}' \mp d\omega_3\omega_{2,1} = b(t)I^{-1}u_{1,2} \pm AL^2\omega_3\omega_{2,1} - BI^{-1}\omega_3^4\omega_{1,2},\\ \omega' = b(t)I_3^{-1}u_3 - BII_3^{-2}\omega_\perp^2\omega_3^3, \qquad \omega_\perp^2 = \omega_1^2 + \omega_2^2,\\ d = 1 - \frac{I_3}{I}, \qquad A = \frac{m\rho^2 I_3}{I^3\Omega^2}, \qquad B = \frac{m\rho^2\nu d I_3^3}{I^3\Omega^4}.\end{gathered} \tag{7.4.20}$$

Here, $\operatorname{diag}(I, I, I_3)$ is the inertia tensor of the unperturbed body, m is the mass of a moving particle, ρ is the distance from the centre of mass to the point at which the particle is fastened do the carrying body; the constants $\Omega^2 = c/m$, $\nu = \delta/m$ characterize the frequency of oscillations and the rate of their decay: c is a stiffness coefficient, and δ is a viscous friction coefficient. Perturbing torques in (20) are small, provided $\Omega^2 \gg \nu\omega \gg \omega^2$; free oscillations of the system decaying long time before the body performs one revolution [1, 21]. For the derivation of Equations (20) and estimations of their error, see [21].

We pose the problem of minimum-time braking of the rotations which set in after the fast initial transient process has been decayed (see (19) and (20)). It is required to find the control law $\mathbf{u}$, the trajectory $\boldsymbol{\omega}(t)$ and the minimal time T. Based on the sufficient optimality conditions of the method of dynamic programming, using the Schwartz inequality for the vector $\mathbf{L}$ we find the control and the quantities L and T:

$$\mathbf{u} = -\frac{\mathbf{L}}{L}, \quad L(t, t_0, L^0) = L^0 - \int_{t_0}^{t} b(\tau)\,d\tau, \quad \int_{t_0}^{T} b(t)\,dt = L_0. \tag{7.4.21}$$

Since, according to (19), $b \geqslant b^* > 0$, the root $T^* = T(t_0, L^0)$ of Equation (21) exists, is unique, and $L^0/b^{**} \leqslant T^* - t_0 \leqslant L^0/b^*$; $T(t, L)$ is Bellman's function for the problem. Substituting $L(t)$ and $I_3\omega_3 = L\cos\theta$, where θ is the nutation angle, into the equation for ω_3 (20) and integrating this one can obtain the relationship between θ and t:

$$\frac{1}{\cos^2\theta} - \frac{1}{\cos^2\theta^0} + \ln\frac{\tan^2\theta}{\tan^2\theta^0} = -\frac{2B}{II_3^4}\int_{t_0}^{t} L^4(\tau, t_0, L^0)\,\mathrm{d}\tau. \tag{7.4.22}$$

In the particular case when $b = \mathrm{const} > 0$, the dependence (22) is reduced to the form

$$\begin{aligned} &\tau = 1 - \left[1 + \sigma\left(\tan^2\theta^0 - \tan^2\theta + \right.\right.\\ &\qquad \left.\left. + \ln\left(\tan^2\theta^0/\tan^2\theta\right)\right)\right]^{1/5}, \qquad 0 \leqslant \tau \leqslant 1,\\ &\tau \equiv \frac{t - t_0}{T^* - t_0}, \qquad T^* = \frac{L^0}{b} + t_0,\\ &\sigma = \frac{5}{2}\frac{bI^4I_3\Omega^4}{m\rho^2\nu dL^{05}}, \qquad |\sigma| < \infty. \end{aligned} \tag{7.4.23}$$

The dependence of the nutation angle according to (23) is shown in Fig. 7.5 for different values of σ and $\theta^0 = \pi/6$. One can draw the following qualitative conclusions: $\theta \to \pi/2$ as $\sigma \downarrow 0$; $\theta \to 0$ as $\sigma \uparrow 0$ and $\tau > 0$. But if $\sigma \to \pm\infty$ then the nutation angle $\theta(\tau) \approx \theta^0$, i.e. during the braking process the quantity θ has not enough time to change significantly. In the limit as $b \to 0$ the time $T^* \to \infty$ and the obtained formulae coincide with those for "passive" motion [1, 21].

Based on the known dependence $\theta(t, t_0, \theta^0)$, we determine the angular velocity vector $\boldsymbol{\omega}(t, t_0, \boldsymbol{\omega}^0)$:

$$\begin{aligned} &\omega_1 = \omega_1^0(L/L^0)\exp\alpha\cos\psi, \qquad \omega_2 = \omega_2^0(L/L^0)\exp\alpha\sin\psi,\\ &\omega_3 = (L/I_3)\cos\theta \qquad \left(\cos\theta^0 = I_3\omega_3^0/L^0\right),\\ &\alpha = \frac{B}{I}\int_{t_0}^{t}\omega_3^4(\tau, t_0, \omega^0)\,\mathrm{d}\tau, \qquad \psi = \int_{t_0}^{t}(d + AL^2)\omega_3\,\mathrm{d}\tau. \end{aligned} \tag{7.4.24}$$

Substitution of expressions (24) into (21) allows the construction of open-loop controls for specified initial conditions. Therefore, the optimal stabilization problem is completely solved.

Now let us consider the case when the particle is connected to the body by a linear spring with stiffness c and a quadratic-friction damper with damping coefficient μ. Then, assuming $\Omega \gg \omega$, where $\Omega^2 = c/m$ and $\mu/m = \Lambda\Omega^3$, we obtain approximate

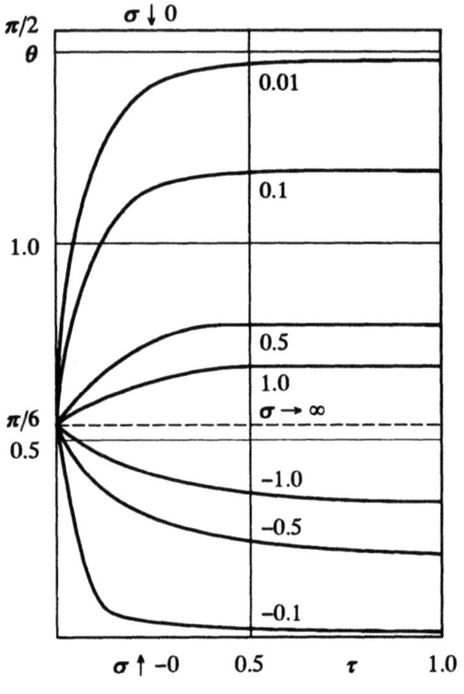

Fig. 7.5.

equations of motion of the form (20). The only difference is in the terms, due to dissipation, which are given by

$$-D\omega_{\perp}\omega_3^6\omega_{1,2}, \quad -DII_3^{-1}\omega_{\perp}^3\omega_3^5; \quad D = m\rho^3 d|d|\Lambda I^{-4} I_3^4 \Omega^{-3}. \tag{7.4.25}$$

It follows from (20) and (25) that the solution of the minimum-time control problem again has the form (21). Further investigation of the optimal motion is conducted analogously to the case considered.

(2) *Braking bodies containing cavities with a strongly viscous fluid.* The aforenamed property of invariance of the uncontrolled system with respect to the quantity of the angular momentum is observed for other mechanical models as well. For example, in the case of a rigid body having axial symmetry while in an unperturbed state, with a spherical cavity containing a strongly viscous fluid (under small Reynolds

numbers), perturbing torques are given by (see [1, 21])

$$N_{1,2} = -P\beta\nu^{-1}I^{-3}I_3(I - I_3)\omega_3^2\omega_{1,2},$$
$$N_3 = -P\beta\nu^{-1}I^{-1}I_3^{-1}(I_3 - I)\omega_\perp^2\omega_3, \qquad P = (8\pi/525)R^7. \tag{7.4.26}$$

Here, β is the density, ν is the kinematic viscosity of the fluid, and R is the cavity radius. Using the technique developed in Sections 7.1 and 7.2 we find the optimal control, the value of the angular momentum and the minimal time in the form (21). If $b = \text{const}$, then the nutation angle θ is determined in the explicit form:

$$\theta = \tan^{-1}\left[\tan\theta^0 \exp\frac{1-(1-\tau)^3}{\sigma}\right], \qquad \sigma = \frac{3\nu b I^3 I_3}{\beta P(I - I_3)L^{03}},$$
$$\tau = (t - t_0)(T^* - t_0)^{-1} \in [0, 1], \qquad T^* = t_0 + L^0/b. \tag{7.4.27}$$

Further investigation is conducted in a way similar to that described above, which does not cause any difficulties (see (24)).

If the body is nonsymmetric while in an unperturbed state ($I_1 > I_2 > I_3$), then the quantities N_i are given by [1, 21]

$$N_1 = +P\frac{\beta}{\nu}\frac{\omega_1}{I_1I_2I_3}\left[I_3(I_1 - I_3)(I_1 + I_3 - I_2)\omega_3^2 + \right.$$
$$\left. + I_2(I_1 - I_2)(I_1 + I_2 - I_3)\omega_2^2\right] \qquad (1, 2, 3).$$

These coincide with (26) when $I_1 = I_2 = I$. For the quantity $k^2 \in [0, 1]$ (6.4.2) the following equation admitting separation of variables is obtained ($L^2 \geqslant 2EI_2$):

$$\frac{dk^2}{d\xi} = (1 - \varkappa)(1 - k^2) - \left[(1 - \varkappa) + (1 + \varkappa)k^2\right]\frac{\mathbf{E}(k)}{\mathbf{K}(k)},$$
$$\xi = \int_{\tau_0}^{\tau} \frac{ds}{n(s)},$$
$$n(\tau) = (3\nu/\beta P)(I_1I_2I_3)^2(I_1 - I_3)^{-1} \times \tag{7.4.28}$$
$$\times \left[I_2(I_1 + I_3 - I_2) + 2I_1I_3\right]^{-1}L^{-2}(\tau),$$
$$\varkappa = 3I_2\left[I_1^2 + I_3^2 - I_2(I_1 + I_3)\right](I_1 - I_3)^{-1} \times$$
$$\times \left[I_2(I_1 + I_3 - I_2) + 2I_1I_3\right]^{-1}.$$

For the region $L^2 \leqslant 2EI_2$, one should permute indices $1 \rightleftarrows 3$ in expressions (28). Further numerical analysis of controlled motion dynamics of the body with a cavity does not cause any difficulty, and the qualitative behaviour of k^2 coincides with that shown in Fig. 7.5.

It is worth noting that by using a similar approach some other optimal stabilization problems can be posed and solved for rigid bodies containing flywheels (gyrostats), viscoelastic elements [91, 97], and so on.

The solution of the stabilization problem can be regarded as part of a more general problem of spacecraft orientation and bringing it to a desirable rotatory state. Changing the orientation of a spacecraft is significantly simpler if the spacecraft is at rest at the initial time instant. This mode is widely applied in practice. Similar problems have been considered in many publications; see, for instance, [1, 46, 91, 97], and Section 7.5.

7.5. Synthesis of a Quasioptimal System of Spacecraft Orientation and Stabilization

7.5.1. INITIAL ASSUMPTIONS AND STATEMENT OF THE PROBLEM

When describing the dynamics of controlled rotation of a spacecraft, we consider the latter as a system of rigid bodies. Equations of motion for the "undeformable part" (the body) of the spacecraft are given by [91]:

$$\begin{gathered} \dot{\mathbf{L}} + [\boldsymbol{\omega} \times \mathbf{L}] = \mathbf{M}, \qquad \mathbf{L} = J\boldsymbol{\omega}, \qquad \boldsymbol{\omega}(t_0) = \boldsymbol{\omega}^0, \\ J = \operatorname{diag}(J_1, J_2, J_3) = \text{const}, \qquad \mathbf{M} = \mathbf{N} + \mathbf{G}. \end{gathered} \tag{7.5.1}$$

Here, $\boldsymbol{\omega}$ and $\mathbf{L}$ are angular velocity and angular momentum vectors, J is the principal central tensor of inertia; and t_0 and $\boldsymbol{\omega}^0$ are initial data. The vector $\mathbf{M}$ is the resultant torque of external and internal forces acting on the carrying body. The torque of the internal forces $\mathbf{G}$ is caused by displacements and rotations of inner parts of the spacecraft. The torque $\mathbf{G}$ can be regarded as a control. Implementation of control torques (jet engines, flywheels, etc.) is discussed in [72, 91]. The external torque $\mathbf{N}$ used in "passive" orientation and stabilization systems is considered as a known function of time and the state vector of the whole system.

Let the torque $\mathbf{N}$ be considered as a perturbation. The control vector $\mathbf{G}$ of system (1) is included in the set of variable parameters p of the control unit. The vector p is an output of the unit described by singularly perturbed equations of the form

$$\begin{gathered} \mu\dot{p} = P(p, \boldsymbol{\omega}, u), \quad p(t_0) = p^0, \quad \mu \in (0, \mu_0], \quad \mu_0 \ll 1, \\ p = (G_1, G_2, G_3, p_4, \ldots, p_n)^T, \quad u = (u_1, \ldots, u_m)^T, \quad u = u(t) \in D^u. \end{gathered} \tag{7.5.2}$$

In Equations (2), a small numerical parameter μ is introduced which characterizes the time constants $\tau_i^{(p)}(\mu) \sim \mu$ of transient processes in the control unit. The components of the m-vector $u(t)$ are considered as control functions. We suppose that $u(t) \in D^u$ where D^u is some admissible domain. If the smoothness conditions are fulfilled for $u(t)$, then the solution $p(t, t_0, p^0, \mu)$ of system (2) for $t \in (t_0, T]$, $T \leqslant \infty$, (outside the boundary layer) is asymptotically close (in terms of μ) to the solution p^* of the finite (not differential) equation (2) for $\mu = 0$ [26, 27, 54]:

$$P(p, \omega, u) = 0, \quad p = p^*(\omega, u). \tag{7.5.3}$$

The vector $\omega = \omega^*(t, t_0, \omega^0)$ is determined from (1) after having substituted the control vector $\mathbf{G} = \mathbf{G}^*(\omega, u)$ found according to (3). The proximity (in terms of μ) of the solutions of original system (1) and (2) and generating system (1) and (3) for the vector ω is uniform. Therefore, the error of the approximate solution is estimated as $O(\mu)$ outside the transient interval (the boundary layer), i.e. for $t \geqslant t_\mu$, $t_\mu - t_0 \sim \mu \ln \mu^{-1}$. In the case when the control $u(t)$ is a piecewise smooth function, the additional analysis is required (see below).

Let us impose more restrictive assumptions on the control system. Let the singularly perturbed system (1) contain a small parameter ε, $\varepsilon \gg \mu$: $\mathbf{G}^*(\omega, u) = \varepsilon^{-1}\mathbf{G}_*(\omega, u)$, the equality of $\mathbf{G}_*(\mathbf{0}, 0) = \mathbf{0}$ being maintained. The parameter ε characterizes the feedback in the loop of the angular velocity control according to (1) as well as time constants $\tau_i^{\omega}(\varepsilon) \sim \varepsilon$ of setting in the required value ω corresponding to a given $u(t) \in D^u$. Thus, neglecting transients in the systems (1) and (2) we obtain the finite relationships

$$\begin{aligned} &\mathbf{G}_*(\omega, u) = 0, \qquad \omega^* = \omega^*(u), \qquad \omega^*(0) = 0; \\ &\omega^* = \omega^*(t) \in D^\omega, \qquad u \in u(t) \in D^u; \qquad \omega^*(t) \equiv \omega^*(u(t)); \\ &\omega^* = \mathbf{0} \in D^\omega, \qquad u = 0 \in D^u. \end{aligned} \tag{7.5.4}$$

We assume in what follows that the domain D^ω is bounded, closed, and, as a rule, strictly convex. Then the error in determining the variables ω and p is estimated by the amounts $O(\varepsilon)$ and $O(\mu/\varepsilon)$, respectively, outside the boundary layer: $t \geqslant t_\varepsilon$, $t_\varepsilon - t_0 \sim \varepsilon \ln \varepsilon^{-1}$. The reasonings given are used when proceeding to the limit system of the "kinematic control" of the spacecraft body orientation. To describe this system, we use the kinematic relationships in convenient variables, the vector $\omega^* \in D^\omega$ being considered as a control.

In applications, one can encounter angular variables (Euler's angles, angles of Euler and Krylov, etc.), direction cosine matrices, quaternions (parameters of Rodrigues and Hamilton), rotation vectors and so on [84, 91, 97, 112, 127]. To be definite, we use below the relationships for quaternions and rotation vectors [84, 91]. Let us pose the problem of the complete orientation of the spacecraft body: to bring all the axes connected with the body to a desired angular position by using parametric control. For control functions, we take projections of the angular velocity vector $\omega^*(t) \in D^\omega$ onto the axes connected with the body. Performance indices are considered below. If the problem is solvable within the class of the piecewise smooth function $\{\omega^*(t)\}$, $\omega^*(t) \in D^\omega$, $u(t) \in D^u$, then the uniform proximity takes place with respect to both the orientation variables and the functional.

Let us give estimates for errors caused by the approximation of system (4) for the control torque $\mathbf{G}^* = \varepsilon^{-1}\mathbf{G}_*$, where $\mathbf{G}_* = -\varkappa J(\omega - \mathbf{u})$. This expression satisfies the conditions of theorems of asymptotic proximity between solutions of the original singularly perturbed system ($\varepsilon > 0$) and the degenerate system ($\varepsilon = 0$). Indeed, using the "gyroscopic property" of the cross product in (1) (i.e. $\mathbf{L}^T[\omega \times \mathbf{L}] = \omega^T[\omega \times \mathbf{L}] \equiv 0$) we obtain the estimates for the modulus $L = |\mathbf{L}|$ of the angular momentum vector as well as for its components L_i, $i = 1, 2, 3$:

$$\begin{aligned} &0 \leqslant L(t) \leqslant L^0 e^{-\varkappa\tau} + u_0^* J_3 (e^{\varkappa\tau} - 1) e^{-\varkappa\tau}, \qquad \tau = t/\varepsilon, \\ &|L_i(t)| \leqslant L(t) \qquad \big(t_0 = 0,\ L^0 = L(0),\ J_3 \geqslant J_2 \geqslant J_1,\ u_0^* = \max_t |u|\big). \end{aligned} \tag{7.5.5}$$

Using (1) and (5), we obtain the estimates for ω_i: $|\omega_i| = |L_i|/J_i \leqslant L/J_i$ ($i = 1, 2, 3$). Approximate integration of system (1), taking into account the estimates, gives

$$\begin{aligned} &\big|\omega(t,\varepsilon) - \omega_0(t,\varepsilon)\big| \leqslant \varepsilon C_0, \qquad t \in [0, T], \\ &C_0 = C(0), \qquad C = C(\tau) = 4J_1^{-2}\big(L^{02} e^{-\varkappa\tau} + u_0^{*2} J_3^2/\varkappa\big), \\ &\omega_0(t,\varepsilon) = \omega^0 e^{-\varkappa\tau} + \frac{\varkappa}{\varepsilon} e^{-\varkappa\tau} \int_0^t \mathbf{u}(\sigma) e^{\varkappa s}\, d\sigma, \qquad s = \frac{\sigma}{\varepsilon}. \end{aligned} \tag{7.5.6}$$

If the control $\mathbf{u}(t)$, $t \in [0, T]$, is a piecewise differentiable function then integration by parts in (6) results in the approximate expression for $\omega(t, \varepsilon)$:

$$\begin{aligned} &\omega_0(t,\varepsilon) = \big[\omega^0 - \mathbf{u}(0)\big] e^{-\varkappa\tau} + \mathbf{u}(t), \qquad t \in [0, T], \\ &\big|\omega(t,\varepsilon) - \omega_0(t,\varepsilon)\big| \leqslant \varepsilon C_*. \end{aligned} \tag{7.5.7}$$

The constant C_* includes C_0 from (6) and the quantity $\dot{u}_0^*/\varkappa$, where $\dot{u}_0^* = \max_t |\dot{u}^*|$. For $t \geqslant (\varepsilon/\varkappa)\ln\varepsilon^{-1}$, the initial difference in (7) disappears, i.e. becomes an amount of the order of the error $O(\varepsilon)$. Integration of the boundary-layer function in the kinematic equations leads to the error $O(\varepsilon)$ for the variables determining the spacecraft orientation. Therefore, the error estimates for the variables to be determined are uniform. Note that, provided $\mathbf{u}(0) \simeq \omega^0$, the transient process $\omega(t) \to \mathbf{u}(t)$ can be ignored and the proximity estimate $|\omega(t,\varepsilon) - \mathbf{u}(t)| = O(\varepsilon)$ is also uniform. The above estimates justify the consideration of the function $\omega(t)$ as a control when solving the orientation problem for sufficiently small $\varepsilon > 0$. If, apart from the initial point $t = 0$, the function $\mathbf{u}(t)$ has a finite number of isolated discontinuity points of the first kind t_j $(0 < t_1 < t_2 < \cdots < t_{n^*} \leqslant T)$ and, between these points, the derivative of the function $\mathbf{u}(t)$ is bounded, then expressions (6) and (7) include additional boundary-layer functions

$$\begin{aligned} &\omega_0(t,\varepsilon) = \mathbf{u}(t) + \left[\omega^0 - \mathbf{u}(0)\right]e^{-\varkappa\tau} + \sum_{j=1}^{n(t)} \Delta\mathbf{u}_j e^{-\varkappa(\tau-\tau_j)}, \\ &1 \leqslant n(t) \leqslant n^* \quad (t > t_1), \\ &\Delta\mathbf{u}_j = \mathbf{u}(t_j - 0) - \mathbf{u}(t_j + 0), \qquad t = t/\varepsilon. \end{aligned} \tag{7.5.8}$$

The total influence of the boundary layers (8) will be of the order of ε in the integral (weak) metric. Note that the estimating procedure can be described transferred to the case $\mathbf{G}_*(\omega,\mathbf{u}) = K(\omega - \mathbf{u})$, where K is a (3×3)-matrix whose eigenvalues have negative real parts. However, in this case, the manipulations will be more cumbersome. Under the additional conditions of smoothness and nonsingularity, the estimates remain valid for the general function $\mathbf{G}_*(\omega,\mathbf{u})$ in (4).

7.5.2. SOLUTION OF THE MINIMUM-TIME PROBLEM OF SPACECRAFT ORIENTATION

Let us consider quaternionic kinematic equations, determining the orientation of a rigid body, when $\omega^*(t)$ is represented in projections onto the axes connected with the body [84, 91, 112]:

$$\begin{aligned} &\dot{\Lambda} = \frac{1}{2}\Omega^{(4)}\Lambda, \quad \Lambda = \left(\lambda_0, \boldsymbol{\lambda}^T\right)^T, \quad \Lambda(t_0) = \Lambda^0, \quad \left|\Lambda^0\right| = 1, \\ &\Omega^{(4)} = \begin{bmatrix} 0 & -\omega^{*T} \\ \omega^* & \Omega^{(3)} \end{bmatrix}, \quad \Omega^{(3)} = \begin{bmatrix} 0 & -\omega_3^* & \omega^* \\ \omega_3^* & 0 & -\omega_1^* \\ -\omega_2^* & \omega_1^* & 0 \end{bmatrix} = -\Omega^{(3)T}. \end{aligned} \tag{7.5.9}$$

Here, λ_0 is the scalar and $\boldsymbol{\lambda}$ is the vector part of the quaternion Λ; vectors are considered as column vectors. The (3×3)-matrix $\Omega^{(3)}$ satisfies the condition $\Omega^{(3)}\mathbf{r} \equiv [\omega^* \times \mathbf{r}]$ for any vector $\mathbf{r}$. System (9) is considered on the manifold $\Lambda^2 = 1$. The last equality is satisfied, since (9) implies that $(\dot{\Lambda^2}) = \Lambda^T\Omega^{(4)}\Lambda \equiv 0$. The vector $\boldsymbol{\lambda}$

defines a finite rotation of the body. The connection between the parameters is given by [127]:

$$\lambda_0 = \cos\frac{\chi}{2}, \qquad \boldsymbol{\lambda} = \mathbf{n}\sin\frac{\chi}{2}, \qquad 0 < \chi < \pi;$$
$$\boldsymbol{\theta} = 2\frac{\boldsymbol{\lambda}}{\lambda_0} = 2\mathbf{n}\tan\frac{\chi}{2}, \qquad |\mathbf{n}| = 1; \qquad \Lambda^2 = \lambda_0^2 + \boldsymbol{\lambda}^2 \equiv 1. \tag{7.5.10}$$

Here, $\boldsymbol{\theta}$ is the rotation vector, χ is the angle, and $\mathbf{n}$ is the unit vector of the rotation axis.

Thus, the parameter-controlled system (9) is the system with an invariant norm: $|\Lambda| \equiv 1$ for any vector $\boldsymbol{\omega}^*(t)$. For this system, we pose the problem of the synthesis of the time-optimal control $\boldsymbol{\omega}^*(\Lambda)$ transferring the system (9) from the state Λ^0 to the required state $\Lambda(t_f) = \Lambda^f$. Due to the invariance mentioned above, this is equivalent to transferring the vector $\boldsymbol{\lambda}$ from an arbitrary initial state $\boldsymbol{\lambda}(t_0) = \boldsymbol{\lambda}^0$, $\lambda^{02} \leqslant 1$ to the required state $\boldsymbol{\lambda}(t_f) = \boldsymbol{\lambda}^f$. Without loss of generality, we assume $\boldsymbol{\lambda}^f = \mathbf{0}$, $\lambda_0^f = 1$ $(\Lambda^f = (1, \mathbf{0}^T)^T)$.

The solution of this problem is constructed by using the method of dynamic programming [32, 43, 55, 65, 144]. For Bellman's function $S = t_f - t_0$, the functional-differential equation of Hamilton–Jacobi and the boundary conditions

$$\frac{1}{2}\min_{\boldsymbol{\omega}^* \in D^\omega}(Q^T \Omega^{(4)} \Lambda) = -1, \quad S = S(\Lambda), \quad \Lambda^2 = 1,$$
$$S(\Lambda^f) = 0, \quad Q = \begin{pmatrix} q_0 \\ \mathbf{q} \end{pmatrix} = \left(\frac{\partial S}{\partial \Lambda}\right)^T, \quad q_0 = \frac{\partial S}{\partial \lambda_0}, \quad \mathbf{q} = \left(\frac{\partial S}{\partial \boldsymbol{\lambda}}\right)^T \tag{7.5.11}$$

are fulfilled. This initial-value problem represents the necessary optimality conditions (the conditions of regular synthesis [40, 65] are supposed to be satisfied).

The minimum of the linear function in (11) over $\boldsymbol{\omega}^* \in D^\omega$ exists. We suppose in what follows that the domain D^ω is convex. In this case, the minimum is determined uniquely. If the solution $S(\Lambda)$ of the functional-differential initial-value problem exists on the manifold $\Lambda^2 = 1$ (a three-dimensional sphere in E^4) and is unique, then the function $S(\Lambda)$ determines the solution of the time-optimal control problem. The quantity $t_f = t_0 + S(\Lambda^0)$ gives the minimal possible time for the control process, and the minimum point $\boldsymbol{\omega}^*(\Lambda, Q(\Lambda)) = \boldsymbol{\omega}_0(\Lambda)$ defines the optimal control in the feedback form; the optimal control $u_0(\Lambda)$ is determined according to (4): $\mathbf{G}_*(\boldsymbol{\omega}_0(\Lambda), u) = 0$, $\boldsymbol{\omega}_0 \in D^\omega$, $u_0 \in D^u$.

This do not lead to solving problem (11) for domains D^ω of any shape. In applied problems, they often approximate admissible domains for the control by the following solids: 1) a parallelepiped (a cube in particular), 2) a cylinder (circular),

3) an ellipsoid (a ball Σ with the centre at $\omega^* = 0$), etc. In what follows, we suppose that there exist two concentric balls $\Sigma_{1,2}$ of radii $\nu_{1,2} > 0$ such that [1]

$$\begin{aligned} &\Sigma_1 \subseteq D^\omega \subseteq \Sigma_2, \qquad \Sigma_{1,2} = \{\omega^*\colon |\omega^*| \leqslant \nu_{1,2}\},\\ &(\nu_2-\nu_1)/\nu_0 = \delta, \qquad \nu_0 \in [\nu_1, \nu_2], \qquad 0 \leqslant \delta \ll 1,\\ &\quad (\delta \in [0, \delta_0],\ \delta_0 \ll 1). \end{aligned} \tag{7.5.12}$$

Consider the problem of the approximate synthesis of control in the case $\omega^* \in \Sigma_0$. Here, Σ_0 is a ball satisfying the inclusion (12); the centre of the ball being placed at $\omega^* = \mathbf{0}$. In particular, the radius of this ball can be chosen as $\nu_0 = (1/2)(\nu_1+\nu_2)$. In this case, the initial-value problem (11) for the nonlinear partial differential equation on the manifold becomes

$$\begin{aligned} &\nu_0\rho = 1, \quad \rho = |\boldsymbol{\rho}|, \quad \Lambda^2 = 1, \quad S(\Lambda^f) = 0, \quad \Lambda^f = \left(1, \mathbf{0}^T\right)^T;\\ &\rho_1 = -\frac{1}{2}(\lambda_1 q_0 - \lambda_0 q_1 - \lambda_3 q_2 + \lambda_2 q_3), \quad Q \equiv \left(q_0, \mathbf{q}^T\right)^T;\\ &\rho_2 = -\frac{1}{2}(\lambda_2 q_0 + \lambda_3 q_1 - \lambda_0 q_2 - \lambda_1 q_3), \quad Q = (\partial S/\partial\Lambda)^T;\\ &\rho_3 = -\frac{1}{2}(\lambda_3 q_0 - \lambda_2 q_1 + \lambda_1 q_2 - \lambda_0 q_3), \quad \omega^* = -\nu_0\boldsymbol{\rho}/\rho. \end{aligned} \tag{7.5.13}$$

The initial-value problem (13) and the optimal control $\omega^*(\Lambda, Q)$ are obtained by minimizing the linear (in ω^*) expression $Q^T\Omega^{(4)}\Lambda \equiv \boldsymbol{\rho}^T\omega^*$ over the ball Σ_0.

The symmetry of the problem implies that the function S depends only on $\lambda = |\boldsymbol{\lambda}|$ or, which is the same, on λ_0. Indeed, according to (10) and (11) we have the following expressions for the components q_i:

$$q_0 \equiv 0, \quad \mathbf{q} = q\boldsymbol{\eta}, \quad q = \frac{\mathrm{d}S}{\mathrm{d}\lambda}, \quad \boldsymbol{\eta} = \frac{\boldsymbol{\lambda}}{\lambda}, \quad \rho = \frac{1}{2}\lambda_0(\lambda)q. \tag{7.5.14}$$

Having substituted expressions (14) into (13) we obtain a simple one-dimensional nonlinear initial-value problem. Its solution, taking into account the mechanical interpretation of Bellman's function $S_0(\lambda)$, is given by

$$\begin{aligned} &\frac{1}{2}\nu_0 q\left(1-\lambda^2\right)^{1/2} = 1, \qquad S_0(0) = 0;\\ &S = S(\Lambda) = S_0(\lambda) = \frac{2}{\nu_0}\sin^{-1}\lambda = \frac{\chi}{\nu_0} \qquad (0 < \chi < \pi). \end{aligned} \tag{7.5.15}$$

Based on (13)–(15), we obtain the synthesis of the time-optimal angular velocity vector whose modulus is bounded:

$$\boldsymbol{\omega}^*(\Lambda, Q_0(\Lambda)) = \boldsymbol{\omega}_0(\boldsymbol{\lambda}) = -\nu_0\boldsymbol{\eta} \qquad (\lambda \not\equiv 0,\ \boldsymbol{\eta} = \mathbf{n}). \tag{7.5.16}$$

The simple result given by (15) and (16) seems to be evident for $\boldsymbol{\omega}^* \in \Sigma_0$. Having supposed $|\chi| < \infty$, we can naturally generalize these expressions. Substituting (16) into (9) we obtain the completely integrable system and the programmed angular velocity $\boldsymbol{\omega}_{0p}$:

$$\begin{aligned}
&\lambda = \lambda_0\big(t - t_0, \lambda^0\big) = \sin\Big(\sin^{-1}\lambda^0 - \frac{1}{2}\nu_0(t - t_0)\Big),\\
&\boldsymbol{\lambda} = \boldsymbol{\lambda}_0\big(t - t_0, \boldsymbol{\lambda}^0\big) = \lambda_0\big(t - t_0, \lambda^0\big)\mathbf{n}_0\big(t - t_0, \mathbf{n}^0\big),\\
&\mathbf{n}_0\big(t - t_0, \mathbf{n}^0\big) \equiv \mathbf{n}^0,\\
&\boldsymbol{\omega}_{0p} = \boldsymbol{\omega}_{0p}\big(t - t_0, \boldsymbol{\lambda}^0\big) = -\nu_0\mathbf{n}^0, \qquad t_f = t_0 + S_0\big(\lambda^0\big) = t_0 + \chi^0/\nu_0.
\end{aligned} \tag{7.5.17}$$

These results are similar to those obtained in Sections 7.1 and 7.2 for controlled dynamic systems with an invariant norm. In such systems, the control is included additively. These results are also similar to those for the "extensive" rotation of a rigid body with respect to the axis (Euler's axis) fixed both in the inertial and the body-connected reference frames [1, 46, 91]. Kinematic (hard) control $\boldsymbol{\omega}^*$ for the considered case of spherical symmetry leads to a rotation with respect to the fixed axis determined by a nonzero initial (as well as current) value of the vector part of the quaternion.

Let the general case occur when the admissible domain D^ω for $\boldsymbol{\omega}^*$, $\boldsymbol{\omega}^* \in D^\omega$, is close, in the sense of (12), to the ball Σ_0. Then, based on the constructed complete solution (15)–(17) for the generating problem (13), we can make the solution of the synthesis problem (11) for the time-optimal control of the spacecraft rotation more precise by using perturbation techniques to take into account higher powers of the parameter δ. After substitution of the feedback control $\boldsymbol{\omega}_0(\boldsymbol{\lambda}, \delta)$ into system (9) the perturbed optimal trajectory (a quaternion) and the programmed (open-loop) control (the angular velocity vector) are constructed analogously. The procedure of the approximate solution of the nonlinear initial-value problem (11) for the functional-differential equation is simplified if the minimization is performed analytically. This can be done, for instance, if the domain D^ω is a polyhedron, a cylinder or an ellipsoid. In the general case, the approximate minimization is carried out to the required accuracy in terms of δ [1]. For an approximate construction of Bellman's function one can use the method of characteristics (Cauchy). However, it does not seem to be expedient to carry out this construction in the general case because this will lead to rather cumbersome manipulations. It seems more reasonable to make these calculations for domains of some specific shapes, in particular for those mentioned above. Consideration of transient processes in the controlled system (1), and (2) causes significant theoretical difficulties. But this consideration is not of substantial practical value, since it is practically impossible to implement the control in the multi-dimensional space of the system parameters. It is more important for applications to have estimates of errors in the manoeuvre implementation and to determine the sensitivity of these errors with respect to structural parameters. The errors can be caused by the dynamics of the control system (in particular by the presence of delay) as well as other factors. Parallel with controls of a time-optimal type, similar to those considered above, they usually envisage stabilization modes in a small vicinity of the desired orientation. These modes must have a high perturbation resistance. In particular, linear feedback controls are widely used. They can be found by using the well-known technique of the analytical construction of regulators.

Consider now the problems of the time-optimal orientation of a spacecraft, based on kinematic equations for the rotation vector $\boldsymbol{\theta}$ (10). According to [91, 112, 127] we have the relationships

$$\begin{aligned}
&\dot{\boldsymbol{\theta}} = \boldsymbol{\omega}^* + \frac{1}{2}[\boldsymbol{\omega}^* \times \boldsymbol{\theta}] + \frac{1}{4}\boldsymbol{\theta}(\boldsymbol{\theta} \cdot \boldsymbol{\omega}^*), \qquad \boldsymbol{\theta}(0) = \boldsymbol{\theta}^0;\\
&\boldsymbol{\theta}^0 = 2\frac{\boldsymbol{\lambda}^0}{\lambda_0} = 2\mathbf{n}^0 \tan\frac{\chi^0}{2}, \qquad 0 < \chi^0 < \pi, \\
&\big(\boldsymbol{\theta}(t_f) = \mathbf{0},\ t_f \to \min_{\boldsymbol{\omega}^*},\ \boldsymbol{\omega}^* \in \Sigma_0;\ t_0 = 0\big).
\end{aligned} \tag{7.5.18}$$

The right-hand side in (18) is linear in the control $\boldsymbol{\omega}^*$ and quadratic in the variable $\boldsymbol{\theta}$. It can be represented both in the matrix-vector and the coordinate forms:

$$\begin{aligned}
&\dot{\boldsymbol{\theta}} = \Big(I_{(3)} - \frac{1}{2}\Theta_{(3)} + \frac{1}{4}\boldsymbol{\theta}\boldsymbol{\theta}^T\Big)\boldsymbol{\omega}^* \equiv \Pi_{(3)}(\boldsymbol{\theta})\boldsymbol{\omega}^*, \qquad \boldsymbol{\theta}(0) = \boldsymbol{\theta}^0;\\
&\Pi_{(3)}(\boldsymbol{\theta}) = \big(\pi_{ij}(\boldsymbol{\theta})\big) \quad (i, j = 1, 2, 3); \qquad \boldsymbol{\theta} = (\theta_1, \theta_2, \theta_3)^T;\\
&\Pi_{(3)}(\boldsymbol{\theta}) = \begin{bmatrix} 1 + \dfrac{\theta_1^2}{4} & \dfrac{\theta_3}{2} + \dfrac{\theta_1\theta_2}{4} & -\dfrac{\theta_2}{2} + \dfrac{\theta_1\theta_3}{4} \\ -\dfrac{\theta_3}{2} + \dfrac{\theta_1\theta_2}{4} & 1 + \dfrac{\theta_2^2}{4} & \dfrac{\theta_1}{2} + \dfrac{\theta_2\theta_3}{4} \\ \dfrac{\theta_2}{2} + \dfrac{\theta_1\theta_3}{4} & -\dfrac{\theta_1}{2} + \dfrac{\theta_2\theta_3}{4} & 1 + \dfrac{\theta_3^2}{4} \end{bmatrix}
\end{aligned} \tag{7.5.19}$$

The skew-symmetric matrix $\Theta_{(3)}$ in (19) is composed of the components of the vector $\boldsymbol{\theta}$: $\Theta_{(3)}\boldsymbol{\omega}^* \equiv [\boldsymbol{\theta} \times \boldsymbol{\omega}^*]$; it is similar to the matrix $\Omega^{(3)}$ composed on the basis of $\boldsymbol{\omega}^*$

in (9). $I_{(3)}$ in (19) is a unit matrix. Therefore, the matrix $\Pi_{(3)}(\boldsymbol{\theta})$ is the sum of the unit, symmetric and skew-symmetric matrices. The elements π_{ij} of this matrix are quadratic forms of $\boldsymbol{\theta}$.

The initial-value problem of the type (13) for the unknown Bellman's function $S(\boldsymbol{\theta})$ and the expression for the optimal control $\boldsymbol{\omega}^*$ (13) are given by

$$\begin{aligned} &\nu_0 \rho = 1, \quad \rho = |\boldsymbol{\rho}|, \quad S = S(\boldsymbol{\theta}), \quad S(\mathbf{0}) = 0; \\ &\boldsymbol{\omega}^* = -\nu_0 \boldsymbol{\rho}/\rho, \quad \boldsymbol{\rho} = (\rho_1, \rho_2, \rho_3)^T = \Pi^T_{(3)}(\boldsymbol{\theta})(\partial S/\partial \boldsymbol{\theta})^T. \end{aligned} \tag{7.5.20}$$

The solution of the nonlinear problem (20) has a singularity of a "cone-point" type at the origin $\boldsymbol{\theta} = \mathbf{0}$. Similarly to (14) and (15), we seek this solution as a function S of $\theta = |\boldsymbol{\theta}| = 2\tan(\chi/2)$. Due to the identity $\Theta_{(3)}\boldsymbol{\theta} \equiv 0$, we obtain the one-dimensional initial-value problem of the type (15)

$$\nu_0 |S'| (1 + \theta^2/4) = 1, \qquad S(0) = 0. \tag{7.5.21}$$

The solution $S_0(\theta)$ of problem (21) is Bellman's function for the time-optimal control problem (18). This function, the feedback control $\boldsymbol{\omega}_0 = \boldsymbol{\omega}_0(\theta)$, the optimal trajectory $\boldsymbol{\theta}_0 = \boldsymbol{\theta}_0(t - t_0, \theta^0)$, $\boldsymbol{\theta}_0(S_0(\theta^0), \theta^0) \equiv \mathbf{0}$, and the open-loop control $\boldsymbol{\omega}_{0p} = \boldsymbol{\omega}_{0p}(t - t_0, \theta^0)$ are obtained by using standard techniques. They are given by

$$\begin{aligned} &S = S_0(\theta) = \frac{2}{\nu_0} \tan^{-1} \frac{\theta}{2} = \frac{\chi}{\nu_0}, \qquad 0 < \chi < \pi; \\ &\boldsymbol{\omega}_0 = \boldsymbol{\omega}_0(\theta) = -\nu_0 \boldsymbol{\eta}, \qquad \boldsymbol{\eta} = \boldsymbol{\theta}/\theta, \qquad 0 < \theta < \infty; \\ &\theta_0 = \theta_0(t - t_0, \theta^0) = 2\tan\left(\tan^{-1}\frac{\theta^0}{2} - \frac{\nu_0}{2}(t - t_0)\right), \\ &\theta_0(t_f - t_0, \theta^0) = \theta_0(S_0(\theta^0), \theta^0) \equiv 0; \\ &\chi = \chi_0(t - t_0, \chi^0) = \chi^0 - \nu_0(t - t_0); \\ &\boldsymbol{\theta}_0 = \boldsymbol{\theta}_0(t - t_0, \boldsymbol{\theta}^0) = \theta_0 \boldsymbol{\eta}_0 = \theta_0(t - t_0, \theta^0)\boldsymbol{\eta}^0, \qquad \boldsymbol{\eta}^0 = \boldsymbol{\theta}^0/\theta^0; \\ &\boldsymbol{\omega}_{0p} = \boldsymbol{\omega}_{0p}(t - t_0, \boldsymbol{\theta}^0) = -\nu_0 \boldsymbol{\eta}_0 = -\nu_0 \boldsymbol{\eta}^0 = \text{const.} \end{aligned} \tag{7.5.22}$$

Expressions (22) correspond to the equalities (16) and (17) and can be obtained from them by the changes (10). These are clear and can be easily implemented in practice. These results can be immediately transferred to the case of an arbitrary rotation angle $|\chi| < \infty$. If the domain D^ω has a general shape and satisfies the conditions (12), the remarks made above are valid. We can obtain the solution for the problem of minimum-time rotation of a spacecraft more precisely by using the approximate procedure developed in Section 7.2. To counter uncontrolled perturbations of a different physical nature, imperfections and delays in servosystems, hierarchical (adaptive) control in a small vicinity of the desired orientation is often utilized. In particular, for stabilization of the angular position, a linear controller is widely used which provides the required characteristics of transient processes.

7.5.3. OTHER STATEMENTS FOR PROBLEMS OF ORIENTATION AND STABILIZATION OF A SPACECRAFT

Here, we consider some problems having an integral quadratic performance index and terminal conditions of the following form [43, 65, 106, 111]:

$$\begin{aligned} &\Phi[\omega^*] = c_f\theta^2(t_f) + \int_0^{t_f} \left(c_\theta\theta^2 + c_\omega\omega^{*2} + c_f\right)dt \longrightarrow \min_{\omega^*\in\Sigma_0}, \\ &1)\ \theta(t_f) = 0 \quad (c_f = 0); \\ &2)\ \theta(t_f) < \infty \quad (c_f \geqslant 0). \end{aligned} \tag{7.5.23}$$

Here, c_f, c_θ, c_ω, and c_t are non-negative constants (weighting factors), the inequality $c_\theta + c_f > 0$ being maintained. These coefficients define the quality of transient processes. The termination instant t_f can be either fixed ($c_t = 0$) or determined as a result of solving the problems ($c_t > 0$). The assumption $t_0 = 0$ does not restrict the statements. Problems similar to (23) are usually considered for a small vicinity of $\theta = 0$ and the control does not attain its boundary $|\omega^*(t)| = \nu_0$. This, for instance, takes place for sufficiently large t_f or sufficiently small c_t and c_ω.

Consider problems (23) with conditions (1) and (2) for fixed t_f, $f_f \leqslant \infty$. In this case, the Hamilton–Jacobi equation is two-dimensional and Bellman's function depends on two arguments: $S = S(t, \theta)$ (more exactly $S = S(t_f - t, \theta)$, see below). Indeed, initial-value problems for Bellman's function have the form

$$\begin{aligned} &\frac{\partial S}{\partial t} + \min_{\omega^*\in\Sigma_0} \left(\rho^T\omega^* + c_\omega\omega^{*2}\right) + c_\theta\theta^2 = 0, \\ &1)\ S = 0, \quad \theta = 0, \quad (t = t_f); \\ &2)\ S = c_f\theta^2, \quad t = t_f. \end{aligned} \tag{7.5.24}$$

Minimization of the quadratic form over ω^* in (24) gives the following expressions:

$$\begin{aligned} &\omega^* = \nu \equiv -\frac{\rho}{2c_\omega}, \quad \nu < \nu_0; \qquad \frac{\partial S}{\partial t} - \frac{\rho^2}{4c_\omega} + c_\theta\theta^2 = 0; \\ &\omega^* = -\nu_0\frac{\rho}{\rho}, \quad \nu \geqslant \nu_0; \qquad \frac{\partial S}{\partial t} - \nu_0\rho + c_\omega\nu_0^2 + c_\theta\theta^2 = 0. \end{aligned} \tag{7.5.25}$$

Equations (25) are symmetrical with respect to the components θ_i of the vector $\boldsymbol{\theta}$. Hence, we can assume $S = S_0(t, \theta)$, see (20). Then, using (19) and (20), we obtain the following expression for the vector $\boldsymbol{\rho}$: $\boldsymbol{\rho} = (\partial S_0/\partial\theta)(1+\theta^2/4)\boldsymbol{\eta}$, $\boldsymbol{\nu} = \nu\boldsymbol{\eta}$, $\boldsymbol{\eta} = \boldsymbol{\theta}/\theta$. Substitution of $\boldsymbol{\rho}$ into (25) results in differential equations depending on two arguments, t and θ:

$$\begin{aligned}
&\frac{\partial S_0}{\partial t} - \frac{1}{4c_\omega}\left(\frac{\partial S_0}{\partial\theta}\right)^2\left(1+\frac{\theta^2}{4}\right)^2 + c_\theta\theta^2 = 0,\\
&\left|\frac{\partial S_0}{\partial\theta}\right|\left(1+\frac{\theta^2}{4}\right) < 2\nu_0 c_\omega;\\
&\frac{\partial S_0}{\partial t} - \nu\left|\frac{\partial S_0}{\partial\theta}\right|\left(1+\frac{\theta^2}{4}\right) + c_\omega\nu_0^2 + c_\theta\theta^2 = 0,\\
&\left|\frac{\partial S_0}{\partial\theta}\right|\left(1+\frac{\theta^2}{4}\right) \geqslant 2\nu_0 c_\omega.\\
&1)\ S_0 = 0,\quad \theta = 0,\quad (t = \tau_f);\qquad 2)\ S_0 = c_f\theta^2,\quad t = t_f.
\end{aligned} \tag{7.5.26}$$

It is required to find the solution $S_0 = S_0(t,\theta)$ of Equations (26) satisfying conditions 1) and 2). The solution of the initial-value problem (26) "in large" (i.e. for $\theta \sim 1$) causes significant difficulties due to the essential nonlinearity of the controlled system (18) or (19). Using the variable χ ($\theta = 2\tan(\chi/2)$, $0 < \chi < \pi$) according to (10) leads to the elementary controlled system $\dot\chi = \nu$, $0 \geqslant \nu \geqslant -\nu_0$ ($\boldsymbol{\nu} = \nu\boldsymbol{\eta}$). But, in this case, the expression under the integral sign and the terminal term in functional (23) become more complicated: $c_{\theta,f}\theta^2 = 4c_{\theta,f}\tan^2(\chi/2)$. Equations of characteristics can be completely integrated, but nevertheless one has not succeeded in determining the explicit analytical solution of the two-point boundary-value problem (see below).

Depending on the initial value $\chi^0 = \chi(0)$ and the parameters t_f, ν_0, c_f, c_θ, and c_ω ($c_t = 0$), the control ν, analogously to (25), can have different analytical representations. To be definite, consider problem (23), 1) using the variable χ. This problem has a solution if $\chi^0 \leqslant \nu_0 t_f$. In the case of equality, the optimal open-loop control is $\nu_{0p}[t] = -\nu_0$ and $\chi_0(t,\chi^0) = \chi^0(1-\nu_0 t)$ (the mode A). Having been presented in feedback form, the mode A is defined by the formulae

$$\chi = \nu_0(t_f - t),\quad \chi > 0;\qquad \nu_s(t_f - t, \chi) = -\nu_0\quad (\dot\chi = \nu_s). \tag{7.5.27}$$

If the strict inequality $\chi^0 < \nu_0 t_f$ takes place, then two control modes are possible: the mode B without attaining the constraint by the control (for sufficiently small χ^0), and the mode C when the constraint is attained. The mode C contains the initial interval $t \in [0, t_*]$ where $\nu = -\nu_0$ (i.e. the control assumes its limiting value) and the terminal one ($t \in (t_*, t_f]$, $t_* < t_f$) where $0 \geqslant \nu > -\nu_0$. Therefore, the mode C

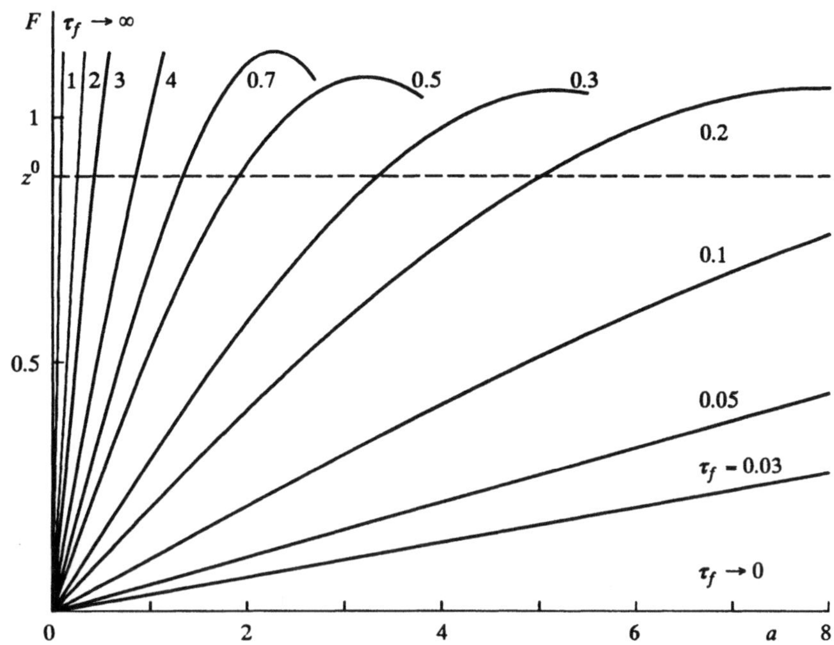

Fig. 7.6.

is a combination of the modes A and B. Optimal open-loop control in mode B takes place if the following relationships are fulfilled

$$\alpha\left(a^{*2}+\tan^2\frac{\chi^0}{2}\right)^{1/2}<\nu_0,\qquad a^*=\min \operatorname{Arg}\left[F(a,\tau_f)-z^0\right],$$

$$F(a,\tau_f)\equiv\begin{cases} a\sinh\left(\sqrt{1-a^2}\tau_f\right)/\sqrt{1-a^2}, & 0<a<1;\\ \tau_f, & a=1;\\ a\sin\left(\sqrt{a^2-1}\tau_f\right)/\sqrt{a^2-1}, & a>1;\end{cases} \tag{7.5.28}$$

$$z^0=\sin\left(\chi^0/2\right),\quad \tau_f=\alpha t_f,\quad \alpha=(c_\theta/c_\omega)^{1/2}\quad (\tau=\alpha t).$$

Since $0<z^0<1$, the root $a^*=a^*(z^0,\tau_f)$ of Equation (28) exists. The diagrammatic solution is given in Fig. 7.6; the curves 1–4 correspond to $\tau_f=0.5$, 1.0, 2.0, 3.0, respectively. For sufficiently large τ_f or small z^0 (or χ^0), the root a^* is determined by the upper relationship (28) for F; the following asymptotic expansion is valid:

$$a^* \simeq \frac{z_0}{\sinh \tau_f}\left[1 + z^{02}\frac{\tau_f}{2}\frac{\operatorname{cotanh}\tau_f}{(\sinh \tau_f)^2}\right].$$

Therefore, in mode B, the open-loop control and the optimal trajectory are given by

$$\begin{aligned}
&\nu_p^* = \nu_p^* \eta_0, \quad \eta_0(t, \theta^0) = \eta^0 = \text{const}; \quad \dot{\chi} = \nu_p^*, \quad \chi(0) = \chi^0;\\
&\nu_p^*[t, \chi^0, a^*] = -\alpha\left[a^{*2} + \tan^2\frac{\chi_0(t)}{2}\right]^{1/2}, \quad \chi_0(t) = \chi_0(\tau, z^0, a^*),\\
&\chi_0(\tau, z^0, a^*) = 2\sin^{-1} z_0(\tau, z^0, a^*), \quad \tau = \alpha t \in [0, \tau_f],\\
&z = z_0(\tau, z^0, a^*(z^0, \tau_f)) \qquad\qquad (7.5.29)\\
&= \begin{cases}
z^0 \cosh(\sqrt{1 - a^{*2}}\tau) - \\
\quad -\dfrac{[a^{*2}(1 - z^{02}) + z^{02}]^{1/2}}{\sqrt{1 - a^{*2}}}\sinh(\sqrt{1 - a^{*2}}), & 0 < a^* < 1;\\
z^0 - \tau, & a^* = 1;\\
z^0 \cos(\sqrt{a^{*2} - 1}\tau) - \\
\quad -\dfrac{[a^{*2}(1 - z^{02}) + z^{02}]^{1/2}}{\sqrt{a^{*2} - 1}}\sin(\sqrt{a^{*2} - 1}\tau), & a^* > 1.
\end{cases}
\end{aligned}$$

To determine the feedback control in the mode in question we have to check the condition

$$-\nu_0 < \nu_s < 0, \quad \nu_s = -\alpha\left(a_s^{*2} + \tan^2\frac{\chi}{2}\right)^{1/2}, \qquad t \in [0, t_f],$$

where $a_s^* = a_s^*(\tau_f - \tau, z)$ is the root of (28) for the current state z; $(t_f - t)$ is the time before the end of the process. Consider the control mode C corresponding to

$$\chi^0 < \nu_0 t_f, \qquad \alpha\left(a^{*2} + \left(\tan\frac{\chi}{2}\right)^2\right)^{1/2} > \nu 0. \tag{7.5.30}$$

This mode consists in the successive application of the scheme A for $t \in [0, t_*]$ and the scheme B for $t \in (t_*, t_f]$, $t_* < t_f$. To construct the open-loop control and the respective trajectory we have to solve two transcendental equations with respect to the unknown parameters t_* and a:

$$\begin{aligned}
&\alpha\left(a^2 + \tan\frac{\chi_*}{2}\right)^{1/2} = \nu_0 \qquad (\chi_* = \chi^0(1 - \nu_0 t)),\\
&F(a, \tau_f - \tau_*) = z_* \qquad \left(t_* = \alpha t_*, \ z_* = \sin\frac{\chi_*}{2}\right).
\end{aligned} \tag{7.5.31}$$

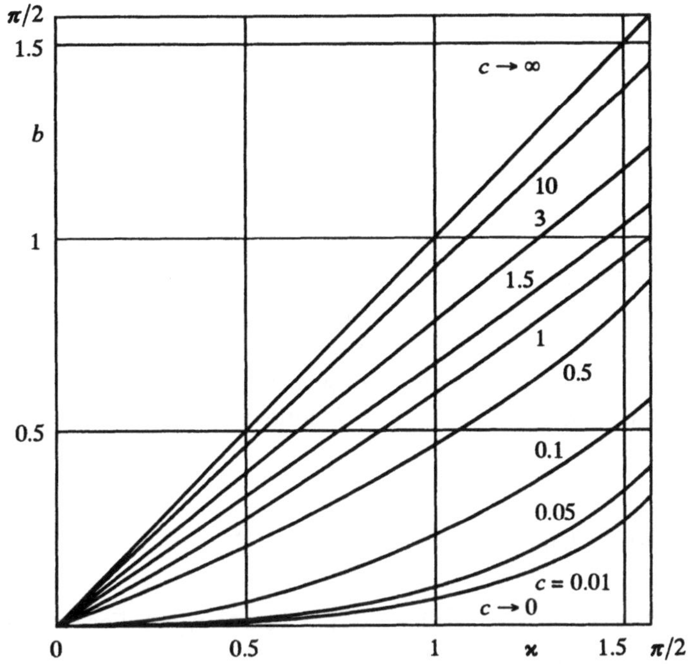

Fig. 7.7.

Using the first equation (31) we express the instant t_* of leaving the constraint through the unknown a. As a result, we obtain the only transcendental equation for determining $a = a^*(\chi^0, \tau_f)$. To obtain the feedback control we have to check the inequalities (30) and to solve the system (31) for current values of the variables: $t \to 0$, $t_f \to t_f - t$, $\chi^0 \to \chi$. Note that the possibility for mode C to appear is accounted for by the significant influence of the "penalty" coefficient c_θ and the constraint on the control. For sufficiently large ν_0 ($\nu_0 \to \infty$), mode B defined by (28) and (29) is optimal; for $c_\theta/c_\omega \to 0$, the constant open-loop control is optimal. Indeed, for $c_\theta = 0$ we have $\nu_{0p} = -\chi^0/\tau_f$ $(0 \geqslant \nu_{0p} \geqslant -\nu_0)$ and the feedback control is $\nu_s = -\chi/(t_f - t)$. It is worth paying attention to a singularity of the feedback control as $t \to t_f$. Such a singularity is inherent in problems having fixed terminal values. It complicates the approximate implementation of the feedback control.

The singularity does not appear if t_f is not fixed, or the right endpoint (for instance, $\theta(t_f)$) is free, or $t_f = \infty$. Consider for purpose of illustration the problem (23), 2) where $c_f > 0$, $c_\theta = 0$, $t_f < \infty$, t_f is fixed. For the mode B, the open-loop control is $\boldsymbol{\nu}_{0p} = \nu_{0p}\boldsymbol{\eta}^0$, $\nu_{0p} = -d = \text{const}$ $(0 < d < \nu_0)$; the modulus of the rotation vector is $\theta_0(t) = 2\tan(\tan^{-1}(\theta^0/2) - td/2)$. The optimal value for the unknown $d = d^*$ is a root of the equation (the transversality condition)

$$g(b,\theta^0) \equiv b\cos^2\left(\tan^{-1}\frac{\theta^0}{2} - b\right) = c\tan\left(\tan^{-1}\frac{\theta^0}{2} - b\right) \equiv ch(b,\theta^0),$$
$$b^* = \tan^{-1}(g - ch) = b^*(\varkappa, c), \qquad \varkappa = \tan^{-1}\frac{\theta^0}{2}, \quad c = c_f t_f. \tag{7.5.32}$$

The diagrammatic solution $b^* = b^*(\varkappa, c)$ of Equation (32) is given in Fig. 7.7. ($d^* = 2b^*/t_f$) where c is the parameter of the family of curves. The feedback control defined through the root $d^*(\theta, t_f - t)$ is a regular (nonsingular) function as $t \to t_f$: $\nu_s \to c_f\theta/[1 + (\theta/2)^2]$ ($\nu_s = \nu_s\eta$). Consider now the case $t_f = \infty$. Then both modes A and B are possible ($\nu_0 t_f > \chi^0$). It follows from (28) that under the mode B, $a^* = 0$ and (29) gives

$$\nu^* = -\alpha\tan(\chi/2) = -(\alpha/2)\theta \quad \text{and} \quad z_0(t) = z^0 e^{-\tau/2}.$$

Under the mode C $a^* = 0$ and $\chi_* = \chi^0(1 - \nu_0 t_*) = 2\tan^{-1}(\nu_0/\alpha)$. For both modes the control function is regular.

7.5.4. SOLUTION OF LINEAR PROBLEMS OF ANGULAR POSITION STABILIZATION FOR A SPACECRAFT

For applied purposes, it seems more preferable to solve control problems of the linearized system (18) or (19). Linearization is possible if the modulus θ of the rotation vector is sufficiently small and one can assume $\Pi_{(3)}(\boldsymbol{\theta}) = I_{(3)} - (1/2)\Theta_{(3)}$. For the variable χ (the rotation angle) this admission means that in the quadratic (in θ) functional (23) θ is changed for χ and appropriate weighting coefficients c_f, c_ω, and c_χ are introduced. Consider the terminal condition 1): $\chi(t_f) = 0$ ($c_f = 0$). Then we obtain the relationships determining the modes A, B, and C:

$$\text{A.} \quad \nu_{0p} = -\nu_0 \quad (\chi^0 = \nu_0 t_f), \quad t \in [0, t_f], \quad \chi_0 = \chi^0 - \nu_0 t; \tag{7.5.33}$$

$$\text{B.} \quad \nu_{0p} = -\alpha\chi^0\varphi(g,\tau)(g^2-1)^{-1}, \quad t \in [0,t_f], \quad \tau = \alpha t,$$
$$\left(\alpha\chi^0(g^2+1)(g^2-1)^{-1} < \nu_0\right), \quad \varphi(g,\tau) \equiv e^\tau + g^2 e^{-\tau},$$
$$g = e^{\tau_f}, \quad \tau_f = \alpha t_f; \quad d\varphi/dt < 0, \quad t \in [0, t_f]; \tag{7.5.34}$$
$$\chi_0 = \chi^0(g^2 e^{-\tau} - e^\tau)(g^2 - 1)^{-1};$$

$$\begin{aligned}
\text{C.}\quad & \nu_{0p} = -\nu_0, \quad t \in [0, t_*], \quad \chi_0 = \chi^0 - \nu_0 t; \\
& \nu_{0p} = -\alpha\chi_*\varphi(g_*, \tau - \tau_*)(g^2 - 1)^{-1}, \quad t \in (t_*, t_f], \\
& g_* = e^{\tau_f - \tau_*}(\tau_f > \sigma = \alpha\chi^0/\nu_0, \ \tanh\tau_f < \sigma), \\
& \chi_0 = \chi_*\left(g_*^2 e^{-(\tau-\tau_*)} - e^{\tau-\tau_*}\right)(g_*^2 - 1)^{-1}.
\end{aligned} \tag{7.5.35}$$

The regions of existence for the modes A, B, and C in the set of parameters χ^0, t_f, ν_0 and α are defined by the two-dimensional plane (τ_f, σ). The region for the mode C (35) is a wedge-shaped domain in the first quadrant between the ray $\sigma = \tau_f \geqslant 0$ and the curve $\sigma = \tanh\tau_f \leqslant \tau_f$ $(\tau_f \geqslant 0)$. The ray itself corresponds to the mode A (33); there is no solution above this ray. Below the curve $\sigma = \tanh\tau_f$ $(\tau_f > 0)$, there are parameters satisfying conditions of the mode B (34). The unknown parameter $\tau_* = \alpha t_*$ which determines the instant of leaving the limiting control $\nu = -\nu_0$ is found uniquely as a root of the equation $y = x - \tanh x$ $(y = \sigma - \tau_f,\ x = \tau_* - \tau_f)$. The diagrammatic solution is given in Fig. 7.8. Let us investigate the dependence of the root $\tau_*(\sigma, \tau_f)$ on σ in the vicinities of the points $\sigma = \tanh\tau_f$ and $\sigma = \tau_f$, τ_f is fixed. These points correspond to $\tau_* = 0$ and $\tau_* = \tau_f$, respectively. If $\sigma = \tanh\tau_f + \gamma$ and $\sigma = \tau_f - \beta$ where $\gamma, \beta > 0$ are small quantities, then we have the following approximate expressions for τ_*:

$$\tau_* \simeq \frac{\gamma}{(\tanh\tau_f)^2} \simeq \frac{\gamma}{\sigma^2}, \qquad \tau_* \simeq \tau_f - (3\beta)^{1/3} \simeq \tanh^{-1}\sigma - (3\beta)^{1/3}.$$

Analogously, we obtain the approximate expressions for the unknown root τ_* if σ is fixed in the expressions $\tau_f = \tanh^{-1}\sigma - \xi$, $\tau_f = \sigma + \zeta$ (ξ and ζ are small quantities):

$$\tau_* \simeq \xi\frac{1 - \sigma^2}{\sigma^2} \simeq \frac{\rho}{(\sinh\tau_f)^2}, \qquad \tau_* \simeq \sigma - (3\zeta)^{1/3} \simeq \tau_f - (3\zeta)^{1/3}.$$

Different orders in expansions in terms of powers of small perturbations can also be discovered from the diagram in Fig. 7.8.

In the limit as $t_f \to \infty$ only the modes B (34) or C (35) can take place, with the singularity of the feedback control disappearing. Then the instant of leaving the constraint is $t_* = \chi^0/\nu_0 - 1/\alpha$. The construction of the feedback control is carried out by changing $t \to 0$, $t_f \to t_f - t$, $\chi^0 \to \chi$ and checking the appropriate conditions for current χ and $(t_f - t)$.

Using similar approaches, one can construct solutions for control problems with a free right endpoint ($\theta(t_f)$ or $\chi(t_f)$), provided t_f is fixed, as well as problems

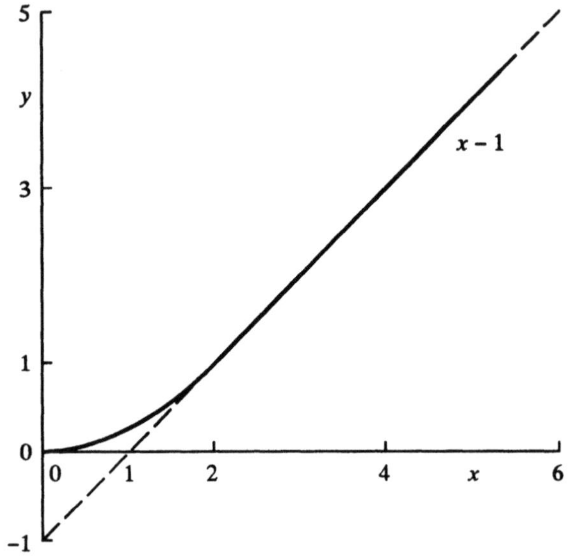

Fig. 7.8.

with a free termination time t_f ($c_t > 0$). When solving practical problems concerning the rotation and stabilization of a spacecraft, one can confine oneself to the modes constructed in Sections 7.5.2–7.5.4. Combination of the time-optimal control (Section 7.5.2) and the linear regulator (Section 7.5.4) in many cases leads to a performance which is close to optimal.

For solutions of a number of orientation and stabilization problems for spacecrafts based on other models, see [1, 46, 90, 91, 97, 111, 132] and others.

CHAPTER 8

Other Prospects for Developing Methods of Optimal Control Synthesis

In this chapter, we consider some constructive perturbation methods for solving various control problems which amplify the techniques described in previous chapters. These methods arise from new problems of precise and efficient control of complex engineering systems, such as manipulation robots, large-scale space structures, etc. Control of these systems requires taking account of numerous perturbations of various physical natures: external force fields, the influence of the environment, dynamics of electrical actuators, distributed elasticity of structural elements, interaction of different parts of the system, imperfections of joints, etc.

8.1. Construction of Approximate Feedback Control for Smooth Systems on the Basis of Dynamic Programming

8.1.1. STATEMENT OF THE SYNTHESIS PROBLEM

Consider the problem of constructing Bellman's function $V(t, x, \varepsilon)$ and the feedback optimal control $u(t, x, \varepsilon)$ for a general regularly perturbed system of the following form:

$$\begin{aligned} &\dot{x} = F(t, x, u, \varepsilon), \quad x(t_0) = x^0; \qquad x, x^0 \in D_x, \quad \varepsilon \in [0, \varepsilon_0], \\ &\Phi(t, x, \varepsilon)\big|_T = 0, \quad J[u] = \varphi(t, x, \varepsilon)\big|_T \longrightarrow \min, \qquad u \in U; \\ &x = (x_1, \dots, x_n), \quad u = (u_1, \dots, u_m), \quad \Phi = (\Phi_1, \dots, \Phi_l). \end{aligned} \tag{8.1.1}$$

The functions F, Φ, and φ are assumed to be sufficiently smooth in the domain of changing the variables t, x, u and the parameter ε. The boundary of the domain U is also assumed to be sufficiently smooth. The smoothness properties are defined by subsequent constructions. The time of terminating the control process T ($T \sim 1$) can be either fixed beforehand or determined as a result of the problem solution. Some special cases of problem (1) are considered in Chapter 7.

Analogously to Chapter 7, we construct the unknown solution by using the perturbation technique based on the dynamic programming method [1, 25, 32, 40, 43, 55, 65, 111, 131, 155]. This method gives the necessary optimality conditions in the form

$$\begin{aligned}
&q + \big(p \cdot F_*(t, x, p, \varepsilon)\big) = 0, \quad q = \frac{\partial V}{\partial t}, \quad p = \frac{\partial V}{\partial x}, \quad F_* = F\big|_{u_*}, \\
&u_* = u(t, x, p, \varepsilon) = \arg\min_{u \in U} \big(p \cdot F(t, x, u, \varepsilon)\big), \\
&V(t, x, \varepsilon)\big|_T = \varphi(t, x, \varepsilon)\big|_T, \quad \Phi(t, x, \varepsilon)\big|_T = 0.
\end{aligned} \tag{8.1.2}$$

Provided some additional conditions (sufficient optimality conditions, for example, in the form [40, 65]) are fulfilled, the solution $V^*(t, x, T_\varepsilon, \varepsilon)$ determines the regular synthesis of the optimal control $u^*(t, x, T_\varepsilon, \varepsilon)$ and, by this, the optimal motion of system (1). Let the function $V_0(t, x, T_0)$, for $\varepsilon = 0$, be known. It is required, analogously to Chapter 7, to find constructive sufficient conditions for the existence and uniqueness of Bellman's function V for $\varepsilon > 0$ and to construct this function with an arbitrary prescribed accuracy, in terms of ε.

8.1.2. MATHEMATICAL BODY, THE CAUCHY METHOD OF CHARACTERISTICS

To solve the initial-value problem (2) by using the method of characteristics we have, as is the case when using the maximum principle technique, to find the solution of the two-point boundary-value problem

$$\begin{aligned}
&\frac{dx}{dt} = F_0, \quad \frac{dp}{dt} = -\frac{\partial}{\partial x}(p \cdot F_*), \\
&\frac{dq}{dt} = -\frac{\partial}{\partial x}(p \cdot F_*), \quad \frac{dV}{dt} = (p \cdot F_*), \\
&x(t_0) = x^0, \quad p(T) = \left[\frac{\partial \varphi}{\partial x} + \frac{\partial}{\partial x}(\lambda \cdot \Phi)\right]_T, \quad \Phi(t, x, \varepsilon)\big|_T = 0, \\
&q(T) = \left[\frac{\partial \varphi}{\partial t} + \frac{\partial}{\partial t}(\lambda \cdot \Phi)\right]_T \big(= -(p \cdot F_*)\big|_T\big), \quad V\big|_T = \varphi(t, x, \varepsilon)\big|_T.
\end{aligned} \tag{8.1.3}$$

Here, λ is a vector of Lagrange's multipliers; the dimension of this vector is the same as that of the vector Φ. When solving the boundary-value problem (3), one has to take into account the relationships for the scalar variable q if T is not fixed. The variables $V(t)$ and $q(t)$ are found by simple quadratures with respect to t after having integrated the Hamiltonian system for x and p:

$$
\begin{aligned}
&x = x_*(t, t_0, x^0, p^0, \varepsilon), \qquad p = p_*(t, t_0, x^0, p^0, \varepsilon), \\
&V = V^0 + \int_{t_0}^{t} (p_* \cdot F_*)\, d\tau, \qquad q = q^0 - \int_{t_0}^{t} \left[p_* \cdot \frac{\partial F_*}{\partial t}\right] d\tau.
\end{aligned}
\tag{8.1.4}
$$

Here, x^0, p^0, V^0, and q^0 are the values of the corresponding variables at $t = t_0$. The unknowns p^0, V^0, q^0 (and T) are determined from the terminal conditions (3):

$$
\begin{aligned}
&\Phi(T, x_*(T, t_0, x^0, p^0, \varepsilon), \varepsilon) \equiv \Phi_*\big|_T = 0, \\
&p_*(T, t_0, x^0, p^0, \varepsilon) = \frac{\partial \varphi_*}{\partial x}\bigg|_T + \frac{\partial}{\partial x}(\lambda \cdot \Phi_*)\bigg|_T, \\
&V^0 + \int_{t_0}^{t} (p_* \cdot F_*)\, d\tau = \varphi(T, x_*(T, t_0, x^0, p^0, \varepsilon), \varepsilon) \equiv \varphi_*|_T, \\
&q^0 - \int_{t_0}^{t} \left(p \cdot \frac{\partial F}{\partial t}\right)_* d\tau = \left(\frac{\partial \varphi}{\partial t}\right)_*\bigg|_T + \left(\lambda \cdot \frac{\partial \Phi}{\partial t}\right)_*\bigg|_T = -(p_* \cdot F_*)|_T.
\end{aligned}
\tag{8.1.5}
$$

Unknown vector parameters p^0 and λ are found from the first two equations (5) as functions of the specified or measured parameters t_0, x^0, T, and ε. However, if the termination instant T is not specified beforehand, it is determined as a function of the parameters t_0, x^0, and ε and then substituted into the expression for p^0, defined earlier. This procedure results in determining the unknown parameters p^0, V^0, q^0 (and T_ε) as functions of known quantities:

$$
\begin{aligned}
&p^0 = p_*^0(t_0, x^0, T_\varepsilon, \varepsilon), \qquad q^0 = q_*^0(t_0, x^0, T_\varepsilon, \varepsilon), \\
&V^0 = V_*^0(t_0, x^0, T_\varepsilon, \varepsilon) \qquad (T_\varepsilon = T_*(t_0, x^0, \varepsilon);\ T_\varepsilon = T).
\end{aligned}
\tag{8.1.6}
$$

The vector Lagrangian multiplier λ is eliminated by solving the linear system, see the second (and the fourth) equation (5). Here, the matrix $(\partial\Phi_*/\partial x)_T$ (and $(\partial\Phi_*/\partial t)_T$) is supposed to have the maximal rank l.

Using the dependences (4)–(6), we obtain the solution

$$
V(t, x, T, \varepsilon) \quad (\text{or } V(t, x, T_*(t, x, \varepsilon), \varepsilon)
$$

of the nonlinear initial-value problem (2), the optimal feedback control u, the optimal state trajectory and, from this, the open-loop control (and the time of terminating the control process)

$$\begin{aligned} &V(t,x,\varepsilon) = V_*^0(t,x,T_\varepsilon,\varepsilon) = \varphi_*\big|_{T_\varepsilon} - \int_t^{T_\varepsilon} (p_* \cdot F_*)\,d\tau, \\ &p(t,x,\varepsilon) = p_*\big(t,t,x,p_*^0(t,x,T_\varepsilon,\varepsilon),T_\varepsilon,\varepsilon\big) = p_*^0(t,x,T_\varepsilon,\varepsilon), \\ &q(t,x,\varepsilon) = q_*^0(t,x,T_\varepsilon,\varepsilon); \qquad T_\varepsilon(t,x,\varepsilon), \quad T_\varepsilon = T, \\ &u(t,x,\varepsilon) = u_*\big(t,x,p(t,x,\varepsilon),\varepsilon\big), \\ &x = x_*\big(t,t_0,x^0,p_*^0\big(t_0,x^0,T_\varepsilon^0,\varepsilon\big)\big); \qquad T_\varepsilon^0 = T_*\big(t_0,x^0,\varepsilon\big), \quad T_\varepsilon^0 = T. \end{aligned} \tag{8.1.7}$$

It is worth paying attention to a singularity of the solution of the synthesis problem as $t \uparrow T$, if T is fixed and the terminal set Φ ($l \geqslant 1$) in (1) is specified. The open-loop control $u^*(t,\varepsilon)$ corresponding to some fixed values (t_0, x^0) is obtained from expressions (7), after having substituted x_* into u_*.

The described procedure of constructing the solution of the initial-value problem (2) is supposed to be correct. It means that the smoothness conditions are fulfilled for right-hand sides of the equations of characteristics, and the sufficient conditions for the existence and uniqueness of the solution of boundary-value problem (3) are satisfied in the admissible domain of changing the variables and parameters t, x, ε (and T) of the control problem (1). Thus, the approach in question is, in essence, equivalent to using the maximum principle technique. It reduces to constructing the solution of the two-point boundary-value problem for any values of t_0 and x^0 from the admissible domain.

8.1.3. CONSTRUCTION OF THE SOLUTION FOR THE PERTURBED PROBLEM

The procedure based on the Cauchy method of characteristics described in Section 8.1.2 provides, together with the technique for the easy approximate construction of Bellman's function, the optimal control and the state trajectory, analogously to Chapter 7. The solution of the unperturbed (for $\varepsilon = 0$) problem of the optimal control synthesis is supposed to be known ($T_\varepsilon = T_{*0}(t,x)$, or $T_\varepsilon = T$):

$$V = V_0(t,x), \qquad u = u_0(t,x), \qquad x = x_0\big(t,t_0,x^0\big). \tag{8.1.8}$$

We also consider the functions $u_*(t,x,p,\varepsilon)$ and $F_*(t,x,p,\varepsilon)$ in (2) to be determined with sufficient accuracy in terms of ε. Some modes for the approximate construction of these functions are developed in Chapter 7.

Then the unknown solution (7) of the initial-value problem (2) is constructed on the basis of the solution of the two-point boundary-value problem (3) in the form

$$V = V_0 + \varepsilon W, \qquad p = p_0 + \varepsilon r, \qquad q = q_0 + \varepsilon s. \tag{8.1.9}$$

Here, $W = W(t, x, \varepsilon)$ is the unknown function to be determined; r and s are partial derivatives of W with respect to x and t, respectively. Substitution of expressions (9) into (2) leads to the weakly nonlinear partial differential equation

$$s + \big(r \cdot F_{*0}(t, x, p_0)\big) + g(t, x) + \varepsilon G(t, x, r) = 0 \tag{8.1.10}$$

and the corresponding terminal (at $t = T$) conditions. The expression F_{*0} in (10) means that the function F_* is taken for $\varepsilon = 0$. The functions g and G are known, and are obtained by using the standard technique similar to that of Chapter 7.

When deriving Equation (10), we use the property of the derivative of the minimum function [46] (see also Section 1.2):

$$\begin{gathered} M = M(z) = \min_u N(z, u), \qquad u \in U, \quad z \in G_z, \\ \frac{\partial M}{\partial z} = \frac{\partial N}{\partial z}\Big|_{u_*}, \qquad u_*(z) = \arg\min_{u \in U} N(z, u). \end{gathered} \tag{8.1.11}$$

Relationships (11) imply that $(p_0 \cdot \partial F_{*0}/\partial p_0) \equiv 0$. Therefore, the equations of characteristics for (10) are reduced to the form

$$\begin{gathered} \dot{x} = F_{*0}\big(t, x, p_0(t, x)\big) + \varepsilon \partial G/\partial r, \qquad x(t_0) = x^0, \\ \dot{r} = -\frac{\partial}{\partial x}\big(r \cdot F_{*0}\big) - \frac{\partial g}{\partial x} - \varepsilon \frac{\partial G}{\partial x}, \qquad r(t_0) = r^0. \end{gathered} \tag{8.1.12}$$

For the approximate construction of the characteristics on the basis of Equations (12) we use the known general solution x_0 of the generating problem. The expressions to be found for W and s are then obtained by quadratures. Integration of system (12) allows to determine the variables x and p in the form

$$\begin{gathered} x = x_0\big(t, t_0, x^0\big) + \varepsilon y\big(t, t_0, x^0, \varepsilon\big), \\ p = p_0\big(t, x_0\big(t, t_0, x^0\big)\big) + \varepsilon r\big(t, t_0, x^0, r^0, \varepsilon\big). \end{gathered} \tag{8.1.13}$$

By using the expressions (13), V and q can be represented by quadratures.

The functions y and r are found as a result of the approximate solution of the initial-value problem for the generalized quasilinear system:

$$\begin{aligned} \dot{y} &= \left(\frac{\partial F_*}{\partial x}\right)_0 y + Y_0(t,r) + \varepsilon f(t,y,r), \qquad y(t_0) = 0, \\ \dot{r} &= -\frac{\partial}{\partial x}(r \cdot (F_*)_0) + B(t) + \varepsilon h(t,y,r), \qquad r(t_0) = r^0. \end{aligned} \tag{8.1.14}$$

The expressions like $(F_*)_0$ mean that the function F_{*0} is taken for $x = x_0(t, t_0, x^0)$. The solution of the initial-value problem (14) is constructed by series expansions or successive approximations in terms of powers of the small parameter. At each step, the equation for r is linear and self-consistent. After integration, we substitute the variable r into the equation for y which is also integrated. It is worth noting that these equations are integrated in an explicit form, since the general solution $x = x_0(t, t_0, x^0)$ of the unperturbed controlled system is known. Fundamental matrices for corresponding systems (14) according to (12) and Section 2.2 (see [46, 150]) are given by

$$Y(t) = \frac{\partial x_0}{\partial x^0}, \qquad R(t) = \left(Y^T(t)\right)^{-1}. \tag{8.1.15}$$

The substantiation of the small parameter technique under appropriate assumptions as to the smoothness of system (14) is conducted in a standard way (by using either the Cauchy method of majorizing functions, or the Banach theorem dealing with a contractive operator [49]). Having constructed the variables x and p as functions of t, t_0, x^0, r^0, and ε to the required accuracy, we determine the variables V and q by simple quadratures according to (4). Then, based on the known generating solution p_0^0, λ_0, q_0^0, T_0, V_0^0 and the boundary condition (5), we calculate the unknown parameters of the boundary-value problem as functions of the known (or measured) variables with a prescribed accuracy. The calculations are performed by using perturbation methods. Here, we suppose that the conditions of the existence of the implicit function are satisfied, i.e. the root of the unperturbed system is nondegenerate for all x^0 from the domain in question. The parameters q^0 and V^0 are easily determined according to (5). Thus, we have constructed the algorithm for the approximate determination of Bellman's function, the optimal feedback control, the optimal state trajectory, the open-loop (programmed) optimal control, the time of the control process (if T is not fixed), as well as other characteristics of the optimal control problem.

As to the accuracy of the obtained expressions, it is appropriate to make the following remark. For problems with a fixed termination instant and terminal set, both Bellman's function and the optimal control have a singularity of the type $V = \widetilde{V}(x,t)(T-t)^{-1}$ and $u = \tilde{u}(x,t)(T-t)^{-1}$. Here, $\widetilde{V}$ and $\tilde{u}$ are smooth functions vanishing on optimal trajectories at $t = T$. Additions to the unperturbed solution will also have such a singularity. Therefore, the coefficient of any power of ε in the expression for the error depends on the proximity of the state point (x,t) to the terminal set, i.e. the error estimate for the approximate solution is not uniform.

For a concrete example of the optimal stabilization of dynamically symmetric rigid body axial rotations see [1, 9].

It is worth noting that the perturbation method for quasilinear systems (the problem with quadratic performance index, the minimum-time problem, etc.) has been previously developed in [25, 30, 33, 43, 85, 92, 95, 106, 109, 111] and other publications.

8.2. Construction of the Perturbed Feedback Control by Using Canonical Transformations

8.2.1. DECOMPOSITION IN OPTIMAL CONTROL PROBLEMS

The perturbation method given in Section 8.1 is efficient if the generating problem (for $\varepsilon = 0$) admits sufficiently full solution. In particular, we can consider a multidimensional controlled system of dimension $n \gg 1$ which separates into a large number (l) of subsystems having fewer dimensions if $\varepsilon = 0$:

$$\begin{aligned}
&\dot{x}_k = F_k(t, x_k, u_k) + \varepsilon f_k(t, x, u), && x_k(t_0) = x_k^0,\\
&x_k = \left(x_k^{(1)}, \ldots, x_k^{(n_k)}\right), && k = 1, \ldots, l, \quad n_1 + \cdots + n_l = n,\\
&u_k = \left(u_k^{(1)}, \ldots, u_k^{(m_k)}\right), && m_1 + \cdots + m_i = m,\\
&u \in U = U_1 \times \cdots \times U_l, && u_k \in U_k.
\end{aligned} \tag{8.2.1}$$

To be determined, we consider the motion of the controlled system (1) on a fixed time interval $t \in [t_0, T]$. The following optimal control problem is stated:

$$\begin{aligned}
&M_k\big(x_k(T)\big) + \varepsilon \widetilde{M}_k\big(x(T)\big) = 0; \qquad J_\varepsilon[u] \longrightarrow \min_{u \in U},\\
&J_\varepsilon[u] = \sum_{k=1}^{l} J_k[u_k] + \varepsilon \widetilde{J}[u], \qquad J_k = \Phi_k(x_k)\big|_T, \quad \widetilde{J} = \widetilde{\Phi}(x)\big|_T.
\end{aligned} \tag{8.2.2}$$

If $\varepsilon = 0$, the optimal control problem for the multidimensional system (1) and (2) is divided into l problems of fewer dimensions. Usually, the solution of each subproblem is substantially simpler than that of the original problem.

Thus, in what follows, we assume that the solutions of all l unperturbed optimal control problems

$$\begin{aligned}&\dot{x}_k = F_k(t, x_k, u_k), \qquad x_k(t_0) = x_k^0, \quad k = 1, \dots, l,\\ &M_k\big(x_k(T)\big) = 0, \qquad J_k[u_k] = \Phi_k\big(x_k(T)\big) \longrightarrow \min_{u_k \in U_k},\\ &u_k = u_{k0}(t, x_k), \qquad x_k = x_{k0}\big(t, t_0, x_k^0\big)\end{aligned} \tag{8.2.3}$$

are known; the optimal control being found in the feedback form.

To construct the optimal solution for the original synthesis problem (1) and (2), taking account of ε, one can use the method of Section 8.1 based on dynamic programming. Below, we give another approach, based on the maximum principle and canonical transformations. The application of the Cauchy method of characteristics is, in essence, equivalent to using the optimality conditions of the maximum principle [131]. For the unperturbed problem (3), these conditions have the form

$$\begin{aligned}&H_k = \big(p_k \cdot F_k(t, x_k, u_k)\big) \longrightarrow \max_{u_k \in U_k}, \qquad k = 1, \dots, l;\\ &u_k^0 = u_k^0(t, x_k, p_k) = \arg\max_{u_k \in U_k} H_k; \qquad H_k^0 = H_k^0(t, x_k, p_k),\\ &\dot{x}_k = \frac{\partial H_k^0}{\partial p_k} = F_k^0(t, x_k, p_k), \qquad x_k(t_0) = x_k^0, \qquad M_k(x_k)\big|_T = 0,\\ &\dot{p}_k = -\frac{\partial H_k^0}{\partial x_k} = -\frac{\partial}{\partial x_k}\big(p_k \cdot F_k^0\big),\\ &p_k(T) = \left[-\frac{\partial \Phi_k}{\partial x_k} + \frac{\partial}{\partial x_k}(\lambda_k \cdot M_k)\right]_T.\end{aligned} \tag{8.2.4}$$

Here, F_k^0 and H_k^0 are the functions F_k and H_k into which the functions $u_k = u_k^0$ have been substituted. The solutions of problems (4) are supposed to be known for any t_0 and x_k^0 from the domains in question, where the motion of system (3) is defined. The Hamiltonian function of the original system (1) and the corresponding optimal control u^* can be expressed as follows:

$$\begin{aligned}&H = \sum_{k=1}^{l} \big[H_k + \varepsilon\big(p_k \cdot f_k(t, x, u)\big)\big] \longrightarrow \max_{u \in U},\\ &u^* = u^0(t, x, p) + \varepsilon v(t, x, p), \qquad u^0 = (u_1, \dots, u_l).\end{aligned} \tag{8.2.5}$$

For $\varepsilon > 0$, the adjoint system can be represented by using the perturbed Hamiltonian function H^*:

$$\begin{aligned}
\dot{x} &= \frac{\partial H^*}{\partial p} = F^0(t,x,p) + \varepsilon \tilde{f}(t,x,p), \\
\dot{p} &= \frac{\partial H^*}{\partial p} = -\frac{\partial}{\partial x}(p \cdot F^0) - \varepsilon \frac{\partial}{\partial x}(p \cdot \tilde{f}(t,x,p)), \\
H^* &= H^*(t,u,p,\varepsilon) = H^0(t,x,p) + \varepsilon \widetilde{H}(t,x,p), \\
h^* &= H\big|_{u^*}, \quad H^0 = H^*\big|_{\varepsilon=0} = \sum_{k=1}^{l} H_k^0, \quad \varepsilon\widetilde{H} \equiv H^* - H^0.
\end{aligned} \tag{8.2.6}$$

The initial and terminal conditions, as well as the transversality conditions, based on (2), are written similarly to (4);

$$\begin{aligned}
&x(t_0) = x^0, \qquad M(x(T)) + \varepsilon\widetilde{M}(x(T)) = 0, \\
&p(T) = -\frac{\partial}{\partial x}(\Phi + \varepsilon\widetilde{\Phi})\Big|_T \left(\lambda \cdot \frac{\partial}{\partial x}(M + \varepsilon\widetilde{M})\right)\Big|_T.
\end{aligned} \tag{8.2.7}$$

We propose to solve the perturbed problem (6) and (7) by separating the variables x_k, $k = 1, \ldots, l$, $l \geqslant 2$ on the basis of canonical transformations [1, 44, 53, 69, 73, 80, 118, 134], see Chapters 2 and 4. Using the perturbation technique we find a univalent canonical transformation close to the identity $(t,x,p) \xrightarrow{(\varepsilon)} (t,y,q)$:

$$y = x + \varepsilon \frac{\partial \sigma}{\partial q}, \qquad p = q + \varepsilon \frac{\partial \sigma}{\partial x}, \qquad \sigma = \sigma(t,x,q,\varepsilon), \tag{8.2.8}$$

such that new variables y and q would be governed by a decoupled set of equations, i.e. the Hamiltonian function for (y,q) would have a form similar to that for the unperturbed case (4):

$$\begin{aligned}
K &= K(t,y,q,\varepsilon) = \sum_{k=1}^{l} K_k(t,y_k,q_k,\varepsilon), \\
K_k &= K_k^0 + \varepsilon K_k^{(1)} + \varepsilon^2 + \cdots.
\end{aligned} \tag{8.2.9}$$

The unknown function σ in (8) satisfies the partial differential equation

$$\varepsilon\frac{\partial\sigma}{\partial t}+H^*\left(t,x,q,+\varepsilon\frac{\partial\sigma}{\partial x},\varepsilon\right)=K\left(t,x,+\varepsilon\frac{\partial\sigma}{\partial q},q,\varepsilon,\right), \tag{8.2.10}$$

which immediately implies that

$$K^0=\sum_{k=1}^{l}K_k^0,\qquad K_k^0(t,x_k,q_k)=H_k^0(t,x_k,q_k). \tag{8.2.11}$$

For subsequent coefficients $K_k^{(j)}$, $j=1,2,\ldots$, one can take arbitrary sufficiently smooth functions. In concrete cases, these are chosen for reasons of convenience, in particular $K_k^{(j)}=0$. In this case, the transformed system has the form of the unperturbed system (4), for which the general solution

$$\begin{aligned} x_k&=x_k^{(0)}(t,t_0,x_k^0,p_k^0), & x_k^{(0)}\big|_{t_0}&=x_k^0,\\ p_k&=p_k^{(0)}(t,t_0,x_k^0,p_k^0), & p^{(0)}\big|_{t_0}&=p_k^0 \end{aligned} \tag{8.2.12}$$

is supposed to be known.

Knowing the functions $K_k^{(j)}$, we construct the unknown $\sigma=\sigma(t,x,q,\varepsilon)$ by a series expansion in terms of powers of ε:

$$\begin{aligned} &\sigma=\sigma_0+\varepsilon\sigma_1+\cdots+\varepsilon^j\sigma_j+\cdots,\qquad j=1,2,\ldots;\\ &\frac{\partial\sigma_j}{\partial t}+\sum_{k=1}^{l}\left(F_k^0(t,x_k,q_k)\cdot\frac{\partial\sigma_j}{\partial x_k}\right)-\\ &\quad-\sum_{k=1}^{l}\left(\frac{\partial}{\partial x_k}(q_k\cdot F_k^0(t,x_k,q_k))\cdot\frac{\partial\sigma}{\partial q}\right)=W_j(t,x,q). \end{aligned} \tag{8.2.13}$$

Here, the functions W_j are known. They are constructed on the basis of H_k^0 and $\widetilde{H}$, given $K_k^{(i)}$, $i=1,2,\ldots,j$, and the functions σ_j calculated on previous steps $(i=1,\ldots,j-1)$; in particular,

$$W_1=\sum_{k=1}^{l}K_k^{(1)}+\widetilde{H}.$$

The solution for the sequence of Equations (13) is found by the method of characteristics on the basis of the general solution of the generating system (4), since the equations of characteristics are integrated in explicit form, by assumption.

Using the above approach, we calculate the unknown function σ with the required accuracy (in terms of ε) defined by the smoothness of the function H^* in x and p. To solve the weakly nonlinear Equation (10) one can use the method of successive approximations. This method requires the extension of the space of characteristics, with the additional variables $X = \partial\sigma/\partial x$ and $Q = \partial\sigma/\partial q$ being introduced [75].

8.2.2. REDUCTION TO OSCULATING CANONICAL VARIABLES

The method of varying integration constants (initial values) $\xi = x^0$ and $\eta = p^0$ in the solution of the "zero" approximation leads to the following Hamiltonian system

$$\dot{\xi} = \varepsilon \frac{\partial \widetilde{H}^0}{\partial \eta}, \qquad \dot{\eta} = -\varepsilon \frac{\partial \widetilde{H}^0}{\partial \xi}, \qquad \xi(t_0) = x^0, \qquad \eta(t_0) = p^0; \tag{8.2.14}$$

$$\widetilde{H}^0 \equiv \widetilde{H}\left(t, x^{(0)}(t, t_0, \xi, \eta), p^{(0)}(t, t_0, \xi, \eta)\right).$$

The general solution of system (14) on the bounded time interval $t \in [t_0, T]$ is constructed by successive approximations or series expansions in terms of powers of ε. The constant vector p^0 is then chosen from the terminal conditions for x and the transversality conditions (7). This vector is expressed as a function of the initial values t_0 and x^0 which are considered as the current ones in the feedback control u^* (5), i.e. $x^0 \to x$, $t_0 \to t$. If $p^0 = p^0(t_0, x^0, T, \varepsilon)$ is determined nonuniquely, then the optimal branch of p^0 is chosen from the minimality condition for the functional $J_\varepsilon[u]$ (2). Subsequent manipulations do not cause any difficulty in the case where the root p^0 is simple, provided the functions $H^*(t, x, p, \varepsilon)$, $\Phi(x)$, $\widetilde{\Phi}(x)$, $M(x)$, and $\widetilde{M}(x)$ are sufficiently smooth. The critical case, when the root p^0 in the generating boundary-value problem is multiple, requires a special investigation. It is connected, as a rule, with approximations in terms of fractional powers of ε (see Sections 2.2, 2.3, 4.2 and 8.1). In the critical case, the corresponding functional determinant with respect to p^0 and λ vanishes for some admissible values of t_0, x^0, T and other parameters ($\varepsilon = 0$). It is worth noting that a weakly controllable system in osculating variables reduces to the form (14) and in the case of a terminal functional leads, to the first approximation in terms of ε, to the so-called local optimal controls [46]. One can also establish that if the maxima of the functions H_k over $u_k \in U_k$ lie strictly within the domains U_k, i.e. $\partial H_k/\partial u_k = 0$ at $u_k = u_k^0$, then the control of the first approximation gives the minimum to the functional $J_\varepsilon[u]$ with the error $O(\varepsilon^2)$. Therefore, the functional is calculated with the error whose power (in terms of ε) exceeds by unity the power of the error in determining the control $u(t, \varepsilon)$ and the trajectory $x(t, \varepsilon)$. More precise calculation of the control to the first approximation is needed for more precise satisfying the boundary conditions, but it does not increase the accuracy in the functional. Analogously, in order to find the minimum of the functional with the error $O(\varepsilon^{k+1})$ it is sufficient to determine the optimal control with the accuracy $O(\varepsilon^k)$. Calculation of the control taking into account the terms of the order of ε^k, i.e. with the error $O(\varepsilon^{k+1})$, can be performed to increase the accuracy in the fulfillment of the terminal conditions.

It is possible to use another scheme for reducing the equations of the maximum principle to canonical osculating variables. This scheme is based on the synthesis of the unperturbed control $u = u_0(t, x)$ and the corresponding unperturbed optimal trajectories $x = x_0(t, t_0, x^0)$ for arbitrary t_0 and x^0. We change the variable x to ξ in system (1) and the terminal conditions in (2) according to the formulae

$$\begin{aligned}
&x = x_0(t, t_0, \xi), \qquad \xi(t_0) = x^0, \\
&\dot{\xi} = \left(\frac{\partial x_0}{\partial \xi}\right)^{-1} \{[F(t, x_0, u) - F(t, x_0, u_0)] + \varepsilon f(t, x_0, u)\} \\
&\quad \equiv \left(\frac{\partial x_0}{\partial \xi}\right)^{-1} \tilde{F}(t, x_0, u, u_0, \varepsilon), \\
&M_\varepsilon(x_0)\big|_T = 0, \qquad J_\varepsilon[u] = \Phi_\varepsilon(x_0)\big|_T \, h \longrightarrow \min_{u \in U}.
\end{aligned} \tag{8.2.15}$$

For $\varepsilon = 0$, the solution of problem (15) is given by $u = u_0(t, t_0, \xi)$, $\xi = x^0 =$ const. Application of the maximum principle to problem (15) for $\varepsilon = 0$ leads to the relationships

$$\begin{aligned}
&u^* = u^*(t, \xi, \eta, \varepsilon) = \arg\max_{u \in U} h, \\
&h \equiv \left(\eta \cdot \frac{\partial x_0}{\partial \xi} \tilde{F}(t, x_0, u, u_0, \varepsilon)\right), \\
&\dot{\eta} = -\frac{\partial h}{\partial \xi}\bigg|_{u^*} = -\frac{\partial h^*}{\partial \xi}, \qquad \eta(T) = \left[-\frac{\partial \Phi}{\partial \xi} + \left(\nu \cdot \frac{\partial M}{\partial \xi}\right)\right]_T.
\end{aligned} \tag{8.2.16}$$

Here, ν is the Lagrange multiplier. Substitution of u^* (16) into the equation for ξ (15) leads to a Hamiltonian system. The solution of the boundary-value problem for this system is constructed by perturbation methods. If the functions h^*, M, and Φ are smooth, one can use series expansions in terms of powers of ε:

$$\xi = x^0 + \varepsilon \xi_1 + \varepsilon^2 + \cdots, \qquad \eta = \eta^0(x^0, t_0) + \varepsilon \eta_1 + \varepsilon^2 + \cdots.$$

Here, η_0 is the known function obtained on the basis of the known solution $x_0(t, t_0, x^0)$ and $u_0(t, x)$ for the unperturbed problem. It is possible as well to apply the method of successive approximations (similar to Picard's method) which uses the quantities $\xi^{(0)} = x^0$ and $\eta^{(0)}(x^0, t^0)$ for the "zero" approximation. The convergence of successive approximations and the proximity estimate for sufficiently small $\varepsilon > 0$ take place, as is the case for the appropriate scheme for system (14), if the solution of the "zero" approximation is nondegenerate. In this case, the corresponding Jacobian with respect to η^0 and ν must not be equal to zero for all t_0 and x^0.

8.3. Switching Curves of Bang-Bang Controls for Perturbed Nonoscillating Systems

Using optimal control theory and perturbation methods we investigate here the time-optimal feedback control for a rather general nonoscillating system with a single degree of freedom. When doing this, we take into account the results of the analysis of such systems obtained on the basis of the maximum principle [40, 131]. For a wide class of controlled systems, the maximum principle leads to constructing the switching curves for bang-bang controls [1, 14, 28, 46]. This section amplifies the picture of approximate synthesis of control for oscillatory systems resulting from the asymptotic methods developed in Chapters 1–6.

8.3.1. STATEMENT OF THE TIME-OPTIMAL FEEDBACK CONTROL PROBLEM FOR PERTURBED NONOSCILLATING SYSTEMS

(1) *Original control problem.* Consider a perturbed controlled dynamical system with a single degree of freedom:

$$\begin{aligned} &\dot{x} = y, \qquad x(0) = x^0, \\ &\dot{Y} = f(x, y, u) + \varepsilon F(x, y, u), \qquad y(0) = y^0. \end{aligned} \tag{8.3.1}$$

Here, x and y are the coordinate and velocity of the system, they constitute the state vector $(x, y) \in G \subseteq R_2$, where R_2 is a state (phase) plane; $t \in [0, T]$ is time ($T < \infty$); initial values $(x^0, y^0) \in G$; u is a scalar piece-wise smooth control function such that $|u(t)| \leqslant 1$; $\varepsilon \in [0, \varepsilon_0]$ is a small numerical parameter ($0 \leqslant \varepsilon \leqslant \varepsilon_0 \ll 1$); f and F are smooth functions of x, y and u in the domain in question, the perturbation F being able to depend continuously on ε. Additional properties (smoothness, growth rate, etc.) of the functions f and F and the domain G are discussed below. It should be noted that the constraints $r^-(x, y, \varepsilon) \leqslant u \leqslant r^+(x, y, \varepsilon)$ on the control can be reduced to those mentioned above by the linear transformation (v is a new control):

$$u = \frac{1}{2}(r^+ + r^-) + \frac{1}{2}(r^+ - r^-)v, \qquad v \in [-1, 1].$$

For the perturbed system (1), we pose the problem of determining the time-optimal feedback control $u(x, y, \varepsilon)$ which brings the phase point $(x, y) \in G$ to the origin (the point $(0, 0) \in G$) for sufficiently small $\varepsilon > 0$. We suppose that the optimal control for the unperturbed system ($\varepsilon = 0$) is of the bang-bang type and that the switching curve for this control is known [1, 14, 28, 40, 131].

Below, we investigate the nonoscillating systems [1, 14, 40] for which unperturbed switching curves have the simplest form: the curve consists of two semitrajectories of the unperturbed system (1) which correspond to $u = \pm 1$ and go to the origin. In [40], the sufficient conditions are given which guarantee the qualitative similarity of the behaviour of system (1) under the optimal control $u(x, y)$ to that of the simplest dynamical system: $\ddot{x} = u$, $|u| \leqslant 1$. Namely, each optimal control has not more than one switch, and the switching curve passes from the second quadrant to the fourth quadrant and is tangential to the axis y at the origin.

Sufficient conditions for such a behaviour are the following [40]. The function f is supposed to be continuously differentiable with respect to all arguments and satisfy the monotonicity conditions in u:

$$f'_u(x, y, u) > 0, \qquad (x, y) \in G, \quad |u| \leqslant 1. \tag{8.3.2}$$

Furthermore, the following inequalities are valid for $u = \pm 1$ at the terminal point $x = 0$ and $y = 0$:

$$f(0, 0, +1) = f^+ > 0, \qquad f(0, 0, -1) = f^- < 0. \tag{8.3.3}$$

Conditions (3) guarantee the stop of the system at the terminal point for some $u(t) = u_0 = \text{const}$, $|u_0| \leqslant 1$, as well as the aforenamed behaviour of the switching curve and optimal trajectories in a sufficiently small vicinity of this point.

It is also supposed that no trajectory of the unperturbed system (1) can go to infinity (or come from infinity) in a finite time (the sufficient conditions can be expressed as global Lipschitz conditions with respect to x and y [40]). If the motion is considered in a bounded domain G, this requirement is unnecessary.

To complete the set of sufficient conditions for the above mentioned behaviour of the system under optimal feedback control, we add the following, more sophisticated, condition. Namely, we suppose that there exists a function $\varphi(x, y, u)$ which is continuously differentiable with respect to x and y and satisfies the inequality

$$y\varphi'_x + f\varphi'_y + \varphi^2 - \varphi f'_y - f'_x < 0, \qquad u = \pm 1, \quad (x, y) \in G. \tag{8.3.4}$$

The maximum principle and the inequality (4) imply [40] that the optimal control $u(x, y)$ or $u(t)$ is of the bang-bang type and has not more than one switch point.

(2) *Construction of the unperturbed switching curve and the synthesis pattern.* For $\varepsilon = 0$, the switching curve $\Pi(x, y)$ for the control $u(x, y)$ is constructed as follows. The upper left (Π^-) and the lower right (Π^+) branches of this curve are defined by the relationships

$$\begin{aligned} &\Pi = \Pi^- \cup \Pi^+, \\ &\Pi^\mp = \{x, y\colon\ x^\mp(\Theta, x, y) = 0,\ \ y^\mp(\Theta, x, y) = 0,\ \ \Theta \geqslant 0\}. \end{aligned} \tag{8.3.5}$$

Here, Θ is a parameter of the curve; the functions $x^\mp(t, x^0, y^0)$ and $y^\mp(t, x^0, y^0)$ are the solutions of the unperturbed initial-value problem (1) which correspond to $u = \mp 1$ ($y \gtrless 0$, $x \lessgtr 0$) and lead to the origin. Under the above assumptions, these solutions can be constructed on the time interval $t \in [0, \Theta]$ during which the motion remains within the domain G. The quantity Θ in (5) is the duration of transferring the state point (x, y) to the origin along the switching curve for fixed $u = -1$ or $u = +1$. Therefore, the relationships (5) give a parametric representation of the unperturbed switching curve $\Pi(x, y)$.

The switching curve $\Pi(x, y)$ can also be written in another form:

$$\Pi(x, y) = \{x, y\colon\ \Psi^\mp(x, y) = 0\}. \tag{8.3.6}$$

Here, $\Psi^\mp$ are partial integrals of the equations

$$f(x, y, \mp 1)\, dx = y\, dy, \qquad \Psi^\mp(0, 0) = 0.$$

Having succeeded in solving the equation $\Psi^\mp(x, y) = 0$ with respect to x ($x = x^\mp(y)$) or y ($y = y^\mp(x)$), one finds the switching curve in the explicit form. Under the assumptions made as to the smoothness of the function f in a small vicinity of the origin, the expression for the switching curve is given by

$$x = (1/2) f^\mp y^2 + O(y^3), \qquad y \gtrless 0.$$

The switching curve $\Pi(x, y)$ divides the domain G into two parts, open the subdomains G^- and G^+: $G = G^- \cup \Pi \cup G^+$. Within the open subdomains $G^\mp$ and on the parts $\Pi^\mp$ of their boundaries, the optimal control is $u = \mp 1$, respectively, i.e. the optimal synthesis (of the feedback control)is given by

$$u(x, y) = \mp 1, \qquad (x, y) \in G^\mp \cup \Pi^\mp. \tag{8.3.7}$$

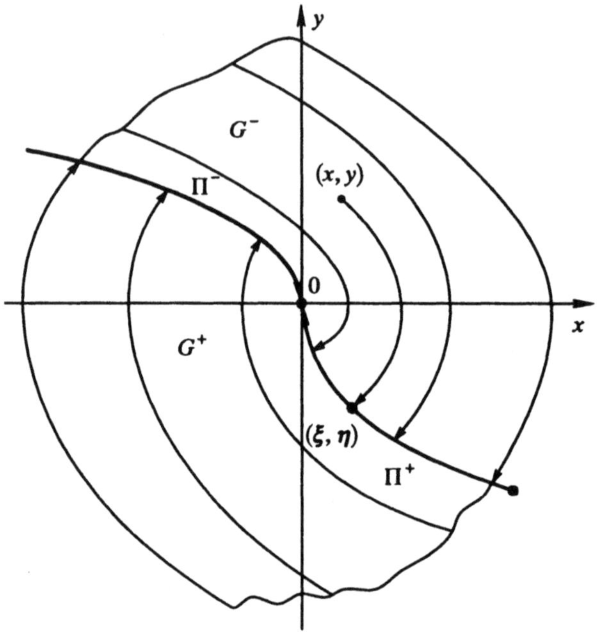

Fig. 8.1.

Typical shapes of the curve Π and the optimal trajectories for $(x, y) \in G$ (qualitative diagram of the synthesis) are shown in Fig. 8.1.

Bellman's function of this problem, i.e the minimal time $T = T(x, y)$ possible for the motion from the point $(x, y) \in G$ to the origin, can be expressed by $T = S + \Theta$. Here, S is the duration of motion from the current point (x, y) to the intersection point (ξ, η) of the trajectory with the corresponding branch of the curve Π; Θ is the duration of motion from the point (ξ, η) to the origin along the switching curve, Θ being defined according to (5), see Fig. 8.1. The unknown quantities T, S, Θ, ξ, and η are uniquely determined from the equations

$$\begin{aligned} &T(x, y) = S(x, y) + \Theta(x, y); \\ &x^{\mp}(S, x, y) = \xi, \qquad y^{\mp}(S, x, y) = \eta, \\ &x^{\pm}(\Theta, \xi, \eta) = 0, \qquad y^{\pm}(\Theta, \xi, \eta) = 0. \end{aligned} \tag{8.3.8}$$

Here, the functions $x^{\mp}$ and $y^{\mp}$ are defined just as in (5); by elementary substitutions, system (8) is reduced to two equations with respect to S and Θ. Concrete examples of constructing the synthesis diagram and Bellman's function are considered in Section 8.3.3.

In complex applied problems, one can use the synthesis given above as a part of a more general quasioptimal synthesis in the vicinity of the origin where, due to (3), the control u determines the motion. For example, if, for large initial deflections, the system's motion is oscillatory, then to quench the oscillations, i.e. to bring the state point into a neighbourhood of the origin, one can use an approximate quasioptimal feedback control of the simple form

$$u = \frac{1}{2}(u^+ + u^-) - \frac{1}{2}(u^+ - u^-)\,\mathrm{sign}\,y,$$

or some other law which corresponds to the model of a weakly controlled oscillatory system (see Chapters 1–6). Then, in the domain where the control action is definitive, the law (7) is used for more accurate stabilization.

8.3.2. CONSTRUCTION OF OPTIMAL FEEDBACK CONTROL FOR PERTURBED SYSTEM

(1) *Sufficient conditions of nonoscillation of a perturbed system.* When investigating a perturbed synthesis problem, usually all conditions, with the exception of the last condition of type (4), are naturally extended to the general case $\varepsilon > 0$, i.e. they should be satisfied uniformly with respect to $\varepsilon \in [0, \varepsilon_0]$, and this is henceforth assumed. A serious difficulty arises with the problem of the existence and construction of a smooth function $\varphi_\varepsilon(x, y, u)$ which possesses property (4), as well as with the estimate of the domain $G_\varepsilon \times [0, \varepsilon_0]$ of the variables x, y, and ε. For this domain, we analogously construct the switching curve $\Pi_\varepsilon = \Pi_\varepsilon(x, y)$ of the optimal control $u_\varepsilon(x, y)$, the optimal trajectory $x_\varepsilon(t, x^0, y^0)$, $y_\varepsilon(t, x^0, y^0)$ for $0 \leqslant t \leqslant T_\varepsilon(x^0, y^0)$, and Bellman's function $T_\varepsilon(x, y)$ of the perturbed system. It should be expected that, for an arbitrary smooth perturbation function F, the estimates lead to the domain G_ε which is more narrow than the original one, i.e. $G_\varepsilon \subseteq G \subseteq R_2$ with $G_\varepsilon \to G$ as $\varepsilon \to 0$. Having solved the above mentioned estimate problem, one can analogously construct the perturbed switching curve, the picture of the synthesis (optimal trajectories) and Bellman's function in the domain $(x, y) \in G_\varepsilon$ by using analytical (in powers of ε) or numerical methods. This does not cause any significant difficulty. Below, in Section 8.3.3 we consider examples of specific systems (1) which are of interest in practice. For these systems, the above mentioned constructions are carried out and justified. For arbitrary smooth perturbations and some special functions f, one may succeed in a constructive investigation regarding the existence of the function φ_ε and the estimate of the domain G_ε, $\varepsilon \in [0, \varepsilon_0]$.

1) For example, let (see [14, 40])

$$f'_x(x, y, u) \geqslant 0, \qquad u = \mp 1, \quad (x, y) \in G. \tag{8.3.9}$$

Then for the unperturbed system, we have $\varphi \equiv 0$ and for $\varepsilon > 0$ one can take the function $\varphi_\varepsilon = -\mu y u$ where the parameter μ is to be determined. Inequalities (3) imply that for sufficiently small positive μ and ε, inequality (4) is valid in a small vicinity $(x, y) \in G_\varepsilon$ of the origin. Indeed, in this case, the dominant term in the left-hand side of inequality (4) is $\mp\mu f(x, y \pm 1)$. It assures the validity of this inequality since, according to (3), $f(0, 0, \pm 1) = f^\pm \gtrless 0$. By an appropriate choice of the parameters $\mu, \varepsilon > 0$, the remaining terms can be made sufficiently small in their absolute value for any $(x, y) \in G_\varepsilon$.

The domain G_ε will be asymptotically large with respect to the parameter ε if we can find a smooth function $\varphi(x, y, \varepsilon)$ such that

$$f^2(\varphi/f)'_y + y\varphi'_x \leqslant h < 0, \qquad (x, y) \in G_\varepsilon, \quad u = \mp 1. \tag{8.3.10}$$

Here, h is a parameter; it is essential that the function φ is constructed on the basis of the function f only, i.e. on the basis of the unperturbed system. But if the function f does not depend on x, then it is natural to seek the function φ in the form $\varphi(y, \varepsilon)$; inequality (10) then takes the form $f^2(\varphi/f)'_y \leqslant h < 0$.

In particular, let $f \equiv u$. Then, assuming as above that $\varphi_\varepsilon = -\mu y u$, we arrive at the inequality

$$-\mu + \max_{(x,y) \in G_\varepsilon} \left[(\mu y)^2 + \varepsilon\left(\mu|F| + \mu\left|yF'_y\right| + \left|F'_x\right|\right)\right] \leqslant 0. \tag{8.3.11}$$

If the quantities $|F'_y|$ and $|F'_x|$ are uniformly bounded for $y \sim 1/\sqrt{\varepsilon}$ and $x \sim 1/\varepsilon$, inequality (11) holds in an asymptotically large domain G_ε with the above mentioned linear dimensions. One can establish the validity of (11) by assuming $\mu = \varepsilon M$.

2) Let the strict inequality

$$\max_{(x,y) \in G_0} \left[(y\varphi - f)'_x + \varphi^2(1 + (f/\varphi)'_y)\right] \leqslant -\delta < 0, \qquad u = \mp 1, \tag{8.3.12}$$

be satisfied for the known function $\varphi(x, y, u)$ in some bounded subdomain $G_0 \subseteq G$. Then, for sufficiently small $\varepsilon > 0$, the inequality of the type (4) holds for the perturbed system (1) if $\varphi_\varepsilon = \varphi$, $(x, y) \in G_0$.

However, if the domain G is unbounded, we must impose additional conditions on the growth of the function F and its derivatives [14, 40]. For instance, for the quasilinear nonoscillating system

$$\ddot{x} + 2k\dot{x} + \omega^2 x = u + \varepsilon F(x, \dot{x}, u), \qquad k > \omega > 0, \tag{8.3.13}$$

we can take the function φ to be constant: $\varphi = \varphi_\varepsilon = -k$. Then according to (12) we have $\delta = k^2 - \omega^2 > 0$. The inequalities of the type (4) or (12) are valid for system (13) in the domain $G_\varepsilon \times [0, \varepsilon_0]$, where

$$\varepsilon\left(kF'_y - F'_x\right) \leqslant \delta, \qquad u = \mp 1. \tag{8.3.14}$$

If the quantities $|F'_y|$ and $|F'_x|$ are uniformly bounded by the constant N for $(x, y) \in R_2$, then inequality (14) is satisfied for $(x, y) \in R_2$ when $\varepsilon \leqslant \delta(k+1)^{-1}N^{-1}$.

(2) *Construction of a perturbed switching curve.* In a domain G_ε, uniformly bounded with respect to ε, the trajectories $(x_\varepsilon^\mp(t, x^0, y^0),\ y_\varepsilon^\mp(t, x^0, y^0))$ which go into the origin and determine the desired curve $\Pi_\varepsilon(x, y)$, in a similar manner to (5), are constructed by simple recursive procedures of successive approximations (by Picard's method [49, 75]) or by expansions in powers of ε [122, 147]. The functions $x^\mp(t, x^0, y^0)$ and $y^\mp(t, x^0, y^0)$, defined in Section 8.3.1 (see (5)), can be taken as initial approximations. The fundamental matrix which is used for the construction of subsequent approximations is obtained from $x^\mp$ and $y^\mp$ by differentiating with respect to the parameters $x^0, y^0 \in G_\varepsilon$. The schemes are substantiated on the basis of the Banach theorem dealing with a contracting operator [49, 75] or by the Cauchy theorem which uses the method of majorant functions [75].

Switching curve Π_ε is constructed in the implicit form of the type (6) by using particular integrals of the perturbed system:

$$(f + \varepsilon F)\, dx = y\, dy, \qquad u = \mp 1, \quad \Psi_\varepsilon^\mp(0, 0) = 0. \tag{8.3.15}$$

The functions $\Psi_\varepsilon^\mp$ can be constructed on the basis of general integrals $\Psi^\mp(x, y) = c$ of the unperturbed problem. Differentiation of the integrals $\Psi^\mp(x, y) = c$ according to (15) gives the equations

$$\begin{aligned} dc &= (\Psi^\mp)'_x\, dx + (\Psi^\mp)'_y\, dy = \varepsilon(\Psi^\mp)'_x (F/f)\, dx \\ &= \varepsilon(\Psi^\mp)'_y \left[f/(f + \varepsilon F)\right] dy. \end{aligned} \tag{8.3.16}$$

When deriving the second and third equalities in (16), we use the relationship $(\Psi^\mp)'_y + (\Psi^\mp)'_y f \equiv 0$ (with respect to x and y); the functions f and F are taken at $u = \mp 1$. Particular integrals of the perturbed system (16) $\Lambda_\varepsilon^\mp(x, y, c) = 0$, satisfying the conditions $\Lambda_\varepsilon^\mp(0, 0, 0) \equiv 0$, together with the general integrals of the unperturbed system $\Psi^\mp = c$ yield the desired switching curve Π_ε in the form (6).

Substantiation of using the small parameter technique for the construction of the switching curve Π_ε in an asymptotically large (with respect to ε) domain G_ε requires uniform boundedness of the derivatives of the functions f and F with respect to x and y, $(x, y) \in R_2$.

8.3.3. INVESTIGATION OF SPECIFIC MECHANICAL SYSTEMS

Using the approach given in Section 2, one can obtain the switching curve for a nonoscillating system of the form

$$\ddot{x} = u + \varepsilon F(x, \dot{x}, u), \qquad u^- \leqslant u \leqslant u^+, \quad u^- < 0, \quad u^+ > 0, \tag{8.3.17}$$

by integrating the equations

$$\frac{dx}{dy} = \frac{y}{u} - \frac{\varepsilon F y}{u(u+\varepsilon F)}, \qquad u = u^{\mp}, \quad y \gtrless 0. \tag{8.3.18}$$

The initial-value problem for Equation (18) is reduced to nonlinear integral equations with respect to $x = x_\varepsilon^{\mp}(y)$:

$$x = \frac{y^2}{2u} - \frac{\varepsilon}{u}\int_0^y \frac{zF(x,z,u)\,dz}{u+\varepsilon F(x,z,u)}, \qquad u = u^{\mp} \quad (y \gtrless 0). \tag{8.3.19}$$

Solutions of Equations (18) or (19) are constructed by using expansions in powers of ε or successive approximations. The expressions $x_0^{\mp} = y^2/(2u^{\mp})$, $y \gtrless 0$, are assumed as the zeroth approximation of the switching curve, i.e. of the functions $x^{\mp}$. The recursive schemes converge uniformly in an asymptotically large domain G_ε ($x \sim 1/\varepsilon$, $y \sim 1/\sqrt{\varepsilon}$) if the function F is differentiable with respect to x and satisfies the uniform Lipschitz condition. Consider some specific examples.

1) Let $|u| \leqslant 1$, $F = F_0 = \text{const}$. Then $\varphi_\varepsilon \equiv 0$, and the following expressions for the switching curve Π_ε and the minimal time T_ε take place when $\varepsilon|F_0| < 1$, $(x,y) \in G_\varepsilon = R_2$:

$$\begin{aligned}
&\Pi_\varepsilon(x,y) = \left\{x, y\colon\ x = \mp\frac{1}{2}y^2/u_*^{\mp} \gtrless 0\right\};\\
&T_\varepsilon(x,y) = S_\varepsilon(x,y) + \Theta_\varepsilon(x,y), \qquad S_\varepsilon = -\left(y + u_*^{\pm}\Theta_\varepsilon\right)/u_*^{\mp},\\
&\Theta_\varepsilon = \left[(y^2 - 2u_*^{\mp}x)/(u_*^{\pm 2} - u_*^{\pm}u_*^{\mp})\right]^{1/2}, \qquad u_*^{\mp} = 1 \mp \varepsilon F_0.
\end{aligned} \tag{8.3.20}$$

2) If the perturbation is a small friction force of the form $\varepsilon F = -\lambda(|y|)y$, where λ is a non-negative function, then $\varphi_\varepsilon \equiv 0$, $(x,y) \in G_\varepsilon = R_2$, $0 \leqslant \varepsilon < \infty$, and the switching curve is found by quadratures:

$$\begin{aligned}
&\Pi_\varepsilon(x,y) = \left\{x, y\colon\ x = \frac{y^2}{2u^{\mp}} + \frac{\varepsilon}{u^{\mp}}\int_0^y \frac{z^2\lambda(|z|)\,dz}{u^{\mp} - \varepsilon z\lambda(|z|)},\ y \gtrless 0\right\},\\
&u^- \leqslant u_\varepsilon(x,y) \leqslant u^+.
\end{aligned} \tag{8.3.21}$$

For particular forms of the function λ, which describe the linear and quadratic friction, the switching curves are given by

$$\begin{aligned}
&\varepsilon\lambda = l = \text{const}, \quad x = \frac{y}{l} - \frac{u^{\mp}}{l^2}\ln\left(1 - \frac{u^{\mp}}{l}y\right);\\
&\varepsilon\lambda = \nu|y|, \quad \nu = \text{const}, \quad x = \mp\frac{1}{2\nu}\ln\left(1 \mp \frac{\nu}{u^{\mp}}y^2\right), \quad y \gtrless 0.
\end{aligned} \tag{8.3.22}$$

The construction of Bellman's function involves the solution of perturbed transcendental equations. The corresponding generating solution is the function $T(x, y)$ defined in (20). For small $\varepsilon > 0$, the additions are constructed by perturbation methods (see below).

3) Consider the linear perturbation $f = Ax$, $A = \text{const}$. The switching curve and the synthesis picture for such a system have been well investigated and constructed for arbitrary $(x, y) \in R_2$, $\alpha = \varepsilon A$, when $-u^- = u^+$. We qualitatively distinguish between oscillating ($\alpha < 0$) and nonoscillating ($\alpha \geqslant 0$) systems. According to Sections 8.3.1 and 8.3.2, we can construct the switching curve and the synthesis picture in an asymptotically large domain G_ε ($x \sim 1/\varepsilon$, $y \sim 1/\sqrt{\varepsilon}$):

$$\begin{aligned}
&\varphi_\varepsilon(x, y, u) = -\varepsilon M y u, \qquad M > 0,\\
&G_\varepsilon = \{x, y\colon -Mu^2 - \alpha M x u + \varepsilon M^2 (uy)^2 - A \leqslant 0\},\\
&\Pi_\varepsilon(x, y) = \left\{x, y\colon -\frac{1}{2}\alpha(x + u/\alpha)^2 + \frac{1}{2}u^2\alpha + \frac{1}{2}y^2 = 0,\right.\\
&\qquad\qquad \left. u = u^\mp,\ y \gtrless 0\right\}.
\end{aligned} \tag{8.3.23}$$

The curve Π_ε in (23) consists of arcs of ellipses when $A < 0$, and hyperbolas when $A > 0$ (for $A = 0$ the curve consists of segments of parabolas; the latter case relates to that discussed above). It should be noted that for $A \geqslant 0$ (see Section 8.3.2, item (1)), condition (4) is fulfilled for $(x, y) \in R_2$ if $M = 0$, i.e. $\varphi_\varepsilon \equiv 0$. To a first approximation in terms of ε we have the explicit expression for the switching curve Π_ε:

$$x = \frac{y^2}{2u^\mp}\left[1 - \alpha\frac{y^2}{4(u^\mp)^2}\right] + O(\alpha^2 y^4), \qquad y \gtrless 0.$$

As has been mentioned in Section 8.3.1, item (2), if the system is oscillatory ($\alpha < 0$) and x and y are large ($|x| \gg \Delta u/\alpha^+$, $|y| \gg \Delta u/\sqrt{\alpha^+}$, $\Delta u = u^+ - u^-$), we can use the quasioptimal feedback control

$$u = \frac{1}{2}(u^- + u^+) - \frac{1}{2}\Delta u \operatorname{sign} y$$

until it becomes $|x| \sim \Delta u/\alpha^+$, $|y| \sim \Delta u/\sqrt{\alpha^+}$, and after that the feedback control given by (7) and (23) ($\alpha^+ = |\alpha|$).

For the analysis of the quasilinear nonoscillating system (13) see [1, 14]. The generalization of the algorithm to the case of perturbations from a wider class and the construction of near-cylindrical switching surfaces for bang-bang controls can be found in [1].

8.4. Investigation of Some Singularly Perturbed Optimal Control Problems

8.4.1. STATEMENT OF THE PROBLEM

The work of a number of engineering systems can be described by models leading to optimal control problems for singularly perturbed systems [1, 17, 18, 22, 26, 27, 54, 68, 85, 89, 93, 96, 102, 103, 114, 126, 139, 140]:

$$\begin{aligned} &\dot{x} = f(t, x, y, u), \qquad x(t_0) = x^0, \quad t \in [t_0, T], \\ &\varepsilon \dot{y} = g(t, x, y, u), \qquad y(t_0) = y^0, \quad \varepsilon \in (0, \varepsilon_0], \\ &M(T, x(T)) = 0, \qquad J[u] = \Phi(T, x(T)) \longrightarrow \min, \quad u \in U. \end{aligned} \tag{8.4.1}$$

Here, $x \in D_x$ and $y \in D_y$ are vectors of arbitrary dimensions (state variables), D_x and D_y are open bounded domains; $u \in U$ is a control vector from an admissible class. The quantities t_0, x^0 and y^0 are specified initial values; T is a limited instant of the process termination, either fixed or to be determined by the solution of the optimal control problem. All functions in (1) are assumed to be sufficiently smooth with respect to the arguments included. The domain U is convex and has a sufficiently smooth boundary.

It is required to construct the approximate solution of problem (1) for sufficiently small $\varepsilon > 0$, i.e. to find the approximate synthesis of the optimal feedback control $u_s(t, x, y, \varepsilon)$, the optimal trajectory $x = x(t, t_0, x^0, y^0, \varepsilon)$, $y = y(t, t_0, x^0, y^0, \varepsilon)$ and, from this, the open-loop (programmed) control $u_p(t, t_0, x^0, y^0, \varepsilon)$, the minimal value of the functional $J_{\min} = \Phi(T, x(T, t_0, x^0, y^0, \varepsilon))$, and the quantity $T = T(t_0, x^0, y^0, \varepsilon)$ if T is not fixed.

8.4.2. THE SCHEME OF THE ASYMPTOTIC SOLUTION

If one formally assumes $\varepsilon = 0$ in system (1), then the equation for y becomes a finite vector relationship. This relationship is supposed to be uniquely solvable with respect to y in the domain of changing arguments in question, i.e.

$$0 = g(t, x, y, u), \qquad y = y_0(t, x, u). \tag{8.4.2}$$

We assume that the root y_0 is isolated, and it is a sufficiently smooth function of t, x, and u. The function $y_0(t, x, u)$ is substituted into Equation (1) for x and then we consider the optimal control problem for the degenerate system having a reduced dimension:

$$\begin{aligned} &\dot{x} = f\big(t, x, y_0(t, x, u), u\big), \qquad x(t_0) = x^0, \quad t \in [t_0, T], \\ &M\big(T, x(T)\big) = 0, \qquad J[u] = \Phi\big(T, x(T)\big) \longrightarrow \min, \quad u \in U. \end{aligned} \tag{8.4.3}$$

Since M and Φ do not depend on $y(T)$, the degenerate problem (3) is closed. We assume that the solution of this problem exists, is unique, and can be constructed in the form

$$\begin{aligned} &u = u_{0s}(t,x), \qquad x = x_0(t,t_0,x^0); \qquad u = u_{0p}(t,t_0,x^0),\\ &J_0 = \Phi\big(T, x_0(T,t_0,x^0)\big); \qquad T = T_0(t_0,x^0);\\ &y = y_0\big(t, x_0(t,t_0,x^0), u_{0p}(t,t_0,x^0)\big). \end{aligned} \tag{8.4.4}$$

Here, y_0 is the solution of the degenerate equation (2), defined from the solution of problem (3). The problem arises as to how to estimate the proximity of the constructed solution (4) of the degenerate problem (2) and (3) and the desired solution of the original singularly perturbed problem (1). The investigation of this problem in the general case is complicated, but the estimate of the proximity of solution (4) and the solution of the singularly perturbed system (1), where the approximate control $u_{0s}(t,x)$, or $u_{0p}(t,t_0,x^0)$, is substituted, is also of practical interest. Let the function $u = u_{0s}(t,x)$ be substituted into (1):

$$\begin{aligned} \dot{x} &= f\big(t,x,y,u_{0s}(t,x)\big) \equiv f_{0s}(t,x,y),\\ \varepsilon\dot{y} &= g\big(t,x,y,u_{0s}(t,x)\big) \equiv g_{0s}(t,x,y). \end{aligned} \tag{8.4.5}$$

Suppose that the stationary point $\bar{y}^*$ of the associated system [27, 148, 152]

$$\bar{y}' = g_{0s}(t,x,\bar{y}), \qquad \bar{y}^* = y_0\big(t,x,u_{0s}(t,x)\big), \tag{8.4.6}$$

is uniformly asymptotically stable with respect to the linear approximation. Here, the prime denotes the derivative with respect to the "fast" time $\tau = t/\varepsilon$; and the arguments t and x are considered to be fixed. Then, under appropriate assumptions regarding the smoothness of the right-hand sides of Equations (5) with respect to $(t,x,y) \in [t_0,T] \times D_x \times D_y$ and the above mentioned isolatedness of the root y_0, the proximity between the unknown solution x_s and y_s of problem (5) and the known solution x_0 and y_0 of problem (4) takes place (see [148]):

$$\begin{aligned} &\big|x_s(t,t_0,x^0,y^0,\varepsilon) - x_0(t,t_0,x^0)\big| \leqslant C\varepsilon, \qquad t \in [t_0,T],\\ &\big|y_s(t,t_0,x^0,y^0,\varepsilon) - y_0(t,x_0,u_{0p})\big| \leqslant C\varepsilon, \qquad t \in I_\varepsilon \subset [t_0,T]. \end{aligned} \tag{8.4.7}$$

Here, I_ε is the segment $[t_0, T]$, excluding some subsets of small measure, the point t_0 in particular. If the control u_{0s} has a finite number of discontinuity points, then the set I_ε does not include these points with their small vicinities.

Instead of the function $u = u_{0s}$, we may substitute the known open-loop control $u = u_{0p}(t, t_0, x^0)$ into system (1):

$$\begin{aligned} \dot{x} &= f(t, x, y, u_{0p}(t, t_0, x^0)) \equiv f_{0p}(t, t_0, x, x^0, y), \\ \varepsilon\dot{y} &= g(t, x, y, u_{0p}(t, t_0, x^0)) = g_{0p}(t, t_0, x, x^0, y). \end{aligned} \tag{8.4.8}$$

For the solution x_p and y_p of system (8) the proximity estimates such as (7) are established analogously. The estimate is violated at the discontinuity points of the control u_{0p} because of the boundary layer due to discontinuities of y_0.

Thus, the control (4) in the form u_{0s} or u_{0p} solves the problem of bringing the variable x to the manifold M (1) at the instant $t = T_0(t_0, x^0)$ with the error $O(\varepsilon)$. The issue of optimality requires an additional analysis, based on the maximum principle, analogously to Sections 2.5 and 4.5. To calculate the controls and trajectories more accurately, we are required to consider boundary layers and construct boundary-layer-type functions [26, 27, 89, 93, 96, 103].

8.4.3. TIME-QUASIOPTIMAL CONTROL IN SINGULARLY PERTURBED LINEAR SYSTEMS

In practice, we may sometimes confine ourselves to the investigation of time-optimal control problem for linear systems of the form

$$\begin{aligned} &\dot{x} = A_{xx}x + A_{xy}y + B_xu + F_x(t), \qquad x(t_0) = x^0, \\ &\varepsilon\dot{y} = A_{yx}x + A_{yy}y + B_yu + F_y(t), \qquad y(t_0) = y^0, \\ &x(T) = x^*, \qquad T \longrightarrow \min, \qquad u \in U. \end{aligned} \tag{8.4.9}$$

Here, x and y are vectors of arbitrary dimensions, x^* is a specified vector, U is a convex domain; in particular, for a number of applied problems $x^* = 0$ $(F_{x,y} \equiv 0)$, U is a convex polyhedron. In accordance with Section 8.4.2, the matrix A is assumed to be negative-definite. Henceforth, we suppose that the degenerate optimal control problem (9) for the variable x

$$\begin{aligned} &\dot{x} = (A_{xx} - A_{yy}^{-1}A_{yx})x + (B_x - A_{yy}^{-1}B_y)u + (F_x - A_{yy}^{-1}F_y), \\ &x(t_0) = x^0, \quad x(T) = x^*, \quad T \longrightarrow \min, \quad u \in U \end{aligned} \tag{8.4.10}$$

has the optimal solution

$$
\begin{aligned}
&u = u_{0s}(t, x), \qquad x = x_0\left(t, t_0, x^0\right);\\
&u = u_{0p}\left(t, t_0, x^0\right), \qquad T = T_0\left(t_0, x^0\right),\\
&y = y_0 = -A_{yy}^{-1}(A_{yx}x_0 + B_y u_{0p} + F).
\end{aligned}
\tag{8.4.11}
$$

The function u_{0p} may be a smooth (continuous) or piecewise smooth (of the bang-bang type) function of time. Having been substituted into the original system (9), the control u_{0p} (11) brings the phase point (x, y) to an ε-vicinity of the hyperplane $x = x^*$, y in the time $(T_0 - t_0)$. If $x^* = 0$, $F_x(t) = F_y(t) \equiv 0$ and $u = 0$ for $t > T_0$, then the variable y will also come to an ε-vicinity of the value $y = 0$ in the time $\Delta t = t - T_0 \sim \varepsilon \ln(1/\varepsilon)$. A more accurate solution of problem (9) requires taking into account boundary layers for the variable y [89, 93, 96, 103].

8.4.4. EXAMPLE

Consider the special system of the second order as a simple model example of problem (9). Let us investigate a single-degree-of-freedom object possessing small inertia. In dimensionless variables, the time-optimal control problem for such an object can be presented in the form

$$
\begin{aligned}
&\varepsilon\ddot{x} + \dot{x} = u,\\
&x(0) = x^0, \quad \dot{x}(0) = \dot{x}(0) = \dot{x}^0,\\
&x(T) = x^* \ (= 0), \quad \dot{x}(T) = 0, \quad T \longrightarrow \min, \quad |u| \leqslant 1.
\end{aligned}
\tag{8.4.12}
$$

The solution of the control problem (12) is easily constructed for any value of $\varepsilon \geqslant 0$ in accordance with [40, 131]. This solution is given in Section 8.3. Figure 8.2 shows the synthesis picture (the phase portrait) for small ε. The switching curve Π of the bang-bang control is close to the ordinate axis $x = x^*$ $(= 0)$; in the limit case, when $\varepsilon = 0$, the control is given by $u = -\operatorname{sign}(x - x^*)$. Phase trajectories in the domains $G^{\mp}$ rapidly (instantaneously when $\varepsilon = 0$) arrive at a small vicinity $(O(e^{-t/\varepsilon}))$ of the straight lines $\dot{x} = \mp 1$. When on these lines, the phase point moves towards the ordinate axis $x = x^* = 0$ with the velocity

$$
\dot{x} = -\operatorname{sign}(x - x^*) + O(e^{-t/\varepsilon})
$$

and attains an ε-vicinity of this axis in the time $T_0 = |x^0|$. If then one assumes $u = \pm 1$, the phase point will come to the origin $x = \dot{x} = 0$ in the time $t - t_0 \sim \varepsilon \ln \varepsilon^{-1}$.

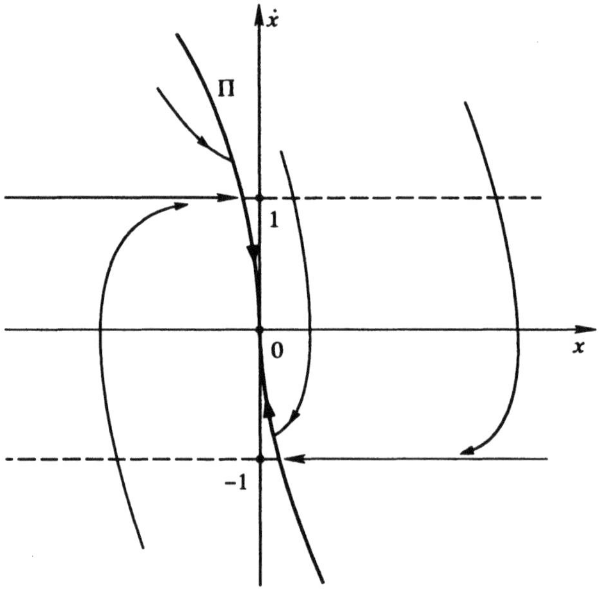

Fig. 8.2.

If we take $u \equiv 0$ for $t \geqslant T(x^0, \dot{x}^0, \varepsilon)$, where T is the known minimal time, then the phase point will remain in the specified terminal state. If we assume $u \equiv 0$ in an ε-vicinity of the ordinate axis, then the phase point will arrive at an ε-vicinity of the terminal state $x = \dot{x} = 0$ in the time $t - T_0 = O(\varepsilon \ln \varepsilon^{-1})$.

8.5. Control of Elastic Parts of Mechanisms by Electromechanical Actuators

Theoretical analysis of the performance characteristics (for example, the accuracy, operation speed, energy consumption, etc.) for complex engineering objects such as manipulation robots, large-scale space structures, etc., often leads to problems concerning the dynamics and control of systems of interconnected rigid and elastic bodies. With the purpose of increasing the system's efficiency, we arrive at the necessity to lighten the structure (especially its terminal link). This can cause a significant elastic compliance or unwanted oscillations worsening the above mentioned performance characteristics [1, 13, 16, 22, 34, 41, 42, 90, 91, 103, 112, 146].

We consider plane rotations of an elastic loaded beam (the link of a manipulator) driven by an electric actuator. The actuator is placed at the end of the beam fixed by a hinge. This model often arises when considering more complex electromechanical systems which must meet great demands of their positioning [1, 13, 17, 22, 34, 37,

41, 42, 52, 67, 90, 103, 117, 142, 146]. Based on the weak bending model for thin rectilinear beams [145], we investigate the rotation dynamics of a mechanism's loaded terminal link driven by an electric actuator. Here we take into account the distributed elasticity and inertia of the beam. We perform a mathematical simulation of the dynamics of both rigid and elastic links and construct the minimum-time control within the framework of the limit model considering the link to be a rigid body and taking account of electromagnetic processes in the electric actuator. We analyse various admissible control laws and consider ways of raising the positioning accuracy. The constructed rational control actions are then applied to the elastic model of the link to obtain estimates of the accuracy and amplitude of the residual oscillations.

8.5.1. STATEMENT OF THE PROBLEM

Consider an elastic beam which can rotate in the plane OX^*Y^* [1]. One endpoint of the beam is fixed in the inertial frame $OX^*Y^*Z^*$, the other endpoint is loaded by a rigid body G whose linear dimensions are small compared with the length l of the beam. The rotation axis is perpendicular to the plane OX^*Y^*. At the point O, an electric actuator D is placed which consists of a d.c. independent excitation electric motor and a reduction gear with absolutely rigid parts (see Fig. 8.3).

To describe the motion we introduce the coordinate frame $OXYZ$ rotating together with the beam. The planes OX^*Y^* and OXY coincide. The axis OX coincides

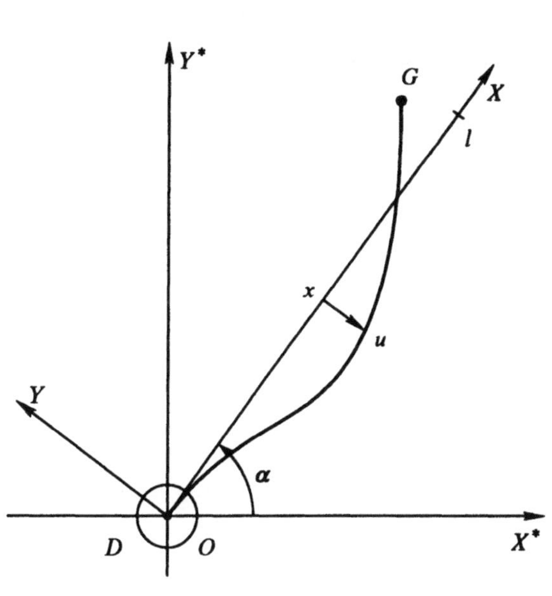

Fig. 8.3.

with the tangent to the neutral line of the beam at the point O. Elastic deformations of the system are supposed to be described within the framework of the linear theory of thin rectilinear beams, the relative elastic displacements $u = u(t, x)$ being small and orthogonal to the straight line OX. The rotation velocity of the beam as a whole is small in comparison with the lowest frequency of free elastic oscillations.

The equation of motion and boundary conditions for a rotating beam are given by

$$\rho(x)\ddot{u} = -\left[EI(x)u''\right]'' - \alpha\rho(x)x, \qquad x \in (0, l); \tag{8.5.1}$$

$$\begin{aligned} &u(t,0) = u'(t,0) = u''(t,l) = 0, \\ &\left[EI(x)u''\right]'\Big|_{x=l} = m\left[\ddot{u}(t,l) + l\ddot{\alpha}\right], \qquad t \in [t_0, T]. \end{aligned} \tag{8.5.2}$$

Here, α is the angle between OX and OX^*, m is the mass of the load (a particle) G on the end of the beam, points and primes denote derivatives with respect to time t and coordinate x, respectively. The linear density ρ of the beam and the bending stiffness EI can be functions of x. Initial conditions are taken in the following general form

$$u(t_0, x) = f(x), \qquad \dot{u}(t_0, x) = g(x), \tag{8.5.3}$$

where $f(x)$ and $g(x)$ are sufficiently smooth functions satisfying conditions (2).

The change of the angular momentum of the whole system with respect to the rotation axis OZ is governed by the integro-differential equation (the equation of the angular momentum) and initial conditions

$$\int_0^l \rho^*(x)(\ddot{\alpha}x + \ddot{u})x\,dx = M, \qquad \alpha(t_0) = \alpha^0, \qquad \dot{\alpha}(t_0) = \omega^0. \tag{8.5.4}$$

Here, M is a torque created by the electric actuator with respect to the axis OZ. The point mass G on the end of the beam is taken into account by adding the delta-function of the intensity m to $\rho(x)$, i.e. we assume $\rho^*(x) = \rho(x) + m\delta(l - x)$ for $x \in [0, l]$.

The requirement of the existence of partial derivatives of the function $u(t, x)$ with respect to x through the fourth order seems to be rather strong, since Equation (1) is derived from the integro-differential equation requiring the derivatives of $u(t, x)$ with respect to x through the second order only.

The equations for the electric actuator have the form

$$L\dot{j} + rj + \Phi\dot{\varphi} = V(t), \qquad j(t_0) = j^0; \tag{8.5.5}$$

$$J_0\ddot{\varphi} = \mu - M/n, \qquad \alpha = \varphi/n, \qquad \mu = \Phi j, \tag{8.5.6}$$

where L is an inductance, Φ is a magnetic flux, R is an ohmic resistance, j is a current in the armature of the electric motor, J_0 is an inertia moment of the armature, φ is the rotation angle, n is a gear ratio, and $V(t)$ is a voltage. Equation (5) describes the voltage balance in the armature circuit, and (6) governs the rotation of the armature.

Eliminating the variables μ (the electromagnetic torque), M, j, and φ from Equations (1), (4)–(6), we arrive at the integro-differential equation

$$\begin{aligned} &LJ_0n^2\dddot{\alpha} + RJ_0n^2\ddot{\alpha} + \Phi^2n^2\dot{\alpha} + \\ &\quad + \int_0^l \rho^*(x)\big[L(x\dddot{\alpha} + \dddot{u}) + R(x\ddot{\alpha} + \ddot{u})\big]\mathrm{d}x = \Phi nV(t). \end{aligned} \tag{8.5.7}$$

Using Equations (1), (4)–(6) we obtain the initial conditions for $\ddot{\alpha}$

$$\ddot{\alpha}(t_0) = \big[EI(0)u''(t_0, 0) + \Phi nj^0\big]/\big(J_0n^2\big), \qquad u''(t_0, 0) = f''(0). \tag{8.5.8}$$

Henceforth, we consider the set of equations (1) and (7), boundary conditions (2) and initial conditions (3), (4), and (8). Given the control $V = V(t)$, one investigates the boundary-value problem with respect to x and the initial-value problem with respect to t using methods of mathematical physics. To be definite, we assume the beam to be uniform ($\rho = \text{const}$, $EI = \text{const}$) and construct the solution of the control problem for the zero initial conditions corresponding to the state of rest for both the beam and the motor.

For the examined system (1) and (7), we state the optimal control problem. We assume that at the initial instant the undeformed beam is at a state of rest. It is required to choose the open-loop control $V(t)$ which turns the beam by a specified angle, brakes the beam and quenches its oscillations by the end of the process, and leads to a minimum for some performance functional $F[V]$ over the admissible set $\{V\}$ of controls, guaranteeing a strong solution of the problem [51, 52, 67, 90, 91, 113, 117, 142, 146]

$$\begin{aligned} &F[V] \longrightarrow \min_V, \quad V \in \{V\}; \qquad \alpha(t_0) = \alpha^0, \quad \alpha(T) = \alpha^T, \\ &\dot{\alpha}(t_0) = \ddot{\alpha}(t_0) = \dot{\alpha}(T) = \ddot{\alpha}(T) = u(t_0, x) = \dot{u}(t_0, x) \\ &\qquad = u(T, x) = \dot{u}(T, x) = 0, \qquad x \in [0, l]. \end{aligned} \tag{8.5.9}$$

8.5.2. ELASTIC OSCILLATIONS OF A LOADED ROTATING BEAM

The solution of Equation (1) with initial and boundary conditions (2) and (3) is constructed by the method of separation of variables (the Fourier method [50]). Having proceeded to the dimensionless variables

$$x_* = x/l, \quad u_* = u/l, \quad t_* = t/\tau, \quad \tau = \left(\rho l^4/EI\right)^{1/2} \tag{8.5.10}$$

and then omitting asterisks in the subscripts, we obtain from (1) and (2)

$$\begin{aligned} &\ddot{u} + u^{IV} = -x\ddot{\alpha}, \qquad u(t,0) = u'(t,0) = u''(t,1) = 0, \\ &u'''(t,1) = \varkappa\left[\ddot{u}(t,1) + \ddot{\alpha}\right], \qquad \varkappa = m/\rho l. \end{aligned} \tag{8.5.11}$$

The elastic displacements can be represented as [50]

$$u(t,x) = \sum_{k=1}^{\infty} X_k(x)\Theta_k(t), \tag{8.5.12}$$

where $X_k(x)$, $x \in [0,1]$, are the eigenfunctions of the following problem

$$\begin{aligned} &X_k^{IV} = \lambda_k^4 X_k, \qquad X_k(0) = X_k'(0) = X_k''(1) = 0, \\ &X_k'''(1) = -\lambda_k^4 \varkappa X_k(1). \end{aligned} \tag{8.5.13}$$

The sequence $\lambda_k = \lambda_k(\varkappa)$, $k = 1, 2, \ldots$, represents the eigenvalues which are the roots of the transcendental characteristic equation

$$1 + \cosh\lambda\cos\lambda + \varkappa\lambda(\sinh\lambda\cos\lambda - \cosh\lambda\sin\lambda) = 0. \tag{8.5.14}$$

The eigenfunctions $X_k(x)$ of the boundary-value problem (13) have the form

$$\begin{aligned} X_k(x) = {} &\cosh\lambda_k x - \cos\lambda_k x - (\cosh\lambda_k + \cos\lambda_k) \times \\ &\times (\sinh\lambda_k x - \sin\lambda_k x)(\sinh\lambda_k + \sin\lambda_k)^{-1}. \end{aligned} \tag{8.5.15}$$

Note that, due to the symmetry of (14) and (15), we may confine ourselves to positive eigenvalues λ_k ($k \geqslant 1$) and the corresponding eigenfunctions $X_k(x)$. The functions $X_k(x)$ are orthogonal in the space $L_2[0,1]$ with the weight $\rho^*(x)$ and the scalar product (δ_{rq} is the Kronecker delta (tensor))

$$\int_0^1 X_r(x)X_q(x)\rho^*(x)\,dx = \delta_{rq}\|X_r\|^2. \tag{8.5.16}$$

Having multiplied the state equations (11) by $\rho^*(x)X_k(x)$, integrated them with respect to x ($x \in [0,1]$), and divided by $\|X_k\|^2$, we obtain, taking into account (12)–(16), the countable set of equations for the Fourier coefficients in the form

$$\begin{aligned}
&\ddot{\Theta}_k + \lambda_k^4\Theta_k = C_k\ddot{\alpha}, \qquad k = 1, 2, \dots,\\
&C_k = -\|X_k\|^{-2}\int_0^1 xX_k(x)\rho^*(x)\,dx = -2\big(\lambda_k\|X_k\|\big)^{-2},\\
&\|X_k\|^2 = 1 + \varkappa(\sinh\lambda_k\cos\lambda_k - \cosh\lambda_k\sin\lambda_k)^2(\sinh\lambda_k + \sin\lambda_k)^{-2}.
\end{aligned} \tag{8.5.17}$$

If $\ddot{\alpha}(t)$ is a known function of time, then, by using (17), we can find the Fourier coefficients $\Theta_k(t)$; these are written as

$$\Theta_k(t) = C_k\lambda_k^{-2}\int_{t_0}^t \sin\lambda_k^2(t-\tau)\ddot{\alpha}(\tau)\,d\tau. \tag{8.5.18}$$

Let us integrate Equation (7), taking into account the simplifications made above, return to the original time $t = \tau t_*$ (see (10)) and substitute the values of $\dddot{u}$ and $\ddot{u}$ into (7), using (1). Having integrated by parts, as a result we obtain

$$\begin{aligned}
&A\dddot{\alpha} + B\ddot{\alpha} + D\dot{\alpha} - \tau^{-2}\big(G\dot{u}''(t,0) + Hu''(t,0)\big) = V^*(t),\\
&A = LJ_0n^2, \qquad B = RJ_0n^2, \qquad D = \Phi^2n^2,\\
&G = \rho l^3L, \qquad H = \rho l^3R, \qquad V^*(t) = \Phi nV(t).
\end{aligned} \tag{8.5.19}$$

Relationship (19) can be considered as an integro-differential equation with respect to α; according to (12) and (18), the quantity $\ddot{\alpha}$ is included in $u(t,x)$ (as well as in $\dot{u}''(t,0)$, $u''(t,0)$) under the integral sign. The kernel of this integro-differential equation is of the difference type. According to (12), one can also consider (19) as a differential equation containing sums of $\dot{\Theta}_k$ and Θ_k. Together with system (17), we obtain the infinite set of ordinary linear differential equations with time-invariant coefficients. Indeed, substituting $\dot{u}''$ and u'' from (12) into (19) and taking account of

(15) and (17), we can write down the countable set of equations, initial and terminal conditions (according to (9)) in the form

$$
\begin{aligned}
&A\dddot{\alpha} + B\ddot{\alpha} + D\dot{\alpha} - \tau^{-2}\sum_{k=1}^{\infty}(G\dot{\Theta}_k + H\Theta_k)\lambda_k^2 = V^*(t),\\
&\ddot{\Theta}_k + \tau^{-2}\lambda_k^4\Theta_k = C_k\ddot{\alpha}, \qquad k = 1, 2, \ldots,\\
&F[V] = \min_v, \qquad V \in \{V\}, \quad \alpha(t_0) = \alpha^0, \quad \alpha(T) = \alpha^T,\\
&\dot{\alpha}(t_0) = \ddot{\alpha}(t_0) = \dot{\alpha}(T) = \Theta_k(t_0) = \dot{\Theta}_k(t_0) = \Theta_k(T) = \dot{\Theta}_k(T) = 0.
\end{aligned}
\tag{8.5.20}
$$

We have not succeeded in finding the exact solution of problem (20). It is of practical interest to develop approximate methods, such as the finite mode approximation ($k = 1, 2, \ldots, k^* < \infty$) and the quasistatic approximation ($\varkappa \gg 1$). It is also important to investigate the case of asymptotically large stiffness $EI \to \infty$ ($\tau/T \ll 1$), the case of an asymptotically small and smooth control actions, etc. The above-mentioned problems have been mostly considered for an ideal actuator.

In what follows, we investigate the case of large stiffness of the elastic link, construct admissible quasioptimal (rational) controls and estimate the control accuracy for the original elastic system.

8.5.3. CONTROL LAWS FOR A THIRD ORDER SYSTEM

Let us introduce the parameter $\mu = \tau/T$ which approaches zero as $EI \to \infty$. According to perturbation theory, we can represent the solution of system (20) as a series in terms of powers of the small parameter μ

$$
\alpha = \alpha_0 + \mu\alpha_1 + \mu^2 + \cdots, \qquad \Theta_k = \Theta_{0,k} + \mu\Theta_{1,k} + \mu^2 + \cdots. \tag{8.5.21}
$$

Substituting (21) into (20) and comparing the coefficients of like powers of μ, we find for the zeroth approximation with respect to μ (the case of a perfectly rigid link) that $\Theta_{0,k} = 0$, and α_0 is found from the following linear differential equation ($J^* = J_0 n^2 + \rho l^3(1/3 + \varkappa)$)

$$
LJ^*\dddot{\alpha}_0 + RJ^*\ddot{\alpha}_0 + D\dot{\alpha}_0 = V^*(t). \tag{8.5.22}
$$

Let us divide Equation (22) by RJ^* and proceed to a new time $t^* = t/\tau_*$, where $\tau^* = RJ^*/(\Phi^2 n^2)$ is the electromechanical time constant. In what follows, the asterisk in t^* is omitted. Then Equation (22) with arbitrary initial conditions is written in the form ($t_0 = 0$):

$$\begin{aligned} &\varepsilon\dddot{\alpha} + \ddot{\alpha} + \dot{\alpha} = v, \qquad \varepsilon = \Phi^2 n^2 L/(R^2 J^*), \\ &v = RJ^* V/(\Phi^3 n^3); \qquad \alpha(0) = \alpha^0, \quad \dot{\alpha}(0) = \omega^0, \quad \ddot{\alpha}(0) = \gamma^0. \end{aligned} \tag{8.5.23}$$

We consider the problem of finding the minimum-time feedback control of the voltage for system (23) with terminal conditions in the following form

$$\alpha(T) = \dot{\alpha}(T) = \ddot{\alpha}(T) = 0, \qquad T \longrightarrow \min_v, \quad |v| \leqslant v_0. \tag{8.5.24}$$

Open-loop controls for problem (23) and (24) are considered, for example, in [110]. Using the maximum principle [40, 131] we find that the optimal control v assumes the values $v = \mp v_0$. For each segment where the control is constant, the variable α is given by

$$\alpha(t) = B_0 + B_1 \exp(\eta_1 t) + B_2 \exp(\eta_2 t) + vt, \tag{8.5.25}$$

where η_1 and η_2 are the roots of the characteristic equation

$$\begin{aligned} &\varepsilon\eta^2 + \eta + 1 = 0, \qquad \eta_{1,2} = \big(-1 \pm \sqrt{1-4\varepsilon}\big)/2\varepsilon, \\ &\eta_1 = -1 + O(\varepsilon), \qquad \eta_2 \sim -1/\varepsilon \longrightarrow -\infty, \qquad \varepsilon \to 0. \end{aligned}$$

For the system $\dot{x} = Ax + bv$, the control has not more than $n-1$ switches, if the eigenvalues of the matrix A are real and the vectors $b, Ab, A^2b, \ldots, A^{n-1}b$ (n is the order of the system) are linearly independent [131]. For system (23), these conditions are just satisfied if $\varepsilon < 1/4$. Using the property of linear time-invariant sets of differential equations, we can find the three-parameter family of points, from which the system arrives at the terminal state in the minimal time so that the control has two switches:

$$\begin{aligned} x^{\mp} = \pm|v|\Big\{ &\exp(-A\tau_1)\Big[\exp(-A\tau_2)\int_0^{\tau_3} \exp(-At')b\,dt' - \\ &- \int_0^{\tau_2} \exp(-At')b\,dt'\Big] + \int_0^{\tau_1} \exp(-At')b\,dt'\Big\}, \end{aligned}$$

$$\tau_1 = t_1, \quad \tau_2 = t_2 - t_1, \quad \tau_3 = -t_1 - t_2 + T.$$

Here, T is the time of the process termination, t_1 and t_2 are the instants of the first and second switch, respectively. Integrating (22) with respect to t from $t = T$ to $t = 0$ we obtain the set of transcendental equations for t_1, t_2, and T, resolved with respect to the initial conditions ($i = 1, 2$):

$$\alpha^0 = \pm v_0\big(2t_1 - 2t_2 + T - P_1\eta_1^{-2} + P_2\eta_2^{-2}\big),$$
$$\omega^0 = \mp v_0(P_1/\eta_1 - P_2/\eta_2), \qquad \gamma^0 = \mp v_0(P_1 - P_2), \tag{8.5.26}$$
$$P_i = \big[1 - 2\exp(-\eta_i t_1) + 2\exp(\eta_i t_2) - \exp(-\eta_i T)\big]/\varepsilon(\eta_2 - \eta_1).$$

By solving, numerically, Equations (26) with respect to the terminal instant T and the switching instants t_1 and t_2 for given initial values of α, $\dot{\alpha}$, $\ddot{\alpha}$, one can construct the open-loop control of the voltage which brings system (22) to the terminal (zero) state with any prescribed accuracy. As an example, we have constructed the function $v(t)$ (Fig. 8.4, the solid line) for the following initial data and parameters of system (22): $\alpha(0) = 1$, $\dot{\alpha}(0) = \ddot{\alpha}(0) = 0$, $|v| \leqslant 0.01$, $\varepsilon = 0.1$.

The optimal values of t_1, t_2, and T_0 and the sign of v just after switching on the voltage ($v(0) > 0$) were found:

$$t_1 = 100.31, \qquad t_2 = 100.69, \qquad T_0 = 100.77. \tag{8.5.27}$$

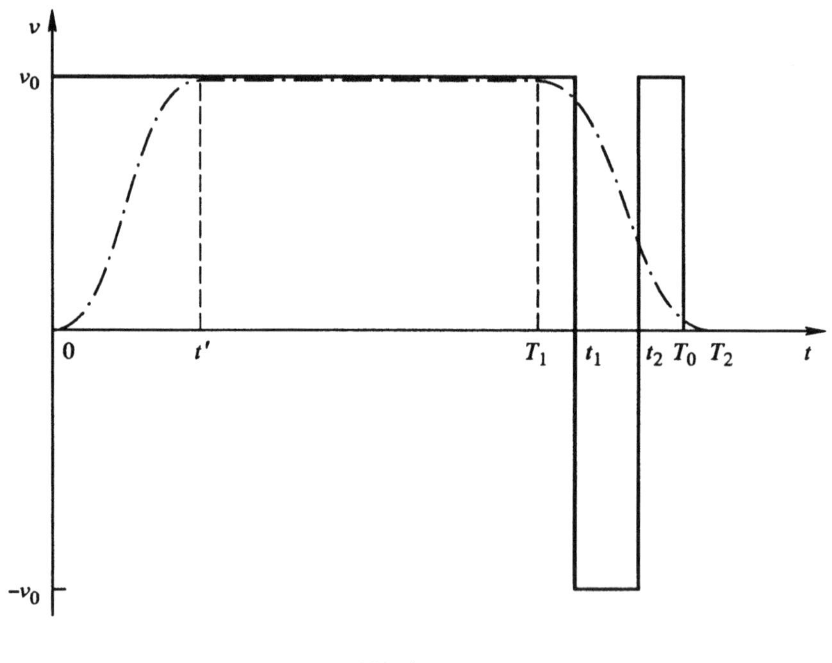

Fig. 8.4.

Assuming $t_1 = 0$ in (26) we obtain the switching surface, specified by Equations (26), from which one can get the zero point involving not more than one switch. This surface, which divides the state space into two regions, can be numerically constructed with any accuracy for any domain of α, $\dot{\alpha}$, and $\ddot{\alpha}$. If $\ddot{\alpha}(t) > \gamma^*$ ($\ddot{\alpha}(t) < \gamma^*$) for the time instant t, then $v < 0$ ($v > 0$). Here, γ^* is the value of the third phase coordinate of the switching surface for current values of $\alpha(t)$ and $\dot{\alpha}(t)$. Assuming that $t_2 = 0$ we obtain the switching line, along which the system can arrive at the zero point without switches. Therefore, having constructed the switching surface and the switching line, we obtain the minimum-time feedback control of the voltage for system (22) with initial and terminal conditions (23) and (24).

Consider now a less general case and consider the problem of finding the control of the voltage which brings system (22) with zero initial conditions for $\dot{\alpha}$ and $\ddot{\alpha}$ to a prescribed δ-vicinity of the zero state in a finite time T_δ. The feature of this is that the system is at rest at the initial time instant. We give here two control laws which solve the problem. The first law is a bang-bang control, piecewise constant on the segment $t \in [0, T]$ and equal to zero outside it:

$$v = \pm v_0, \quad t \in [0, T], \qquad v \equiv 0, \quad t \notin [0, T]. \tag{8.5.28}$$

This control is shown by the dashed line in Fig. 8.4. For such a control, the phase coordinates at time instants $t > T$ have the form (25):

$$\begin{aligned} &\alpha(t) = C_0 + C_1 \exp(\eta_1 t) = C_2 \exp(\eta_2 t), \\ &\dot{\alpha}(t) = d\alpha/dt, \quad \ddot{\alpha}(t) = d^2\alpha/dt^2, \\ &C_0 = vT + \alpha^0, \quad C_i = \eta_i^{-2} v\big[\exp(-\eta_i T) - 1\big]/\varepsilon(\eta_2 - \eta_1), \quad i = 1, 2. \end{aligned} \tag{8.5.29}$$

It is seen from (29) that if one chooses $T = T_1 \equiv |\alpha^0/v_0|$ and $v = -v_0 \operatorname{sign} \alpha^0$, then the control (28) drives system (23) to the δ-vicinity of the point $(0, 0, 0)$ in the time $T_{1,\delta} = T_1 + (1/\eta_1) O(\ln \delta / v_0)$, since $\varepsilon \ll 1$. If $T_1/\tau^* \gg 1$, then it is possible to replace the bang-bang minimum-time control with a simpler control (28), the loss of optimality being relatively small ($(T_{1\delta} - T_1)/T_1 \ll 1$).

The second, "smoothed", control law is described by the function twice differentiable on the segment $t \in [0, T]$. It has the form (dashed-dotted line in Fig. 8.4):

$$v = \begin{cases} \pm v_0 \sin^2 \xi t, & 0 \leqslant t \leqslant \pi/2\xi; \\ \pm v_0, & \pi/2\xi < t \leqslant \Delta = T - \pi/2\xi; \\ \pm v_0 \cos^2 \xi(\Delta - t), & \Delta < t \leqslant T. \end{cases} \tag{8.5.30}$$

Here, ξ is the parameter defining the time necessary for the control to reach its extremal value. Having integrated system (22) with the initial conditions $\dot{\alpha}(0) =$

$\dot{\alpha}(0) = 0,\ \alpha(0) = \alpha^0$, we obtain that at $t \geqslant T$ the phase coordinates are given by expressions (29), where

$$\begin{aligned} C_0 &= \alpha^0 + v\Delta, \\ C_i &= \left[1 - \exp(-\eta_i \Delta)\right] \times \\ &\quad \times \left[\eta_1\eta_2(B + v_0/2) - 2\xi\eta_i A\right] / \left[\eta_i^2(\eta_2 - \eta_1)\right], \qquad i = 1, 2, \\ A &= \frac{v_0\xi}{d}, \quad B = -\frac{v_0(1 - 4\varepsilon\xi^2)}{2d}, \quad d = \left(1 - 4\varepsilon\xi^2\right)^2 + 4\xi^2. \end{aligned} \tag{8.5.31}$$

Having chosen $\Delta = |\alpha^0/v^0|$ and $v = -v_0 \operatorname{sign} \alpha^0$, we find that, for $t > T$, the phase point approaches zero as $v_0\xi^2 O(\exp(\eta_1(t - T)))$, for $t \to \infty$. As is shown below, the application of the law (30) is justified for the control of the countable system (20).

8.5.4. ESTIMATES OF THE AMPLITUDE OF ELASTIC OSCILLATIONS FOR DIFFERENT CONTROL LAWS

In Section 8.5.3, we proposed three control schemes for system (23) (Fig. 8.4). These controls satisfy definite criteria of positioning accuracy and operation speed. Let us estimate the elastic displacements arising in system (20) when using the above mentioned control laws. Considering the influence of the elastic compliance of the structure as a small perturbation, we will determine the acceleration $\ddot{\alpha}(t)$ from the model considering the rotating link as an absolutely rigid body. Therefore, given a specified control, $\ddot{\alpha}(t)$ is a known function of time. After switching on the voltage obeying the bang-bang law (28), the variable $\ddot{\alpha}$ has the form

$$\ddot{\alpha}(t) = v\left[\exp(\eta_1 t) - \exp(\eta_2 t)\right] / \varepsilon(\eta_1 - \eta_2). \tag{8.5.32}$$

Let us find now the value of $\Theta_k(t)$ in (18) by substituting, for $\dot{\alpha}(t)$, expression (32). Integrating the right-hand side of (18) gives

$$\begin{aligned} \Theta_k(t) &= \frac{C_k v}{\varepsilon\nu_k(\eta_1 - \eta_2)}(\chi_k \cos\nu_k t + \zeta_k \sin\nu_k t) + O\left(\exp(\eta_1 t)\right), \\ \chi_k &= \frac{\nu_k}{d_{k_2}} - \frac{\nu_k}{d_{k_1}}, \qquad \zeta_k = \frac{\eta_2}{d_{k_2}} - \frac{\eta_1}{d_{k_1}}, \\ d_{k_{1,2}}^{-1} &= \nu_k^2 + \eta_{1,2}^2, \qquad \nu_k = \frac{\lambda_k^2 \tau^*}{\tau}. \end{aligned} \tag{8.5.33}$$

Residual oscillations for $t \geqslant T_1$ are estimated by an expression similar to (33). Note that by choosing the parameters T and ξ one can significantly reduce the residual vibrations ($t \geqslant T$) for the chosen (for example, the lowest, $k = 1$) oscillation mode k.

Having supposed the above assumptions to be valid and applying the "smoothed" control (30) to system (20) we obtain

$$\ddot{\alpha}(t) = -\eta_1^2 C_1 \exp(\eta_1 t) + \eta_2^2 C_2 \exp(\eta_2 t) + 2\xi(A\cos 2\xi t + B\sin 2\xi t), \qquad (8.5.34)$$

where the coefficients C_1, C_2, A, and B are determined from (31). Assuming $\eta_1 t' \gg 1$, we find the expressions for $\Theta_k(t)$ at the time instant $t' = \pi/2\xi$:

$$\begin{aligned}\Theta_k(t') = \frac{vC_k}{\nu_k}\Big\{\Gamma\Big[\frac{\mu_2}{d_{k_1}}(\eta_1 \sin\nu_k t' - \nu_k\cos\varphi_k t') - \\ - \frac{\mu_1}{d_{k_2}}(\eta_2\sin\nu_k t' - \nu_k\cos\nu_k t')\Big] + \\ + \frac{2\xi^2}{\nu_k^2 - 4\xi^2}(\nu_k - \nu_k\cos\nu_k t' + \sin\nu_k t')\Big\} + O\Big(\frac{1}{\eta_1^2 t'^2}\Big),\end{aligned} \qquad (8.5.35)$$

$$\mu_{1,2} = 1 - \eta_{1,2}^{-1} - \varepsilon, \qquad \Gamma = -4\xi^2(1-4\varepsilon)^{-1/2}.$$

Let us compare estimates (33) and (35) for $\Theta_k(t)$. For the control $\ddot{\alpha}(t)$ from (32), the estimate of $\Theta_k(t)$ is given by

$$\Theta_k(t) \simeq C_k v_0/\varepsilon(\nu_k^2 + \eta_2^2)(\eta_1 - \eta_2)$$

and, for the "smoothed" control (34), we have

$$\Theta_k(t) \simeq C_k v_0 \xi^2/\varepsilon(\nu_k^2 + \eta_2^2)(\eta_1 - \eta_2).$$

For k fixed (for instance, $k = 1$), the estimates can be improved.

8.5.5. NUMERICAL SIMULATION OF THE CONTROL PROCESS

We have integrated the differential equations pertaining to a rigid beam and an elastic beam (taking into account the first oscillation mode $\Theta_1(t)$). The orders of these equations were 3 and 5, respectively. Integration was performed numerically by the Runge–Kutta method with automatic selection of the step, and accuracy control. Figure 8.5 shows the variables α and $\nu = \dot{\alpha}$ versus time, and Fig. 8.6

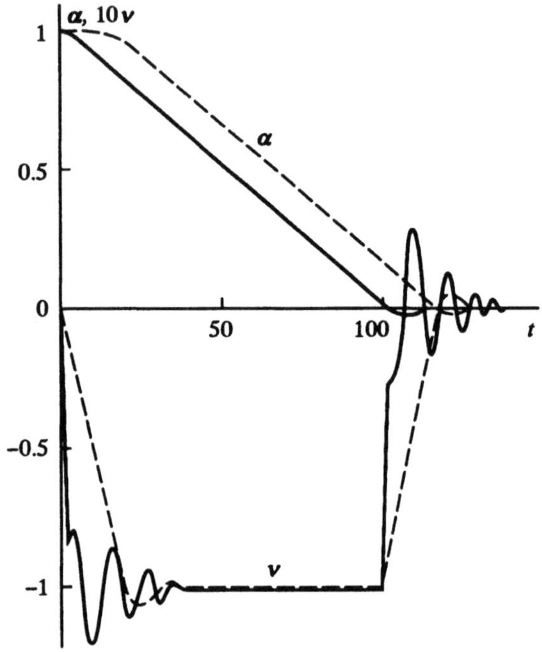

Fig. 8.5.

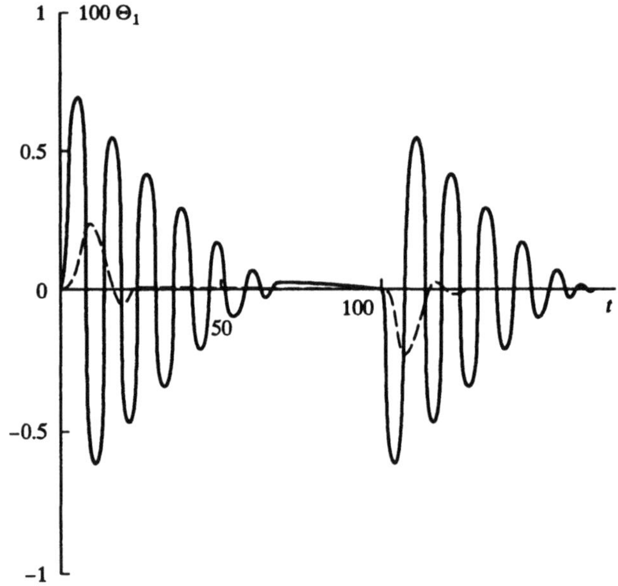

Fig. 8.6.

gives the dependence of the coefficient Θ_1 on time t for the system (20) with $k = 1$. Solid lines correspond to the bang-bang control with three switches according to (26) and (27); dashed lines show the response to the "smoothed" control (30). The parameters of the model have been taken as follows: $\alpha^0 = 1$, $\varepsilon = 0.1$, $|v| \leqslant 0.01$, $\varkappa = 100$, $\tau = 0.26$, $\xi = 0.1$, $\tau^* = 1$.

Let us draw some conclusions. In the system considered, elastic decaying oscillations are excited. The decay is due to the ohmic resistance of the armature winding. For the chosen model, the maximal amplitude of elastic displacements is given by $|u/l| \cong 0.02$, when using the bang-bang control. The current magnitude $j(t)$ at the switching instants is rather significant, which can complicate the implementation of this control. When the control voltage is engaged smoothly, we have the estimate $|u/l| \leqslant 4 \cdot 10^{-3}$. Having increased the process duration by 14%, the amplitude of the elastic displacements becomes five times less when the control is smooth. At the instant of terminating the control, the ratio α_1/α_2 satisfies the inequality $|\alpha_1(T_0)/\alpha_2(\Delta + t')| \leqslant 0.01$ (the angle α_1 corresponds to the bang-bang control, while α_2 corresponds to the "smoothed" control). The minimum-time control for the system (22) brings the phase point of the "rigid" model to the vicinity of the zero point with a high accuracy δ ($\delta \ll 0.01$) at the instant T.

Problems of the beam rotation control taking into account reduction gear elasticity are studied in [1, 22], see also Section 8.6.

8.6. Analysis of the Dynamic Equations and Construction of the Control for an Elastic Space Manipulator with Electromechanical Actuators

Studies of manipulation robot dynamics usually deal with models where links and other manipulator elements are assumed to be perfectly rigid bodies. When the links are relatively long, and the load carried is massive, the elastic compliance of the manipulator structure may affect the precision of the operations. The elasticity of the links should be taken into account in the analysis of the dynamics and the design of robot control systems. Disturbance factors such as the elastic compliance of reduction gears in actuators, coupling flanges, and other components should also be considered in practical applications.

The dynamics of manipulators with elastic links have been studied, for example, in [1, 13, 22, 34, 41, 42, 51, 67, 103, 112, 117, 142, 146]. The dynamics of an anthropomorphic loaded manipulator with elastic weightless links in the approximation of perfect actuators has been discussed in [13, 34, 112, 146]. Here, we analyse the equations of motion of an elastic manipulator with electromechanical actuators which contain d.c. motors with independent excitation and inertialess elastic reduction gears. The methods of singular perturbations is used to estimate the influence of the elastic compliance of the structure on the dynamics of the controlled motions of an elastic manipulator, corrected for the actual characteristics of the electromechanical actuators. Under various assumptions concerning the parameters of the perturbations, simplified equations of the transport motions of a robot are derived. The approach allows computing approximate open-loop (programmed) control voltages and correcting them on the basis of feedback (synthesis). The open-loop control problem, for bringing the manipulator approximately from the initial state of rest to a desired state, is solved.

8.6.1. DESCRIPTION OF THE MODEL

We consider an electromechanical model of an elastic two-link manipulator and study its controlled three-dimensional motions. The mechanical model of an elastic anthropomorphic manipulator considered here is defined by the following assumptions (see Fig. 8.7).

1. The manipulator $OD_0D_1D_2G$ is a two-link structure. The length of the first link is l_1 and the length of the second link is l_2; we assume that the link lengths are of a similar order of magnitude ($l_1 \sim l_2$), which is typical for such manipulators.

2. At a fixed point O, two cylindrical joints are combined: the axis $\mathbf{e}_0$ of the first joint is fixed, while the axis $\mathbf{e}_1$ of the second joint rotates and is perpendicular to $\mathbf{e}_0$. At the point D_2 there is the third (elbow) cylindrical joint; its axis $\mathbf{e}_2$ in the absence of elastic deformations of the rod (D_1D_2) is parallel to the axis $\mathbf{e}_1$.

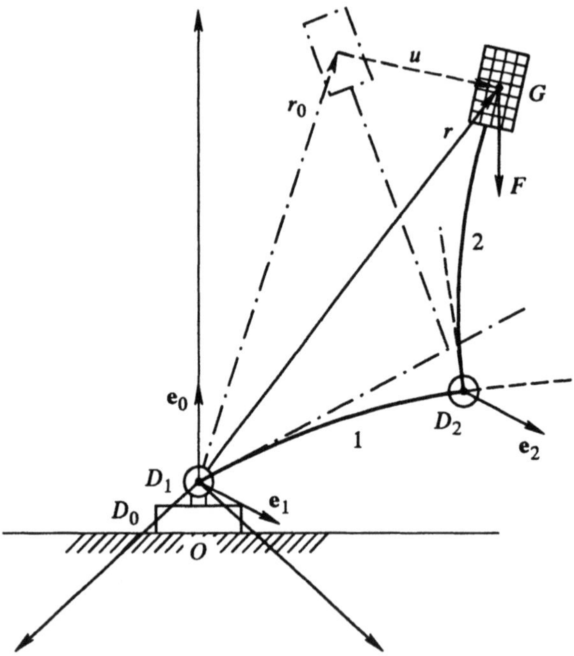

Fig. 8.7.

3. The load is considered to be a particle of mass m and is placed at the end of the second link (at the point G).

4. The links of the manipulator $(D_{0,1}D_2)$ and (D_2G) are homogeneous rectilinear elastic beams with a circular cross section, subjected to bending and tortional deformations.

5. Elastic displacements of the load and the manipulator during the motion are assumed to be small, compared with the link length.

6. The mass of the manipulator is small compared with the mass of the load.

7. The manipulator with the load is acted upon by the torques $M = (M_0, M_1, M_2)$ applied at the joints D_0, D_1, and D_2; it is also affected by the force F applied to the load and the reaction of the support at the point O.

We assume that at the manipulator joints D_0, D_1, D_2 there are d.c. electric motors with independent excitation. We introduce additional assumptions concerning the dynamic properties of the electric actuators.

8. The kinetic energy of the reduction gear and the motor is determined by a rapid rotation of the rotor (the armature of the motor) only, its moment of inertia is J_i, where $i = 0, 1, 2$ is the motor number.

9. The potential energy of the system of electric actuators is reduced to the potential energy of elastic deformations of the reduction gears; the elasticity coefficient of the reduction gear of the ith actuator is denoted by k_i $(i = 0, 1, 2)$.

Assumptions 8 and 9 are based on the fact that the gear ratios of reduction gears are usually large (about 10^3–10^4). The bulk of the kinetic energy of the motor and the joint therefore arises from the fast-rotating rotor. The kinetic energy of the motion of the centre of mass of the motor D_2 in the inertial space is usually negligibly small, compared with the kinetic energy of the load transported and the fast-rotating rotors of the actuators. Force loads at the output of each reduction gear are greater by several orders of magnitude than those at the input, because of the very large gear ratios. The potential energy is therefore mainly determined by elastic deformations of the gear mechanism. The energy of elastic deformations of the rotors of the electric motors can be neglected.

8.6.2. DERIVATION OF THE EQUATIONS OF MOTION

We will derive the equations of motion of a manipulator with a load by using the Lagrange principle [112, 127]. From the above assumptions, the kinetic energy of the system is

$$K = \frac{m}{2}\dot{\mathbf{r}}^2 + \frac{1}{2}\sum_{i=0}^{2} J_i\dot{\varphi}_i^2. \tag{8.6.1}$$

Here, $\mathbf{r}$ is the position vector of the load (the point G) and φ_i are the rotation angles of the rotors $(i = 0, 1, 2)$.

The potential energy of the system is

$$\Pi = \frac{1}{2}(C(\alpha)\mathbf{u}, \mathbf{u}) + \sum_{i=0}^{2} k_i(\varphi_i - n_i\alpha_i)^2, \quad \mathbf{r} = \mathbf{r}_0(\alpha) + \mathbf{u} \tag{8.6.2}$$

$$(\mathbf{u} = \mathbf{r} - \mathbf{r}_0).$$

Here, $\mathbf{u}$ is the elastic displacement of the load due to the elastic compliance of the structure (links) of the manipulator; n_i are the gear ratios ($i = 0, 1, 2$); the stiffness matrix of the elastic two-link structure is $C(\alpha)$; and the combination of the angles $\alpha = (\alpha_0, \alpha_1, \alpha_2)$ defines the configuration of a manipulator with perfectly rigid links and, by this, the position vector $\mathbf{r}_0(\alpha)$. We should note that the matrix $C(\alpha)$ has a singularity due to the loss of a degree of freedom by the system when the manipulator configuration is fully extended ($\alpha_2 = 0$) or fully folded ($\alpha_2 = \pi$). This matrix has the form

$$\begin{aligned}
&C = W^T C^* W, \qquad C^* = \varkappa(c^*_{ij}), \qquad i, j = 0, 1, 2;\\
&c^*_{00} = (4\xi + 3\lambda)(3\xi\zeta^{-1}\sin^2\lambda_2 + \xi + 3\lambda^2\cos^2\alpha_2 +\\
&\qquad + 3\lambda\cos\alpha_2 + 1)^{-1},\\
&c^*_{11} = 4(\xi + 3\lambda^2 + 3\lambda\cos\alpha_2 + \cos^2\alpha_2)/\sin\alpha_2, \qquad c^*_{22} = 4,\\
&c^*_{12} = c^*_{21} = -2(3\lambda + 2\cos\alpha_2)/\sin\alpha_2,\\
&c^*_{01} = c^*_{10} = c^*_{02} = c^*_{20} = 0;\\
&\varkappa = 3E_1I_1\left(l_1^3(4\xi + 3\lambda^2)\right)^{-1}, \qquad \lambda = l_2/l_1 \sim 1,\\
&\xi = E_1I_1\lambda^3/E_2I_2, \qquad \chi = \sigma\lambda/E_2I_2.
\end{aligned} \tag{8.6.3}$$

Here, $E_iI_i = \text{const}$ ($i = 1, 2$) are bending stiffnesses, σ is the torsional stiffness of the first link.

The auxiliary matrix W in (3) is ($\beta = \alpha_1 + \alpha_2$):

$$W = \begin{bmatrix} \cos\alpha_0 & \sin\alpha_0 & 0 \\ -\sin\alpha_0\cos\beta & \cos\alpha_0\cos\beta & \sin\beta \\ \sin\alpha_0\sin\beta & -\cos\alpha_0\sin\beta & \cos\beta \end{bmatrix}. \tag{8.6.4}$$

The expressions for the work of the external forces on virtual displacements are

$$\delta A = (\mathbf{F}, \delta\mathbf{r}), \qquad \delta A_i = \mu_i\delta\varphi_i \qquad (i = 0, 1, 2). \tag{8.6.5}$$

Here, **F** is the vector of the external forces applied to the load and μ_i are the torques of the electromagnetic forces defined below. We assume that the function **F** is known and satisfies the necessary conditions of smoothness and boundedness.

The equations of motion of the load and the armatures are defined in a standard fashion with expressions (1)–(5). By taking into account the kinematic relation (2) we can assume for the generalized coordinates the variables $\mathbf{r}, \alpha_1$ or $\mathbf{r}, \mathbf{u}$, or $\alpha_i, \mathbf{u}$ $(i = 0, 1, 2)$. The infinitesimal increments of these variables are connected by the expression

$$\delta\mathbf{r} = \sum_{i=0}^{2} \frac{\partial \mathbf{r}_0(\alpha)}{\partial \alpha_i} \delta\alpha_i + \delta\mathbf{u}. \tag{8.6.6}$$

If necessary, we can assume that the linear system (6) is solvable uniquely for $\delta\alpha_i$ in the given variation range of α, i.e. the matrix $(\partial\mathbf{r}_0/\partial\alpha)$ is nonsingular. For describing the motion, we will subsequenty use the variables $\mathbf{r}$, α, and φ; the Lagrange equations appear as

$$\begin{aligned}
&m\ddot{\mathbf{r}} = \mathbf{F} - C(\alpha)\big(\mathbf{r} - \mathbf{r}_0(\alpha)\big) \qquad (\mathbf{F} = \mathbf{F}(t, \mathbf{r}, \dot{\mathbf{r}}))\\
&\big(C(\alpha)(\mathbf{r} - \mathbf{r}_0(\alpha)), \partial\mathbf{r}_0(\alpha)/\partial\alpha_i\big) - \\
&\quad - \frac{1}{2}\big(\partial C(\alpha)/\partial\alpha_i(\mathbf{r} - \mathbf{r}_0(\alpha)), (\mathbf{r} - \mathbf{r}_0(\alpha))\big) - k_i n_i(\varphi_i - \nu_i\alpha_i) = 0,\\
&J_i\ddot{\varphi}_i + k_i(\varphi_i - n_i\alpha_i) = \mu_i \qquad (i = 0, 1, 2).
\end{aligned} \tag{8.6.7}$$

The first vector equation (7) describes the motion of a load; the next three equations in (7) are equations of the equilibrium of the elastic elements; the last three equations describe the rotations of the motor armatures.

The equations of the electric circuits for the armatures and the excitation windings determine the electromagnetic torgues μ_i [1, 34]:

$$L_i\dot{I}_i + R_iI_i + \Phi_i\dot{\varphi}_i = U_i, \qquad \mu_i = \Phi_iI_i \qquad (i = 0, 1, 2). \tag{8.6.8}$$

Here, I_i are currents; U_i are voltages; L_i are inductances; R_i are ohmic resistances; $\Phi_i = L_i^*I_i^*$ are magnetic fluxes generated by excitation currents I_i^*, L_i^* are cross-induction coefficients. The quantities Φ_i are assumed to be constant, i.e. the influence of the currents I_i in the windings of armatures upon the excitation currents I_i^* is neglected, and the values I_i^* are assumed to be steady. Mechanical losses in the rotation of the axles and joints are disregarded. If the coefficients of (7) and (8), the functions $\mathbf{F}$ and U_i, and the initial conditions for the variables $\mathbf{r}$, $\dot{\mathbf{r}}$, φ, and $\dot{\varphi}$ are given, the motion of the electromechanical system is defined completely.

The torques M_i at the outputs of the joints are

$$M_i = \partial\Pi/\partial\alpha_i = -(C(\alpha)(\mathbf{r} - \mathbf{r}_0(\alpha)), \partial\mathbf{r}_0(\alpha)/\partial\alpha_i) + \\ + \frac{1}{2}(\partial C(\alpha)/\partial\alpha_i(\mathbf{r} - \mathbf{r}_0(\alpha)), (\mathbf{r} - \mathbf{r}_0(\alpha))) \quad (8.6.9)$$

$$(i = 0, 1, 2).$$

Based on the equilibrium conditions (7) and the expressions (9), we can write alternative expressions for M_i:

$$M_i = k_i n_i(\varphi_i - n_i\alpha_i) \qquad (i = 0, 1, 2). \quad (8.6.10)$$

By using the physically obvious expression (10), Equations (7) for the rotation of the armatures can be rewritten as

$$J_i\ddot{\varphi} = \mu_i - M_i/n_i \qquad (i = 0, 1, 2). \quad (8.6.11)$$

Equations of motion in the form of (7)–(11) are convenient for studying various limit cases which often occur in practice: perfectly rigid reduction gears ($k_i \to \infty$), perfectly rigid links ($||C(\alpha)|| \to \infty$), inertialess electric circuit ($L_i \downarrow 0$), and others. The system of differential equations for the generalized coordinates $\mathbf{r}$ and $\mathbf{u}$ or α and $\mathbf{u}$ is singular, because $\mathbf{r}$, $\mathbf{u}$, and α are linked by finite (nondifferential) connections (2) and (7). The order of system (7)–(11) describing the electromechanical model of an anthropomorphic manipulator with inertialess links and reduction gears is 15; it contains three control actions: voltages U_i, $U_i(t) \in U^*$. It can readily be verified that the order of the system of differential equations for a manipulation robot with rigid links in the more general case of inertial links is smaller (equal to 9). This is due to the finite relation linking α and φ ($\varphi_i = n_i\alpha_i$), besides $\mathbf{u} = 0$ (see below).

The problem of the control of the motion of system (7) and (8) describing the transport operations of a manipulation robot is formulated as follows: at the instant $t = t_0$ the initial values of the variables $\mathbf{r}$, $\dot{\mathbf{r}}$, φ_i, $\dot{\varphi}_i$ and I_i ($i = 0, 1, 2$) are given; the objective is to bring the phase point of the system at the instant $t = T$ to a certain admissible set of states and to do so in an optimal fashion:

$$\begin{aligned}
&\mathbf{r}(t_0) = \mathbf{r}^0, \qquad \dot{\mathbf{r}}(t_0) = \mathbf{v}^0,\\
&\varphi_i(t_0) = \varphi_i^0, \qquad \dot{\varphi}_i(t_0) = \omega_i^0, \qquad I_i(t_0) = I_i^0;\\
&Q_i(t, \mathbf{r}, \dot{\mathbf{r}}, \varphi, \dot{\varphi}, I)\big|_{t=T} = 0, \qquad S[U] \longrightarrow \min;\\
&(i = 0, 1, 2;\ j = 0, 1, 2, \ldots,\ j^* \leqslant 15),\\
&\varphi = (\varphi_0, \varphi_1, \varphi_2), \qquad I = (I_0, I_1, I_2), \qquad U = (U_0, U_1, U_2).
\end{aligned} \tag{8.6.12}$$

Here, T is a time instant which is fixed or determined in the course of the solution of the optimal control problem, and S is a functional characterizing the control quality of the motion of the system (7) and (8). In particular, the functions Q_j can define a fixed point, for example, the point of rest, and S the operation speed. The controlled system can be subjected to phase constraints such as geometric, thermal, "power level", and others.

The system of equations of motion (7) is essentially nonlinear because of geometric nonlinearity. For r, the equation has a meaning similar to the equations for oscillatory systems with a controlled equilibrium position $\mathbf{r}_0(\alpha)$, which has been studied approximately by averaging methods in Chapter 5. Under a number of simplifying assumptions, the techniques of the asymptotic solution of optimal control problems of type (12) can be applied to the system (17) and (18), as developed in the previous chapters. Generally, numerical methods of optimal control should be used.

8.6.3. ANALYSIS OF THE EQUATIONS

We examine in this section several limit cases of the system (7) and (8) that are relevant to practical applications.

1) Suppose that the reduction gears of the actuators are close to absolute rigidity, i.e. the coefficients $k_i \to \infty$. In the limit, the variables φ_i and α_i are then linked by a simple finite relationship: $\varphi_i = n_i\alpha_i$ $(i = 0, 1, 2)$. If k_i are asymptotically large, i.e. $1/k_i$ are small parameters, then, formally, system (7) is singularly perturbed. Equations of motion in the first approximation with respect to the parameters $1/k_i$ can be reduced, by taking into account (7)–(11), to

$$\begin{aligned}
&m\ddot{\mathbf{r}} = \mathbf{F} - C(\alpha)(\mathbf{r} - \mathbf{r}_0(\alpha)), \qquad \alpha = \beta,\\
&J_i n_i^2 \ddot{\beta}_i = n_i\mu_i - M_i(\mathbf{r}, \alpha), \qquad \varphi_i = n_i\beta_i,\\
&L_i \dot{I}_i + R_i I_i + n_i \Phi_i \dot{\beta}_i = U_i \qquad (i = 0, 1, 2).
\end{aligned} \tag{8.6.13}$$

For the variables M_i, we take expressions (9) at $\alpha = \beta$. The following term of an asymptotic series is obtained from the finite equations (7) ($\beta_i = \varphi_i/n_i$):

$$\begin{aligned}\alpha_i = \beta_i + (k_i n_i)^{-1}\big[(C(\beta)(\mathbf{r} - \mathbf{r}_0(\beta)), \partial\mathbf{r}_0(\beta)/\partial\beta_i) - \\ - (1/2)\big(\partial C(\beta)/\partial\beta_i(\mathbf{r} - \mathbf{r}_0(\beta)), (\mathbf{r} - \mathbf{r}_0(\beta))\big)\big] + O(k_i^{-2}n_i^{-2}),\end{aligned} \tag{8.6.14}$$

and substituted into the first equation of (13) and expressions (9). For the system (7), regular expansions are thus constructed; the order of system (13) is maximal and equal to 15.

2) Suppose now that the manipulator links are very rigid; in the limit for perfectly rigid links as $\|C(\alpha)\| \to \infty$, the following relations then take place:

$$\begin{aligned}&\mathbf{u} = 0, \quad \mathbf{r} = \mathbf{r}_0(\alpha); \qquad \alpha_i = h_i(\mathbf{r}) \qquad (i = 0, 1, 2), \\ &C(\alpha)\mathbf{u} = -\big(\partial\mathbf{r}_0(\alpha)/\partial\alpha^T\big)^{-1}M = -(R)^T M.\end{aligned} \tag{8.6.15}$$

The equations of motion (7) and (8) in that case are written as

$$\begin{aligned}&m\ddot{\mathbf{r}} = \mathbf{F} + (R^T)^{-1}M, \quad \alpha_i = h_i(\mathbf{r}); \\ &L_i\dot{I}_i + R_iI_i + \Phi_i\dot{\varphi}_i = U_i, \quad n_ik_i(\varphi_i - n_i\alpha_i) = M_i \quad (i = 0, 1, 2).\end{aligned} \tag{8.6.16}$$

Consider the case of a high but finite stiffness of the manipulator links $C(\alpha) = \varepsilon^{-2}H(\alpha)$, where $\varepsilon < 1$, $\|H\| \sim 1$ ($k_i \sim 1$). Assume also that $M_i \sim 1$. From relations (15), we obtain $\mathbf{u} = \varepsilon^2\mathbf{v}$, $|\mathbf{v}| \sim 1$. Substituting the resultant representations for the matrix C and the vector $\mathbf{u}$ into the equations of motion (7), we obtain

$$m\ddot{\mathbf{r}} = \mathbf{F}(t, \mathbf{r}, \dot{\mathbf{r}}) - H(\alpha)\mathbf{v}, \tag{8.6.17}$$

$$M + R^T(\alpha)H(\alpha)\mathbf{v} = (1/2)\varepsilon^2\partial\big(H(\alpha)\mathbf{v}, \mathbf{v}\big)/\partial\alpha. \tag{8.6.18}$$

We can also write down the kinematic relation which follows from expressions (2) and (15):

$$\alpha = h(\mathbf{r} - \varepsilon^2\mathbf{v}). \tag{8.6.19}$$

In order to obtain a closed set of equations for the variables $\mathbf{r}$, φ, and I, we eliminate from system (17) the vectors $\mathbf{v}$ and α by using the relations (18) and (19). Solving (18) for $\mathbf{v}$ to a first approximation with respect to ε^2, we obtain

$$\begin{aligned} &\mathbf{v} = \mathbf{v}_0 + O(\varepsilon^2), \qquad \mathbf{v}_0 = -\left[R^T(\alpha)H(\alpha)\right]^{-1}M = \mathbf{v}_0(\alpha, M), \\ &M = (M_0, M_1, M_2), \qquad M_i = n_i k_i(\varphi_i - n_i\alpha_i). \end{aligned} \tag{8.6.20}$$

Let us substitute the vector $\mathbf{v}_0$ from (20) into the right-hand side of (18); to the next approximation in terms of ε^2 we obtain

$$\begin{aligned} &\mathbf{v} = \mathbf{v}_0 + \varepsilon^2\mathbf{v}_1 + O(\varepsilon^4), \qquad \mathbf{v}_0 = \mathbf{v}_0(\alpha, M), \\ &\mathbf{v}_1 = \mathbf{v}_1(\alpha, M) = (1/2)\left[R^T H\right]^{-1}\partial(H\mathbf{v}_0, \mathbf{v}_0)/\partial\alpha. \end{aligned} \tag{8.6.21}$$

Substituting relations (20) and (21) into (17) and (19), and considering only the terms through $O(\varepsilon^2)$, we can obtain the desired approximate equations with the error $O(\varepsilon^4)$:

$$m\ddot{\mathbf{r}} = \mathbf{F}(t, \mathbf{r}, \dot{\mathbf{r}}) - H(\alpha)(\mathbf{v}_0 + \varepsilon^2\mathbf{v}_1), \tag{8.6.22}$$

$$\alpha = h\big(\mathbf{r} - \varepsilon^2\mathbf{v}_0(\alpha, M)\big). \tag{8.6.23}$$

Relations (23), where the vector $\mathbf{v}_0$ is defined by the equality (20) is a vector equation for defining the joint angles $\alpha(\mathbf{r}, \varphi)$ at each time instant t. This equation can be solved by the method of successive approximations, according to the following scheme, ($n = 1, 2, \ldots$):

$$\alpha^{(0)} = h(\mathbf{r}), \qquad \alpha^{(n)} = h\big(\mathbf{r} - \varepsilon^2\mathbf{v}_0(\alpha^{(n-1)}, M)\big). \tag{8.6.24}$$

The superscripts specify the iteration number. The derivative, with respect to α, of the right-hand side of (24) is a quantity of $O(\varepsilon^2)$. Therefore, iterations (24) converge rapidly to a unique solution $\alpha = \alpha^*(\mathbf{r}, \varphi, \varepsilon^2)$ of (20) and $|\alpha^* - \alpha^{(n)}| = O(\varepsilon^{2n+2})$. For obtaining α with the desired error of $O(\varepsilon^4)$, it is sufficient therefore to perform a single iteration

$$\alpha = h\big(\mathbf{r} - \varepsilon^2\mathbf{v}_0(h(\mathbf{r}), M)\big) + O(\varepsilon^4). \tag{8.6.25}$$

If the manipulator links are rigid, regular expansions are again constructed for the system (7); the order of the system (16) and (17) is also maximal (equal to 15).

3) System (7) is, in fact, singularly perturbed if the quantities $\|C\|$, k_i are asymptotically large. In the limit, we obtain a system of a lower, ninth order [1, 17, 22]:

$$\begin{aligned}
&m\ddot{\mathbf{r}}_0(\alpha) = \mathbf{F}(t, \mathbf{r}_0(\alpha), \dot{\mathbf{r}}_0(\alpha)) + R^T(\alpha)^{-1}M, \qquad t \in \Theta \subset [t_0, T],\\
&\varphi_i = h_i\alpha_i, \qquad M_i = n_i\mu_i - J_i n_i^2\ddot{\alpha}_i, \qquad \mu_i = \Phi_i I_i,\\
&L_i\dot{I}_i + R_i I_i + n_i\Phi_i\dot{\alpha}_i = U_i \qquad (i = 0, 1, 2).
\end{aligned} \tag{8.6.26}$$

For more exact expansions, boundary layers at the beginning and the end of the control process and at the points of bang-bang change of controls $U_i(t)$ must be included into the equation.

4) The system (7) and (8) can be perturbed singularly with respect to the parameters L_i in the equations for the variables I_i. In the limit at $L_i = 0$ its order is reduced by 3.

The expressions for the torques μ_i in that case are given by

$$\mu_i = R_i^{-1}\Phi_i(U_i - \Phi_i\dot{\varphi}_i), \qquad U(t) \in U^* \quad (i = 0, 1, 2) \tag{8.6.27}$$

and are substituted in Equations (7) and (11). With this approach, the initial conditions for the currents I_i are disregarded because, in practice, transient processes in the electric circuits of armatures are rapid. Approximation (27) is often used when studying the motions of manipulation robots with electromechanical actuators.

8.6.4. APPROXIMATE SOLUTION OF THE LOAD TRANSFERRING PROBLEM

Consider the problem of the transport of a load from an initial position to a desired position at rest. The control of the system (7) and (8) is a process that can be subdivided into two stages: 1) motion to a small neighbourhood of the given state of rest; and 2) the adjustment of the load position within the small neighbourhood of the desired position. The initial and terminal states of rest, according to (12), are

$$\begin{aligned}
&\mathbf{r}\big|_{0,T} = \mathbf{r}^{0,T}, \qquad \dot{\mathbf{r}}\big|_{0,T} = 0,\\
&\varphi_i\big|_{0,T} = \varphi_i^{0,T}, \qquad \dot{\varphi}_i\big|_{0,T} = 0,\\
&I_i\big|_{0,T} = 0 \qquad (i = 0, 1, 2,\ t \in [0, T]).
\end{aligned} \tag{8.6.28}$$

In addition, we assume that there are no external forces ($\mathbf{F} = 0$) and that the control is such that the motion of the load G is quasistationary, which results in small elastic displacements $\mathbf{u}$. This assumption is satisfied if the controls U_i are sufficiently small, i.e. $U_i = \varepsilon V_i$, where ε is a small parameter and $V_i(t) \sim 1$, where V_i are piecewise smooth functions. The time T of the control process is asymptotically large: $T \sim \varepsilon^{-1}$. Accurate to within $u^2 \sim \varepsilon^2$, system (7), taking into account (27), can be rewritten as

$$\begin{aligned}
&m\ddot{\mathbf{r}} = \left(R^T(\alpha)\right)^{-1} M, \qquad M_i = k_i n_i(\varphi_i - \nu_i \alpha_i) \quad (i = 0, 1, 2), \\
&-k_i n_i(\varphi_i - n_i \alpha_i) = \left(C(\alpha)(\mathbf{r} - \mathbf{r}_0(\alpha)), \partial \mathbf{r}_0(\alpha)/\partial \alpha_i\right), \\
&J_i \ddot{\varphi}_i + k_i(\varphi_i - n_i \alpha_i) = \mu_i, \qquad \mu_i = R_i^{-1} \Phi_i(\varepsilon V_i - \Phi_i \dot{\varphi}_i).
\end{aligned} \tag{8.6.29}$$

Since in a real situation, manipulator links and reduction gears are stiff, small and smooth (piecewise smooth) control actions will produce small elastic deviations of the load from the path described by the simplified system (26) and (27); these deviations will be of the same order of magnitude as the control action. Under these assumptions, the motion of system (28) and (29) to a first approximation is described by the following equations and constraints:

$$\begin{aligned}
&\sum_{j=0}^{2} m(\partial \mathbf{r}_0/\partial \alpha_i)^T (\partial \mathbf{r}_0/\partial \alpha_j)\ddot{\alpha}_j + J_i n_i^2 \ddot{\alpha}_i \\
&\qquad = \eta_i R_i^{-1} \Phi_i(\varepsilon V_i - n_i \Phi_i \dot{\alpha}_i), \qquad t \in [0, T]; \\
&\alpha_i\big|_{0,T} = \alpha_i^{0,T}, \quad \dot{\alpha}_i\big|_{0,T} = 0, \quad \left|V_i(t)\right| \leqslant V_i^0 \quad (i = 0, 1, 2).
\end{aligned} \tag{8.6.30}$$

The solution of the control problem (30) that is approximate with respect to ε and close to the time-optimal one can be obtained by assuming the control functions V_i to be constant. With the error $O(\varepsilon)$ with respect to the path and $O(\ln \varepsilon^{-1})$ with respect to the functional (the time) T, $T = O(\varepsilon^{-1})$, the controls V_i can be taken in the following simple form:

$$\begin{aligned}
&V_i(t) = V_i^* = \text{const}, \qquad \left|V_i^*\right| \leqslant V_i^0, \quad t \in [0, T], \quad T = \max_i T_i, \\
&T_i = \varepsilon^{-1}\left|\alpha_i^T - \alpha_i^0\right|\left(n_i \Phi_i / V_i^0\right), \qquad V_i^* = n_i \Phi_i\left(\alpha_i^T - \alpha_i^0\right)/(\varepsilon T), \\
&V_i(t) = 0, \qquad t \notin [0, T] \quad (i = 0, 1, 2).
\end{aligned} \tag{8.6.31}$$

In the time $T \sim 1/\varepsilon$ the phase point of system (30) will come into the ε-neighbourhood of the given value. The phase point of the original system (7) and (8) will also be, in this case, in the ε-neighbourhood of a desired terminal point, according to (28).

8.6.5. NUMERICAL SIMULATION OF THE DYNAMICS OF AN ANTHROPOMORPHIC MANIPULATOR WITH AN ELECTRIC ACTUATOR, AND BASIC CONCLUSIONS

Having used the Runge–Kutta method with an automatic choice of the step and accuracy control, we integrated the twelfth order system (13) under the initial conditions (28) for $\mathbf{F} = 0$, $L_i = 0$, $k_i \to \infty$. To this system, we applied the control of the form (31). For the model in question, the following values of dimensionless parameters were chosen:

$$\varkappa = 200, \qquad \xi = 1.5, \qquad \zeta = 1, \qquad l_1 = l_2 = 10,$$

$$J_i n_i^2 = 3000, \qquad i = 0, 1, 2; \qquad m = 1000,$$

$$\Phi_j^2 n_j^2 R_j^{-1} = 140, \quad j = 0, 1; \qquad \Phi_2^2 n_2^2 R_2^{-1} = 70.$$

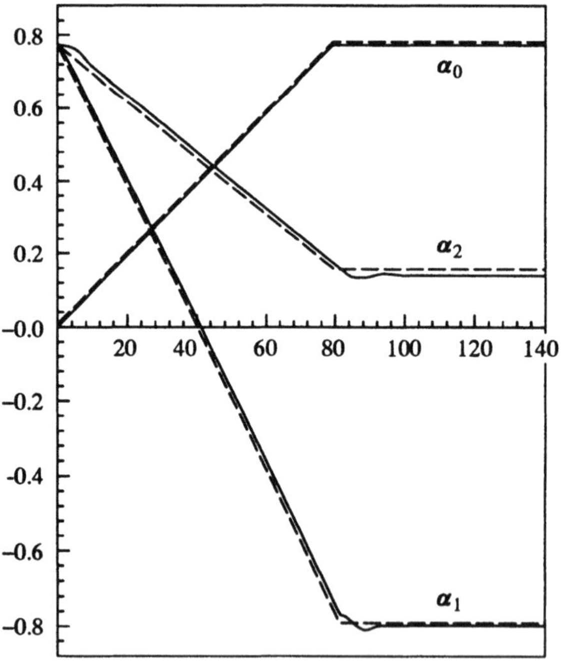

Fig. 8.8.

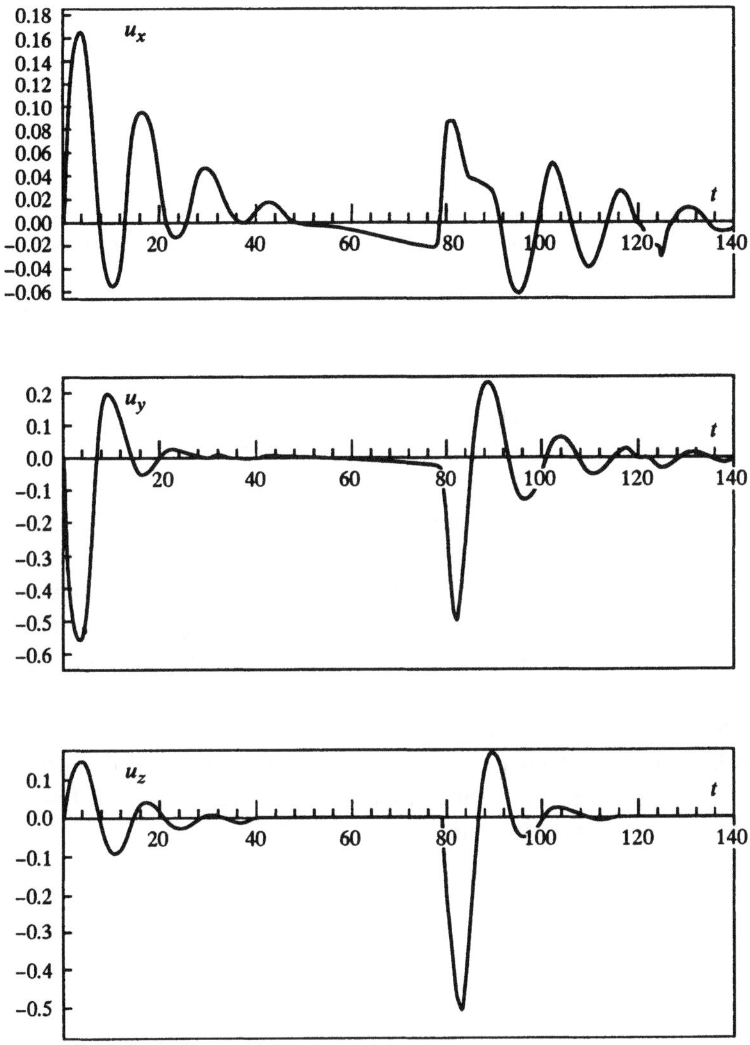

Fig. 8.9.

which corresponded to a hypothetical space-based manipulator. For initial and terminal conditions we have taken the values of joint angles corresponding to a significant change of the manipulator configuration

$$\alpha_0(0) = 0, \quad \alpha_1(0) = \pi/4, \quad \alpha_2(0) = \pi/4;$$

$$\alpha_0(T) = \pi/4, \quad \alpha_1(T) = -\pi/4, \quad \alpha_2(T) = 0.16.$$

The corresponding velocity values have been assumed to be equal to zero and the maximal admissible voltages have been taken as follows ($v_i^* = U_i^*/\Phi_i n_i$): $v_0^* = 10^{-2}$, $v_1^* = 10^{-2}$, $v_2^* = 5 \cdot 10^{-3}$.

Figure 8.8 shows the time-dependence of the angles α_i, $i = 0, 1, 2$ having used the given control (31). Figure 8.9 gives the values of the elastic deviations u_i, $i = x, y, z$, where $u_x = (x - x_0(\alpha))$, $u_y = (y - y_0(\alpha))$, and $u_z = (z - z_0(\alpha))$. During the motion, the elastic bend appears in system (13), with the amplitude $|u| < 0.6$. As is seen from Fig. 8.8, the system comes to the ε-neigbourhood (here $\varepsilon = 0.7$) of the terminal phase point in the time T; the positioning error being $|\Delta r/r| \leqslant 0.5\%$. The derived accuracy estimates for the proposed control (31), as well as for the elastic oscillations which appear, are in good agreement with the values obtained as a result of the numerical simulation of the dynamics of the anthropomorphic manipulator with elastic links. For the system with the given parameters, elastic deviations correspond to quasistatic bending of the manipulator links. The increase of the transported load mass m results in increasing the duration of transient electromechanical processes in the manipulator actuators, when control constraints remain unchanged. The asymptotic estimates obtained above fail to be valid if the time constants of the transient processes are comparable with the total time of the load transfer.

The further correction of the phase state of the system can be conducted on the basis of regular methods of the analytical design of regulators, or by using smoothed controls, as in Section 8.5. With this approach, an optimal regulator can be constructed for a fixed terminal configuration of the manipulator, i.e. $\alpha = \alpha^T$.

The detailed study of this control problem should be based on mathematical simulation. The foregoing methods can be used for estimates and initial approximations. A strategy based on a separate control by different actuators for fixed angles in the remaining joints, is another alternative.

References

1. Akulenko, L. D.: *Asymptotic Methods in Optimal Control*, Nauka, Moscow, 1987 (in Russian).
2. Akulenko, L. D.: 'Investigation of Certain Optimal Systems by the Averaging Method', *J. of Appl. Math. and Mech.* **38**, No. 3 (1974), 392–401.
3. Akulenko, L. D.: 'Asymptotic Solution of Certain Problems of Time-Optimal Type', *J. of Appl. Math. and Mech.* **39**, No. 4 (1975), 565–578.
4. Akulenko, L. D.: 'Approximate Solution of Nonlinear Problems of Optimal Control of Oscillatory Processes Using the Method of Canonical Separation of Motions', *J. of Appl. Math. and Mech.* **40**, No. 6 (1976), 970–978.
5. Akulenko, L. D.: 'Asymptotic Solution of Certain Nonlinear Problems of Time-Optimal Response', *J. of Appl. Math. and Mech.* **42**, No. 1 (1978), 75–87.
6. Akulenko, L. D.: 'Approximate Synthesis of Optimal Response Controls in Nearly Spherically Symmetric Problems', *Mech. of Solids* **13**, No. 2 (1978), 1.
7. Akulenko, L. D.: 'Time-Optimal Stabilization of a Perturbed System with Invariant Norm', *J. of Appl. Math and Mech.* **42**, No. 4 (1978), 646–658.
8. Akulenko, L. D.: 'Control of Motion of a Nonlinear Vibratory System by Displacement of the Equilibrium Point', *Mech. of Solids* **13**, No. 4 (1978), 1–9.
9. Akulenko, L. D.: 'Optimal Synthesis in Some Problems of Terminal Control', *J. of Appl. Math. and Mech.* **42**, No. 6 (1978).
10. Akulenko, L. D.: 'Asymptotic Solution of Two-Point Boundary Value Problems', *J. of Appl. Math. and Mech.* **44**, No. 4 (1980), 446.
11. Akulenko, L. D.: 'Approximate Synthesis of Optimal Control of Motion with Respect to Part of the Variables', *Mech. of Solids* **15**, No. 5 (1980), 1–10.
12. Akulenko, L. D.: 'Applications of the Methods of Averaging and Successive Approximations for Studying Nonlinear Oscillations', *J. of Appl. Math. and Mech.* **45**, No. 5 (1981), 575.
13. Akulenko, L. D.: 'Probleme der Optimalen Steuerung von Manipulierenden Robotern unter Beachtung der Elastizität der Konstruktionen', *Wissenschaftliche Berichte der Technischen Hochschule Leipzig* (Leipzig) H. 8 (1982), 4–8.
14. Akulenko, L. D.: 'The Approximate Synthesis of Perturbed Non-Vibrating Systems with One Degree of Freedom', *J. of Appl. Math and Mech.* **49**, No. 5 (1985), 561–567.
15. Akulenko, L. D.: 'Averaging in a Quasilinear System with a Strongly Varying Frequency', *J. of Appl. Math and Mech.* **51**, No. 2 (1987), 189–196.
16. Akulenko, L. D.: 'Constructive Control of the Motion of Oscillating Systems with Discrete and Distributed Parameters', *J. of Appl. Math. and Mech.* **53**, No. 4 (1989), 464–573.
17. Akulenko, L., Bojadzhiev, G., and Lilov, L.: 'Kinematic Structures of a Hand and Dynamics of Orientational Motions for Electromechanical Manipulation Robot', *Theor. and Appl. Mech.* **21**, No. 3 (1990), 24–36 (in Russian).
18. Akulenko, L. D. and Bolotnik, N. N.: 'Synthesis of Optimal Control of Transport Motions of Manipulation Robots', *Mech. of Solids* **21**, No. 4 (1986), 18–26.
19. Akulenko, L. D., Bolotnik, N. N. et. al.: *Dynamik und Optimale Schwingungstilgung in Unwuchterregten Schwingsystemen*, Report R-03/80, ZIMM AdW DDR, Berlin, 1980.
20. Akulenko, L. D. and Chernousko, F. L.: 'The Method of Averaging in Problems of Optimal Control', *Zh. Vychisl. Mat. Mat. Fiz.* **15**, No. 4 (1985) (English translation).
21. Akulenko, L. D. and Leschenko, D. D.: 'Some Problems of the Motion of a Solid with a Moving Mass', *Mech. of Solids* **13**, No. 5 (1978), 24–28.

22. Akulenko, L. D. and Mikhailov, S. A.: 'Analysis of the Dynamic Equations of an Elastic Manipulator with Electromechanical Actuators', *Mech. of Solids* **23**, No. 1 (1988), 72–78.
23. Akulenko, L. D. and Roshchin, Yu. R.: 'Optimal Control of 'Pendulum-Type' System Motion by Support Point Displacement', *Mech. of Solids* **11**, No. 1 (1976), 12–18.
24. Akulenko, L. D. and Zaremba, A. T.: 'Synthesis of Control of a Pendulum of Variable Length', *Mech. of Solids* **16**, No. 2 (1981), 123–129.
25. Al'brekht, E. G.: 'On the Convergence of Quasilinear Objects', *J. of Appl. Math. and Mech.* **34**, No. 4 (1970), 551–559.
26. Ardema, M. D. (ed.): *Singular Perturbations in Systems and Control*, Springer-Verlag, Wien, New York, 1983.
27. Ardema, M. D.: 'Singular Perturbations and Asymptotic Expansions in Nonlinear Optimal Control', *Lect. Notes Contr. and Inf. Sci.* **95** (1987), 3–18.
28. Athans, M. and Falb, P. L.: *Optimal Control*, Mc Graw-Hill Book Co., New York, 1966.
29. Bahls, D. L. and Paris, S. W.: 'A Simplified Method for Obtaining Near Minimum Time Low Thrust Transfers', *Advances in the Astronautical Sciences: Astrodynamics* (Univ. San Diego, 1980) **40** (1979), 749–764.
30. Baldwin, J. F. and Williams, J. H.: 'The Use of a Method of Perturbations in the Synthesis of Closed-Loop Optimal Control Laws for Nonlinear Systems', *Automatica* **5**, No. 3, 1357–1367.
31. Bell, D. J. and Jacobson, D. H.: *Singular Optimal Control Problems*, Academic Press, New York, 1975.
32. Bellman, R.: *Dynamic Programming*, Princeton Univ. Press, Princeton, NJ, 1957.
33. Bensoussan, A.: 'Regular Perturbations in Optimal Control', *Cours. and Lect. CISM Int. Cent. Mech. Sci.* No. 280 (1983), 169–183.
34. Berbyuk, V. E.: *Dynamics and Optimization of Robotic Systems*, Naukova Dumka, Kiev, 1989 (in Russian).
35. Besjes, J. G.: 'On the Asymptotic Methods for Non-Linear Differential Equations', *J. de Mecanique* **8**, No. 3 (1969), 357–372.
36. Blaquière, A.: *Nonlinear System Analysis*, Academic Press, New York, 1966.
37. Bobrow, J. E., Dubowsky, S., and Gibson, J. S.: 'Time-Optimal Control of Robotic Manipulators Along Specified Paths', *The Intern. J. of Robotic Research* **4**, No. 3 (1985), 3–17.
38. Bogoliubov, N. N. and Mitropolskii, Yu. A.: *Asymptotic Methods in the Theory of Nonlinear Oscillations*, Gordon and Breach, New York, 1961.
39. Bohr, H.: *Almost Periodic Functions*, Chelsea Publ. Co., New York, 1947.
40. Boltyanskii, V. G.: *Mathematical Methods of Optimal Control*, Nauka, Moscow, 1969 (in Russian).
41. Book, W. J.: 'Analysis of Massless Elastic Chains with Servo Controlled Joints', *J. Dynamic Syst. Measur. and Control* **101**, No. 3 (1979), 187–192.
42. Book, W. J., Maizza-Neto, W., and Whitney, D. E.: 'Feedback Control of Two Beam, Two Joint Systems with Distributed Flexibility', *J. of Dynamic Syst., Measur. and Control* **97**, No. 4, 424–431.
43. Bryson, A. E. and Ho, Y. C.: *Applied Optimal Control*, John Wiley and Sons, New York, 1975.
44. Campbell, J. E. M. A.: *Introductory Treatise on Lie's Theory of Finite Continuous Transformation Groups*, Clarendon Press, Oxford, 1903.
45. Cesari, L.: *Asymptotic Behavior and Stability Problems in Ordinary Differential Equations*, Springer-Verlag, Berlin, 1959.
46. Chernousko, F. L., Akulenko, L. D., and Sokolov B. N.: *Control of Oscillations*, Nauka, Moscow, 1980 (in Russian).
47. Coddington, E. A. and Levinson, N.: *Theory of Ordinary Differential Equations*, McGraw-Hill, New York, 1955.
48. Collatz, L.: *The Numerical Treatment of Differential Equations*, Springer-Verlag, Berlin, 1960.
49. Collatz, L.: *Funktionalanalysis und Numerische Mathematik*, Springer-Verlag, Berlin, 1964.
50. Courant, R. and Hilbert, D.: *Methoden der Mathematischen Physik*, Bd. I, Springer-Verlag, Berlin, 1968.
51. Curtain, R. F. and Glover, K.: 'Robust Stabilization of Infinite Dimensional Systems by Finite Dimensional Controllers', *Systems and Control Letters* **7**, No. 1 (1986), 41–47.
52. Curtain, R. F. and Pritchard, A. J.: *Infinite Dimensional Systems Theory*, Springer-Verlag, Berlin, 1978 (LNCIS, Vol. 8).

53. Deprit, A.: 'Canonical Transformations Depending on a Small Parameter', *Celest. Mech.* **1**, No. 1 (1969), 12–30.
54. Dontchev, A. L.: *Perturbations, Approximations and Sensitivity Analysis of Optimal Control Systems*, Springer-Verlag, Berlin, 1983.
55. Dreyfus, S. E.: *Dynamic Programming and Calculus of Variations*, Academic Press, New York, 1965.
56. Eckel, K. G. and Vinh, N. X.: 'Optimal Switching Conditions for Minimum Fuel Fixed Time Transfer between Non-Coplanar Elliptical Orbits', *Acta Astronautica* **11**, No. 10 (1984), 621–631.
57. Edelbaum, T. N.: 'Optimal Low-Thrust Transfer between Circular and Elliptic Orbits', in *The 4th US National Congress of Appl. Mech.*, ASME, New York, 1962, pp. 137–147.
58. Edelbaum, T. N.: 'Optimal Low-Thrust Rendezvous and Station Keeping', *AIAA J.* **2**, No. 7 (1964).
59. Edelbaum, T. N., Sackott, L. L., and Malchow, H. L.: 'Optimal Low Thrust Geocentric Transfer', AIAA paper, No. 1074, 1973, p. 19.
60. Evtushenko, Yu. G.: 'Approximate Calculation of Optimal Control Problems', *J. of Appl. Math. and Mech.* **34**, No. 1 (1970), 86–94.
61. Falb, P. L. and Jong, J. L.: *Some Successive Approximation Methods on Control and Oscillation Theory*, Academic Press, New York, 1969.
62. Fedoryenko, R. P.: *Approximate Solution of Optimal Control Problems*, Nauka, Moscow, 1978 (in Russian).
63. Fiacco, A. V. and McCormick, G. P.: *Nonlinear Programming: Sequential Unconstrained Minimization Techniques*, John Wiley and Sons, New York, 1968.
64. Filippov, A. F.: *Differential Equations with Discontinuous Right-Hand Side*, Nauka, Moscow, 1985 (in Russian).
65. Fleming, W. H. and Rishel, R. W.: *Deterministic and Stochatic Optimal Control*, Springer-Verlag, Berlin, 1975.
66. Formal'sky, A. M. and Osipov, S. N.: 'On the Problem of the Time-Optimal Manipulator Arm Turning', *IEEE Trans. on Automatic Control* **35**, No. 6 (1990), pp. 714–719.
67. Fukuda, T.: 'Control of Flexible Robotic Arms', *Bull. JSME* **29**, No. 250 (1986), 1269–1273.
68. Gaitsgory, V. G.: *Systems with Divided Motions*, Nauka, Moscow, 1991 (in Russian).
69. Giacaglia, G. E. O.: *Perturbation Methods in Non-Linear Systems*, Springer-Verlag, New York, 1972.
70. Golubev, Yu. F. and Demidov, V. N.: Optimal Control Law upon Stopping of Rotation', *Mech. of Solids*, **21**, No. 2 (1986), 15–21.
71. Grishin, S. D., Zakharov, Yu. A., and Odelyevski, V. K.: *Design of Space Vehicles with Low-Thrust Engines*, Mashinostroyeniye, Moscow, 1990 (in Russian).
72. Grodzovskii, G. L., Ivanov, Yu. N., and Tokarev, V. V.: *Mechanics of Space Flight: Problems of Optimization*, Nauka, Moscow, 1975 (in Russian).
73. Hagedorn, P.: 'Canonical Transformation in the Optimal Control of Mechanical Systems', *Int. J. of Non-Linear Mech.* **13**, No. 2 (1978), 103–116.
74. Hale, J. K.: *Oscillations in Nonlinear Systems*, McGraw-Hill Book Co., New York, 1963.
75. Hartman, P.: *Ordinary Differential Equations*, Wiley, New York, 1964.
76. Hay, D. H.: 'An Energy Approach for Orbital Transfers', *J. Guidance* **9**, No. 1 (1986), 23–26.
77. Hayashi, Ch.: *Nonlinear Oscillations in Physical Systems*, McGraw-Hill Book Co., New York, 1964.
78. Henrard, J.: 'On a Perturbation Theory Using Lie Transforms', *Celest. Mech.* No. **3** (1970), 120.
79. Henrard, J. and Roels, J.: 'Equivalence for Lie Transforms', *Celest. Mech.* No. 10 (1974), 512.
80. Hori, G.: 'Theory of General Perturbations with Unspecified Canonical Variables', *Publ. Astron. Soc. Jap.* No. 18 (1966), 296.
81. Hori, G.: 'Theory of General Perturbations for Noncanonical Systems', *J. Japan. Astron. Soc.* **23** (1971), 567–587.
82. Hunt, L. R., Su, R., and Meyer, G.: 'Global Transformation of Nonlinear Systems', *IEEE Transactions on Automatic Control* **28**, No. 1 (1983), 24–31.
83. Isaacs, R.: *Differential Games*, Wiley, New York, 1965.
84. Ishlinskii, A. Yu.: *Orientation, Gyroscopes and Inertial Navigation*, Nauka, Moscow, 1976 (in Russian).
85. Jacobson, D. H.: *Extensions of Linear-Quadratic Control Theory*, Springer-Verlag, Berlin, 1980.

86. Jacobson, R. A. and Powers, W. F.: 'Asymptotic Solution to the Problem of Optimal Low-Thrust Energy Increase', *AIAA J.* **10**, No. 12 (1972), 1679–1680.
87. Janke, E., Emde, F., und Lösch, F.: *Tafelnherer Funktionen*, B.G. Teubner Verlagsgesellschaft, Stuttgart, 1960.
88. Jasper, T. P.: 'Low-Thrust Trajectory Analysis for the Geosynchronous Mission', AIAA paper, No. 73-1072, 1973, pp. 1–8.
89. Javid, S. M.: 'The Time Optimal Control of a Class of Nonlinear Singularly Perturbed Systems', *Int J. Control* **28**, No. 6 (1978), 831–836.
90. Junkins, J. L. (ed.): *Mechanics and Control of Large Flexible Structures*, AIAA, Washington, 1990.
91. Junkins, J. L. and Turner, J. D.: *Optimal Spacecraft Rotational Maneuvers*, Elsevier, Amsterdam, 1986.
92. Kalinin, A. I.: Optimization of Quasilinear Control Systems', *Zh. Vychisl. Mat. Mat. Fiz.* **28**, No. 3 (1988) (English translation).
93. Kalinin, A. I.: 'An Algorithm for the Asymptotic Solution of a Singularly Perturbed Linear Time-Optimal Control Problem', *J. of Appl. Math. and Mech.* **53**, No. 6 (1989), 695–703.
94. Kamel, A.: 'A Perturbations Methods in the Theory of Nonlinear Oscillations', *Celest. Mech.* **3**, No. 1 (1970), 90–106.
95. Kamel, A. A. and Hassan, S. D.: 'A Perturbation Treatment for Optimal Slightly Nonlinear Systems with Linear Control and Quadratic Criteria', *J. Optimiz. Theory and Appl.* **11**, No. 4 (1973), 386–404.
96. Kao, Y. K. and Bankoff, S. G.: 'Singular Perturbation Analysis of Free-Time Optimal Control Problems', *Int. J. Syst. Sci.* **5**, No. 4 (1974), 335–350.
97. Kaplan, M.: *Modern Spacecraft Dynamics and Control*, Wiley, New York, 1976.
98. Kauderer, H.: *Nichtlineare Mechanik*, Springer-Verlag, Berlin, 1958.
99. Kelley, H. J. and Edelbaum, T. N., 'Energy Climbs, Energy Turns and Asymptotic Expansions', *J. Aircraft* **7**, No. 1 (1970), 93–95.
100. Kirchgraber, U. und Stiefel, E.: *Methoden der Analytischen Strungsrechung und ihre Anwedungen*, Teubner, Stuttgart, 1978.
101. Kiselyov, Yu. N.: *Linear Perturbed Theory of Minimum-Time Motion*, Moscow State University Publishing House, Moscow, 1986 (in Russian).
102. Kokotovic, P. V.: 'Application of Singular Perturbations Techniques to Control Problems', *SIAM Review* **26**, No. 4 (1984), 501–550.
103. Kostin, G. V.: 'Dynamics of Controlled Rotation of a Loaded Elastic Link in a Manipulation System with Electromechanical Drive', *Soviet J. of Computer and Systems Sci.* No. 3 (1990), 120–128.
104. Kovaleva, A. S.: *Optimal Control in Mechanical Systems: Methods and Applications*, Nauka, Moscow, 1990 (in Russian).
105. Krasovskii, N. N. and Subbotin, A. I.: *Game-Theoretical Control Problems*, Springer-Verlag, New York, 1987.
106. Kwakernaak, H. and Sivan, R.: *Linear Optimal Control Systems*, John Wiley, New York, 1972.
107. Lass, H. and Lorell, J.: 'Low Acceleration Takeoff from a Satellite Orbit', *ARS J.* **31**, No. 1 (1961), 24–28.
108. Lawden, D. F.: *Optimal Trajectories for Space Navigation*, Butterworths, London, 1963.
109. Lee, E. B. and Markus, L.: *Foundations of Optimal Control Theory*, John Wiley, New York, 1967.
110. Leitmann, G.: *An Introduction to Optimal Control*, McGraw-Hill Book Co., New York, 1966.
111. Letov, A. M.: *Flight Dynamics and Control*, Nauka, Moscow, 1969 (in Russian).
112. Lilov, L. K.: *Modelling of Systems of Connected Bodies*, Nauka, Moscow, 1992 (to appear in Russian).
113. Lions, J.-L.: *Optimal Control of Systems Governed by Partial Differential Equations*, Springer-Verlag, Berlin, 1971.
114. Lions, J.-L.: *Perturbation Singulières dans le Problems aux Limites et en Controle Optimale*, Lect. Notes Math. Vol. 323, Springer-Verlag, Berlin, 1973.
115. Malkin, I. G.: *Some Problems of the Theory of Non-Linear Oscillations*, Gostekhizdat, Moscow, 1956 (in Russian).

116. Marec, J.-P.: *Optimal Space Trajectories*, Elsevier, Amsterdam, 1979.
117. Matsuno, F. et. al.: 'Feedback Control of a Flexible Manipulator with a Parallel Drive Mechanism', *The Intern. J. of Robotics Research* **6**, No. 4 (1987), 76–83.
118. Mersman, W.: 'Explicit Recursive Algorithms for the Construction of Equivalent Canonical Transformation', *Celest. Mech.* No. 3 (1971), 384–389.
119. Mitropolskii, Yu. A. and Lopatin, A. K.: *The Group Theory Approach in Asymptotic Methods of Nonlinear Mehanics*, Naukova Dumka, Kiev, 1988 (in Russian).
120. Mitropolskii, Yu. A. and Lykova, O. B.: *Integral Manifolds in Mechanics*, Nauka, Moscow, 1973 (in Russian).
121. Moiseyev, N. N. and Shmidt, A. G.: 'Asymptotic Methods in the Theory of Optimum Correction for Systems with Slowly Varying Parameters', *J. Optimizat. Theory and Applic.* **3**, No. 3 (1969), 141–152.
122. Nayfeh, A. H.: *Introduction to Perturbation Techniques*, Wiley, New York, 1981.
123. Neishtadt, A. I.: 'On Passage through Resonances in a Two-Frequency Problem', *Dokl. AN SSSR* **221**, No. 2 (1975) (in Russian).
124. Neishtadt, A. I.: 'Passage through a Separatrix in a Resonance Problem with a Slowly-Varying Parameter', *J. of Appl. Math. and Mech.* **39**, No. 4 (1975), 594–605.
125. O'Malley, R. E., Jr.: *Introduction to Singular Perturbations*, Academic Press, New York, 1974.
126. O'Malley, R. E., Jr.: 'Singular Perturbations and Optimal Control', *Lect. Notes Math.* **680** (1978), 171–218.
127. Pars, L. A.: *A Treatise on Analytical Dynamics*, Oxbow Press, Woodbridge, Connecticut, 1979.
128. Plotnikov, V. A.: 'The Averaging Method in Control Problems', *Differential Equations* **21**, No. 10 (1985), 1713–1717 (in Russian).
129. Pnevmaticos, S. (ed.): *Singularities and Dynamical Systems*, North-Holland, Amsterdam, 1985.
130. Poincaré, H.: *Les Méthodes Neuvelles de la Mécanique Céleste*, Gauthier-Villars, Paris, Vol. I – 1892, Vol. II – 1893, Vol. III – 1899.
131. Pontriagin, L. S., Boltyanskii, V. G., Gamkrelidze, R. V., and Mishchenko, E. F.: *The Mathematical Theory of Optimal Processes*, Pergamon Press, 1964.
132. Potter, J. E. and Vandervelde, W. E.: 'Fuel Optimal Reorientation of Axisymmetric Spacecraft', *J. of Spacecraft*, Nov. 1980, 177–181.
133. Povzner, A.: 'Linear Methods in Problems of Nonlinear Differential Equations with Small Parameter', *Int. J. Nonlinear Mech.* **9**, No. 4 (1974), 279–323.
134. Powers, W. F. and Tapley, B. D.: 'Canonical Transformation Applications to Optimal Trajectory Analysis', *AIAA J.* **7**, No. 3 (1969).
135. Roberts, S. M. and Shipman, J. S.: *Two-Point Boundary Value Problems: Shooting Methods*, American Elsevier Publishing Company, New York, 1972.
136. Rosenwasser, E. N. and Yusupov, P. M.: *Sensitivity of Control Systems*, Nauka, Moscow, 1981 (in Russian).
137. Roseu, M.: *Vibrations non Linéares et Théorie de la Stabilite*, Springer-Verlag, Berlin, Heidelberg, 1966.
138. Routh, E. J.: *Dynamics of System of Rigid Bodies*, Parts 1, 2, Macmillan and Co., London, 1905.
139. Sannuti, P. and Wason, H. S.: 'Multiple Time-Scale Decomposition in Cheap Control Problems – Singular Control', *IEEE Trans. Automatic Control* **AC-30**, No. 7 (1985), 633–642.
140. Shinar, J.: 'On Applications of Singular Perturbation Techniques in Non-Linear Optimal Control', *Automatica* **19**, No. 2 (1983), 203–211.
141. Shniad, H.: 'The Equivalence of von Zeipel Mapping and Lie Transforms', *Celest. Mech.* No. 2 (1970), 114–120.
142. Singh, S. N. and Schy, A. A.: 'Elastic Robot Control: Nonlinear Inversion and Linear Stabilization', *IEEE Trans. Aerosp. and Electron. Syst.* **22**, No. 4 (1986), 340–348.
143. Sternberg, S.: 'Infinite Lie Groups and Formal Aspects of Dynamical Systems', *J. Math. and Mech.* **10**, No. 3 (1961), 451–471.
144. Tabak, D. and Kuo, B. C.: *Optimal Control by Mathematical Programming*, Prentice-Hall Inc. Englewood Cliffs, NJ, 1971.
145. Timoshenko, S.: *Vibration Problems in Engineering*, McGraw-Hill, New York, 1955.
146. Truckenbrodt, A.: *Bevegungverhalten und Regelung Hybride Merkorpersystem mit Anvendung auf Industrieroboter*, VDI-Forschrittsbericht, VDI-Verlag, Dösseldorf, 1980, R. 8, No. 33.

147. Vainberg, M. M. and Trenogin, V. A.: *The Theory of Branching of Solutions of Nonlinear Equations*, Nauka, Moscow, 1969 (in Russian).
148. Vasil'eva, A. B. and Butuzov, V. F.: *Asymptotic Expansions of Solutions of Singularly Perturbed Equations*, Nauka, Moscow, 1973 (in Russian).
149. Volosov, V. M.: 'The Method of Averaging', *Dokl. Akad. Nauk SSSR* **137**, No. 1 (1961), 21–24 (in Russian); *Soviet Math. Dokl.* **2**, 221–224 (English translation).
150. Volosov, V. M. and Morgunov, B. I.: *Method of Averaging in the Theory of Nonlinear Oscillatory Systems*, Moscow State University Publishing House, Moscow, 1971 (in Russian).
151. Wacker, H. (ed.): *Continuation Methods*, Academic Press, New York, 1978.
152. Wasow, W.: *Asymptotic Expansion for Ordinary Differential Equations*, John Wiley, New York 1965.
153. Wiesel, W. E. and Alfano, S.: 'Optimal Many-Revolution Orbit Transfer', *J. Guidance* **8**, No. 1 (1985), 155–157.
154. Wittenburg, J.: *Dynamics of Systems of Rigid Bodies*, B.G. Teubner, Stuttgart, 1977.
155. Young, L. C.: *Lectures on the Calculus of Variations and Optimal Control Theory*, W.B. Saunders Company, London, 1969.
156. Zaremba, A. T.: 'Synthesis of Control of a Pendulum', *Mech. of Solids* **16**, No. 1 (1981), 52–59.

Key Index

Other *Mathematics and Its Applications* titles of interest:

B.S. Razumikhin: *Physical Models and Equilibrium Methods in Programming and Economics.* 1984, 368 pp. ISBN 90-277-1644-7

N.K. Bose (ed.): *Multidimensional Systems Theory. Progress, Directions and Open Problems in Multidimensional Systems.* 1985, 280 pp. ISBN 90-277-1764-8

J. Szep and F. Forgo: *Introduction to the Theory of Games.* 1985, 412 pp. ISBN 90-277-1404-5

V. Komkov: *Variational Principles of Continuum Mechanics with Engineering Applications. Volume 1: Critical Points Theory.* 1986, 398 pp. ISBN 90-277-2157-2

V. Barbu and Th. Precupanu: *Convexity and Optimization in Banach Spaces.* 1986, 416 pp. ISBN 90-277-1761-3

M. Fliess and M. Hazewinkel (eds.): *Algebraic and Geometric Methods in Nonlinear Control Theory.* 1986, 658 pp. ISBN 90-277-2286-2

P.J.M. van Laarhoven and E.H.L. Aarts: *Simulated Annealing: Theory and Applications.* 1987, 198 pp. ISBN 90-277-2513-6

B.S. Razumikhin: *Classical Principles and Optimization Problems.* 1987, 528 pp. ISBN 90-277-2605-1

S. Rolewicz: *Functional Analysis and Control Theory. Linear Systems.* 1987, 544 pp. ISBN 90-277-2186-6

V. Komkov: *Variational Principles of Continuum Mechanics with Engineering Applications. Volume 2: Introduction to Optimal Design Theory.* 1988, 288 pp. ISBN 90-277-2639-6

A.A. Pervozvanskii and V.G. Gaitsgori: *Theory of Suboptimal Decisions. Decomposition and Aggregation.* 1988, 404 pp. *out of print,* ISBN 90-277-2401-6

J. Mockus: *Bayesian Approach to Global Optimization. Theory and Applications.* 1989, 272 pp. ISBN 0-7923-0115-3

Du Dingzhu and Hu Guoding (eds.): *Combinatorics, Computing and Complexity.* 1989, 248 pp. ISBN 0-7923-0308-3

M. Iri and K. Tanabe: *Mathematical Programming. Recent Developments and Applications.* 1989, 392 pp. ISBN 0-7923-0490-X

A.T. Fomenko: *Variational Principles in Topology. Multidimensional Minimal Surface Theory.* 1990, 388 pp. ISBN 0-7923-0230-3

A.G. Butkovskiy and Yu.I. Samoilenko: *Control of Quantum-Mechanical Processes and Systems.* 1990, 246 pp. ISBN 0-7923-0689-9

A.V. Gheorghe: *Decision Processes in Dynamic Probabilistic Systems.* 1990, 372 pp. ISBN 0-7923-0544-2

Other *Mathematics and Its Applications* titles of interest:

A.G. Butkovskiy: *Phase Portraits of Control Dynamical Systems.* 1991, 180 pp.
ISBN 0-7923-1057-8

A.A. Zhigljavsky: *Theory of the Global Random Search.* 1991, 360 pp.
ISBN 0-7923-1122-1

G. Ruhe: *Algorithmic Aspects of Flows in Networks.* 1991, 220 pp.
ISBN 0-7923-1151-5

S. Walukuwiecz: *Integer Programming.* 1991, 196 pp. ISBN 0-7923-0726-7

M. Kisielewicz: *Differential Inclusions and Optimal Control.* 1991, 320 pp.
ISBN 0-7923-0675-9

J. Klamka: *Controllability of Dynamical Systems.* 1991, 260 pp.
ISBN 0-7923-0822-0

V.N. Fomin: *Discrete Linear Control Systems.* 1991, 302 pp. ISBN 0-7923-1248-1

L. Xiao-Xin: *Absolute Stability of Nonlinear Control Systems.* 1992, 180 pp.
ISBN 0-7923-1988-5

A. Halanay and V. Rasvan: *Applications of Liapunov Methods in Stability.* 1993, 238 pp.
ISBN 0-7923-2120-0

D. den Hertog: *Interior Point Approach to Linear, Quadratic and Convex Programming.* 1994, 208 pp.
ISBN 0-7923-2734-9

V.S. Tanaev, V.S. Gordon and Y.M. Shafranksy: *Scheduling Theory. Single-Stage Systems.* 1994, 380 pp.
ISBN 0-7923-2853-1

V.S. Tanaev, Y.N. Sotskov and V.A. Strusevich: *Scheduling Theory. Multi-Stage Systems.* 1994, 412 pp.
ISBN 0-7923-2854-X

L.D. Akulenko: *Problems and Methods of Optimal Control.* 1994, 356 pp.
ISBN 0-7923-2855-8

MIX
Papier aus verantwortungsvollen Quellen
Paper from responsible sources
FSC® C105338

If you have any concerns about our products,
you can contact us on
ProductSafety@springernature.com

In case Publisher is established outside the EU,
the EU authorized representative is:
Springer Nature Customer Service Center GmbH
Europaplatz 3, 69115 Heidelberg, Germany

Printed by Libri Plureos GmbH
in Hamburg, Germany